PHILOSOPHY AND TECHNOLOGY II

BOSTON STUDIES IN THE PHILOSOPHY OF SCIENCE

EDITED BY ROBERT S. COHEN AND MARX W. WARTOFSKY

VOLUME 90

PHILOSOPHY

AND

TECHNOLOGY II

Information Technology and Computers
in Theory and Practice

Edited by

CARL MITCHAM

Philosophy & Technology Studies Center,
Polytechnic Institute of New York

and

ALOIS HUNING

University of Düsseldorf

D. REIDEL PUBLISHING COMPANY

A MEMBER OF THE KLUWER  ACADEMIC PUBLISHERS GROUP

DORDRECHT / BOSTON / LANCASTER / TOKYO

Library of Congress Cataloging-in-Publication Data

C̄IP

Main entry under title:

Philosophy and technology II.

(Boston studies in the philosophy of science ; 90)
Selected proceedings of an international conference held in New York,
September 3–7, 1983, and organized by the Philosophy & Technology Studies
Center of the Polytechnic Institute of New York in conjunction with the
Society for Philosophy and Technology.
"A German-language version has appeared under the title: Technik-
philosophie im Zeitalter der Informationstechnik (Braunschweig : Vieweg,
1985)" — Pref.
Bibliography: p.
Includes indexes.
1. Electronic data processing—Congresses. 2. Computers—Con-
gresses. I. Mitcham, Carl J. II. Huning, Alois. III. Polytechnic
Institute of New York. Philosophy & Technology Studies Center. IV.
Society for Philosophy & Technology (U.S.) V. Title: Philosophy and
technology 2. VI. Title: Philosophy and technology two. VII. Series.
Q174.B67   vol. 90   001'.01 s   85–28345
[QA75.5]   [004]
ISBN-13: 978-94-010-8510-6   e-ISBN-13: 978-94-009-4512-8
DOI: 10.1007/978-94-009-4512-8

Published by D. Reidel Publishing Company,
P.O. Box 17, 3300 AA Dordrecht, Holland.

Sold and distributed in the U.S.A. and Canada
by Kluwer Academic Publishers,
190 Old Derby Street, Hingham, MA 02043, U.S.A.

In all other countries, sold and distributed
by Kluwer Academic Publishers Group,
P.O. Box 322, 3300 AH Dordrecht, Holland.

EDITORIAL PREFACE

Until recently, the philosophy and history of science proceeded in a separate way from the philosophy and history of technology, and indeed with respect to both science and technology, philosophical and historical inquiries were also following their separate ways. Now we see in the past quarter-century how the philosophy of science has been profoundly influenced by historical studies of the sciences, and no longer concerned so single-mindedly with the analysis of theory and explanation, with the relation between hypotheses and experimental observation. Now also we see the traditional historical studies of technology supplemented by philosophical questions, and no longer so plainly focussed upon contexts of application, on invention and practical engineering, and on the mutually stimulating relations between technology and society. Further, alas, the neat division of intellectual labor, those clearly drawn distinctions between science and technology, between the theoretical and the applied, between discovery and justification, between internalist and externalist approaches . . . all, all have become muddled!

Partly, this is due to internal revolutions within the philosophy and history of science (the first result being recognition of their mutual relevance). Partly, however, this state of 'muddle' is due to external factors: science, at the least in the last half-century, has become so intimately connected with technology, and technological developments have created so many new fields of scientific (and philosophical) inquiry that any critical reflection on scientific and technological endeavors must henceforth take their interaction into account.

This has been especially and vividly true in the domain of the (so-called) information sciences and computer science. These are, to be sure, 'sciences' proper, in that there is a body of pure theory, largely mathematical (but also physical, e.g. electronics and solid state physics), which have developed as the foundation of information and control processing and of computer science. But what is perhaps more important than this striking and rapid interaction of science and technology, in these contexts, is the fact that fundamental philosophical questions have arisen (or revived) which become of central importance for our time, momentous in their significance for our Western and 'third world' cultures alike, and for our self-understanding. This volume of the *Boston Studies*, consisting of selected papers from the 1983 International Conference on the Philosophy of Technology, held in New York, presents some leading contributions of contemporary thought on these questions. What, then, are they?

At the inception of the contemporary information sciences, Shannon and Weaver (1948) developed their theory based on a mathematical characterization of 'information' in the transmission or communication of a 'message'. What exactly *is* 'information'? Does it have the character imputed to it, or defined, by the theory? Again, from the earliest developments of automatic computational devices, they have been characterized as substituting for, or duplicating, what human beings do in the course of what appear as 'intelligent', i.e. 'mental', operations. Is such computational procedure therefore a sort of artificial intelligence? Is it *thinking*? Turing, one of the admirably creative founders of modern computer science, posed this question in terms of a test to mark the distinction (if there is one) between 'artificial intelligence' and human thinking, between 'artificial' and 'natural' intelligence.

These two questions are at the heart of philosophical discussions of modern information technologies and computers: What is information? and, What is the relation between computer calculation and human reasoning? In a special sense these are basic: to answer them, we need not only consider the technologies (and their theorizations) but also what we take to be human reasoning, and the nature of meaning in communication . . . in short the fundamental questions about ourselves and our language. But another issue is to be confronted. With the computer revolution and the proliferation of information and control technologies in nearly every aspect of our social, political, and economic lives, the question also arises as to how human beings interact with these information and computation systems, and what the social effects of the technology are, or will be, or might be.

These are the questions which the essays in this volume address. They do so from a number of standpoints, but all are *critical*, often strongly so. They present fresh analyses and sharp attacks on some favorite myths and dogmas of the new sciences of the artificial, and they often oppose each other. Professor Carl Mitcham introduces the essays with his customary enlightening overview of the issues and their history. As he also notes in the preface, this is a successor volume to *Philosophy and Technology*, ed. P. T. Durbin and F. Rapp (*Boston Studies* **80**, 1984).

February 1986   ROBERT S. COHEN
                              Center for Philosophy and History of Science
                              Boston University

                              MARX W. WARTOFSKY
                              Department of Philosophy
                              Baruch College, CUNY

PREFACE

These papers constitute the selected proceedings of an international conference on the philosophy of technology held in New York, September 3–7, 1983. The conference was organized by the then nascent Philosophy & Technology Studies Center of the Polytechnic Institute of New York, in conjunction with the Society for Philosophy and Technology.

The idea for such a conference originated at the conclusion of a previous West German-North American meeting on the philosophy of technology held at the Werner-Reimers-Stiftung, Bad Homburg, in 1981. The proceedings of that conference have already appeared in Paul T. Durbin and Friedrich Rapp (eds.), *Philosophy and Technology* (Boston Studies in the Philosophy of Science, vol. 80, 1983), with a parallel German version, *Technikphilosophie in der Diskussion* (Braunschweig: Vieweg, 1982). It thus seemed appropriate to entitle the present book *Philosophy and Technology II* to indicate continuity with that previous work. As with the first series of proceedings, a German-language version has appeared, under the title *Technikphilosophie im Zeitalter der Informationstechnik* (Braunschweig: Vieweg, 1985).

However, unlike in 1981, it was decided that this conference should take as a theme, information technology and computers in theory and practice – hence the descriptive subtitle. Yet papers on other topics were welcome, and sessions were organized on engineering ethics as well as on technology and democracy. Some papers originally planned to deal with the theme turned out to be more directly addressed to other aspects of technology or technology in general. To facilitate dissemination, it was decided to publish papers not primarily concerned with computer-related issues independently in *Research in Philosophy and Technology*, vol. 8 (1985). Conference papers to be found there are as follows:

> Michael Black (State Univ. of New York, Plattsburgh) and Richard Worthington (Rensselaer Polytechnic Institute), "Democracy and Reindustrialization: The Politics of Technology in New York State"
>
> Stanley R. Carpenter (Georgia Tech), "Scale in Technology: A Critique of Design Assumptions"

Alois Huning (Düsseldorf Univ.), *"Homo Mensura*: Human Beings are Their Technology – Technology Is Human"

Don Ihde (State Univ. of New York, Stony Brook), "Technology and Cultural Variations"

C. Thomas Rogers (Montana Tech), "The Ethical End-Use Problem in Engineering Ethics"

Kristin Shrader-Frechette (Univ. of California, Santa Barbara, and Univ. of Florida), "Technology Assessment, Expert Disagreement, and Democratic Procedures"

With this kind of diverse participation, the conference itself was a rich interaction between pro- and anti-technology partisans (e.g., P. Levinson and W. Schirmacher vs. H. Dreyfus and W. Zimmerli, respectively), well-established (J. Margolis and H. Beck) and relatively younger scholars (S. Kramer-Friedrich and D. Cérézuelle), some persons long associated with the field (C. Mitcham and F. Rapp) and some more newly engaged (P. Heelan and E. MacCormac), and some with backgrounds other than professional philosophy. In addition to the Germans and Americans, there were participants or participant observers from Switzerland, The Netherlands, France, Canada, and South Africa. For three days the conference was conducted at a resort hotel north of New York, then for two more days at the United Engineering Center in midtown Manhattan, again to encourage different levels and a wide spectrum of discussion and involvement.

The original conception of this conference was generously supported by the Franklin J. Matchette Foundation, by Goethe House New York, and by the Department of Humanities and Communications of the Polytechnic Institute of New York. Indeed, without the strong encouragement and active involvement of Polytechnic President George Bugliarello, an engineer of exceptional philosophic interests; Arts and Sciences Dean Eli Pearce, a scientist of equally pronounced humanistic sympathies; and Donald Hockney, Head of the Department of Humanities and Communications, and a philosopher of science who recognizes the importance of the philosophy of technology, this conference would not have taken place.

It should also be acknowledged that editorial preparation of these proceedings has been facilitated in part by an Exxon Education Foundation grant to the Philosophy & Technology Studies Center to support course development in philosophy and technology studies. It is expected that this book will serve as a good text for advanced courses in this emerging and important field.

Finally, persons who have contributed to proof reading and index preparation include Robert Mackey, Doahn Nguyen, and Yvonne Williams. Their work has been greatly appreciated.

C.M.

TABLE OF CONTENTS

xi

SELECT ANNOTATED BIBLIOGRAPHY ON PHILOSOPHICAL STUDIES OF INFORMATION TECHNOLOGY AND COMPUTERS

ANALYTIC TABLE OF CONTENTS

Note: Divisions or titles not provided by an author are placed in brackets.

INTRODUCTION: INFORMATION TECHNOLOGY AND COMPUTERS AS THEMES IN THE PHILOSOPHY OF TECHNOLOGY

Philosophical interest in computers and information technology has closely paralleled scientific and technological developments in these fields. Indeed, philosophical reflection on computers and information technology has from its origins constituted one of the primary aspects of the philosophy of technology, particularly in the Anglo-American philosophical community. What follows is an attempt to substantiate these two claims by a review of some historical moments in the computer-philosophy encounter and by an analytic survey of the issues thus raised. A conclusion places the present proceedings in this historico-analytic framework.

I

The early history of the computer has been admirably (if somewhat parochially) traced by Herman Goldstine in *The Computer: From Pascal to von Neumann* (1972). As the subtitle indicates, philosophers have from the beginning been involved with the creation of computers. Pascal (1623–1662), Leibniz (1646–1716), Charles Babbage (1791–1871), and John von Neumann (1903–1957) were all logicians and mathematicians attracted by the idea of mechanizing the process of calculation, which was perceived as a fundamental but subordinate form of thought, and to reflecting on the philosophical implications of that mechanization. Why, for instance, do mathematics and logic lend themselves to mechanization and automation? What is the relation between mathematics and logic?

With von Neumann's formulation of automata theory, philosophical interest broadened to include concerns regarding the ontological status of certain artifacts. Here the work of the British logician, A.M. Turing (1912–1954), on the criteria for distinguishing artificial from human intelligence, raised challenging questions, and in the early 1950s introduced the theory of computers into two discussions prominent in analytic philosophy: the mind/body problem and the problem of other minds. Is the relation between mind and brain in any way similar to that between a computer program (software) and the computer itself (hardware)? Can

1

Carl Mitcham and Alois Huning (eds.), Philosophy and Technology II, 1–14.

the ambiguities of communicating with another person be illuminated by the operations of communicating with a computer?

Particularly during the initial three generations of computer development (from vacuum tubes through transistors to integrated circuits), it was not uncommon for even technical books to include philosophical commentary. Often this entailed a reference to information theory because, as Goldstine remarks in his own concluding reflections on why the mechanization of calculation should have had such a major impact on society,

It is . . . better to recognize that what a computer really deals with is not just numbers alone but rather with *information* broadly. It does not just operate on numbers; rather, it transforms information and communicates it (p. 345).

As with all discussions of information theory in even the most technological works, however, the concept of "information" is left as an undefined primitive, and thus as a natural subject for philosophical analysis and debate.

In fact, there are two quite distinct concepts of information, semantic and mathematical. Information technologies in the semantic sense are those electronic means of communication – such as the telephone (invented in the 1870s), the radio (early 1900s), and television (1930s) – which are able to transmit human verbal and visual behavior over large distances in real time. Interestingly enough, the histories of these technologies reveal little philosophical reflection on the part of their inventors. But the fact that different kinds of communications media – from speech and writing to print and electronics – influence culture and society in a multitude of ways, has aroused the concern of others and even created specialized fields of communication studies and media theory. Colin Cherry's *On Human Communication: A Review, A Survey, and A Criticism* (1957) and Marshall McLuhan's *Understanding Media: The Extensions of Man* (1964) are two formative texts in these areas.

Issues in the philosophical understanding of an information-technology culture have been supplemented by studies on the professional ethics of news reporting, advertizing, and mass media entertainment. The proliferation of information by means of information technologies has also given rise to "information science," a kind of extension of library science dealing with how to store, manipulate, and access enormous amounts of data. Ethical issues of access and privacy have naturally been associated with information science developments.

The mathematical concept of information points in a quite different direction, toward the science of information theory. Unlike as is the case with computers, the telephone, radio, and television, there exists no history of information theory. There is, however, a kind of symbiotic relationship between developments in electronic hardware and in theory. Certain technical problems with early electronic technologies (telephone and radio) gave rise to early information theory (R. Hartley, 1928); this theory in turn contributed to the new generation of information technologies (TV and computers), which themselves stimulated the development of more advanced information theory (C. Shannon and W. Weaver, 1949). Such a relationship is already instructive for eliminating a common misconception of technology as simply applied science. In some instances it might be more exact to speak of science as theoretical technology than of technology as applied science.

Be that as it may, among the earliest philosophical reflections associated with these technical developments were those of Norbert Wiener (1894–1964) on cybernetics. Precisely because cybernetics proposes to enlarge on information theory to create a comprehensive understanding of behavior in both organism and artifact, it constitutes a philosophy of technology. Traditional philosophy aspires to comprehend the whole as a unity of human beings and nature, with artifice implicitly understood as an aspect of the human. When this implicit unity becomes problematic it engenders a new approach to philosophy.

Recognition of the inherent philosophical character of cybernetics is attested not only by the philosophical reflections that formed part of the writings of major cybernetic theorists throughout the world – N. Wiener and Warren S. McCulloch in the U.S.; W.R. Ashby, F.H. George, Gordon Pask, W. Grey Walter, and D.M. MacKay in Britain; L. Brillouin and L. Couffignal in France; K. Steinbuch and G. Klaus in Germany; I.A. Berg and V.M. Glushkov in the Soviet Union – but also by the fact that professional philosophers were early attracted to fruitful engagement with this new field. When Kenneth Sayre and Frederick Crosson ventured an appraisal of *Philosophy and Cybernetics* (1967) they identified the focus as theoretical issues associated with the relation between semantic and mathematical concepts of information, the possibility of artificial intelligence, the adequacy of the cognitive simulation, and the nature of mechanization itself.

For philosophers the primary question was, to put it simply, "Can computers think?" Sayre's *Recognition: A Study in the Philosophy of Artificial Intelligence* (1965) and even more strongly Hubert Dreyfus' *What Computers Can't Do* (1972) argued against the ability of algorithmic programs to imitate human thinking in any global sense. Thus, by the early 1970s there existed a philosophical consensus regarding the limitations of artificial intelligence. Computer scientists, however, continued (and continue) to hold a more affirmative position, at least with regard to future possibilities. The work of Herbert Simon in decision theory and heuristic programming has been particularly important in transcending the limits of so-called logic machines. And Simon's related proposal for creating *The Sciences of the Artificial* (1969) to supplement the sciences of nature challenges standard philosophy of science to recognize the legitimacy of the philosophy of technology. It can also be read as a confirmation of Martin Heidegger's argument that cybernetics is "the last stage of metaphysics," that is, a culmination of the typically Western attempt to objectify and control the world.

In *The Human Use of Human Beings: Cybernetics and Society* (1950), the founder of cybernetics had, barely two years after his classic *Cybernetics*, already pointed to another whole dimension of the cybernetics and philosophy question, one having to do with social philosophy and ethics. (Thirteen years later Wiener also initiated discussion of the ways cybernetics impinges on religion.) But it was not until the early 1960s that practical issues became prominent in the philosophical discussions of computers, perhaps because, as far as philosophers were concerned, theoretical questions had been provisionally exhausted, and because the late 1960s and early 1970s witnessed a revival of interest in ethical and political issues in the Anglo-American philosophical tradition.

In the practical realm discussion has centered around the possibilities and impact of automation. "Automation" is a condensation of "automatization" coined independently by D.S. Hander at Ford Motor Company in 1946 and by John Diebold at the Harvard Business School a few years later, the latter of whom used it as the title for an influential 1952 book. Indeed, initially it was economists and social theorists who drew attention to and analyzed the problems of historical change, unemployment, leisure, consumerism, education, and secrecy inherent in the commercial development and utilization of computers and information systems – first in record keeping (payroll and billing), and only later in manufacturing (robotics). Yet two key aspects of computer utilization, military and scientific, both of which preceded commercial applications, have continued to be slighted even in subsequent philosophical discussion and analysis.

As a result of these various historical encounters between the computer and philosophy, by 1973, when there appeared the first systematic *Bibliography of the Philosophy of Technology*, reflection on information technology and computers constituted the single most well-defined aspect of the field. Questions dealing with artificial intelligence predominated among metaphysical and epistemological concerns; media issues and automation occupied a prime place in the ethical-political domain. And although there were a number of significant anthologies devoted to each subject area, it is significant that only one (Z. Pylyshyn, 1970) explicitly tried to bridge the theory-practice gap.

II

The history of the philosophy of technology can conveniently be divided into three phases. Phase one constitutes a prehistory of the subject – and stretches from explicit discussions of *techne* in Plato and Aristotle, together with the implicit criticisms of technology to be found in the Middle Ages, to a radical rejection of this tradition by the Renaissance and Enlightenment, and the subsequent Romantic critique of modern science and industrialization. The prehistory comes to a close, and a second phase opens, with the publication in 1877 of Ernst Kapp's *Grundlinien einer Philosophie der Technik*. This formative period continues for the next hundred years, through the life and work of Friedrich Dessauer (1881–1963), the father of the philosophy of technology as a recognized academic discipline, and the later Heidegger's attention to "The Question Concerning Technology." The formative phase can in its turn be said to terminate with the previously mentioned publication of the first systematic bibliography of the field.

Since the early 1970s, in a third phase, the questions and issues broached during the formative years have been re-examined and deepened – especially with regard to computers and information technology. The expansion of computers over the last three decades beyond the confines of specialized scientific and military applications to general scientific and engineering work, telecommunications, government and commercial record keeping, white collar office management systems, blue collar production operations, games, personal computers, and dedicated micro-processors in everything from cars to kitchen appliances – to the point where *Time* magazine in January, 1983, replaced the "man of the year" with the computer as "machine of the year" – could not help but encourage the philosophical community to extend and develop its own initial ventures. What follows, then is an analytic outline of the major issues in the philosophy of technology as these have come to the fore in the last decade with respect to information technology and computers.

A. Conceptual Issues

Two basic conceptual issues concern the definition of "information" and the uniqueness of information technology in the spectrum of technologies. Although "information" in the mathematical sense is apparently related in some way to "information" in the ordinary language or semantic sense, the exact character of this relation is not at all clear. For instance, some (e.g., Douglas Hofstadter, 1979) have argued that mathematical information theory can contribute to a theory of meaning, whereas others (e.g., Fred Dretske, 1981) have strongly criticized this notion. The relation of electronic technologies to tools and machines as traditionally conceived, and their place in a possible evolutionary sequence internal to technology itself, are further topics which have been raised especially by philosophically inclined members of the technical community, but have implications as well for historico-philosophical interpretations. There may, for instance, be a sense in which all technologies are information technologies embodying different kinds of information.

B. Ethical Issues

Here a key question concerns the ethical responsibilities of computer professionals and others who regularly use computers and information technology. The salutary fact is that computer scientists in the United States, like nuclear scientists before them, have from an early period seriously debated their ethical obligations. Two issues which have been prominent in these debates are warfare and privacy. Information technology professionals themselves have recognized that computers achieved their greatest developmental stimulus from military demands and that nuclear missiles are not possible without computers. There is also a natural temptation to be guided primarily by what is "technically sweet" (Oppenheimer) at the expense of more fundamental perceptions. Yet it is important to remember, as well, that computer professionals have been among the strongest opponents to the idea of a national data bank from the time that this was proposed in the United States in 1967.

   A second ethical issue concerns the relation between computers and human dignity, as vigorously expressed by Joseph Weizenbaum in *Computer Power and Human Reason* (1976). Against the background of a growing concern about the depersonalization wrought by IBM punch cards and computer programs that mimic human behavior, Weizenbaum argues for the placing of moral limits on two kinds of computer applications: (a) "all projects that propose to substitute a computer system for a human function that involves interpersonal respect, understanding, and love" (p. 269); and (b) those with unforeseeable and irreversible consequences, especially when the needs being met can be dealt with in other ways. One objection to limit (a) is that there are cases (e.g., autistic children) in which persons respond to computers and other electronic technologies better than to human beings; indeed, B.F. Skinner argues that educational technologies built around computers are inherently more humane than human teachers precisely because they can be more "attentive" to the students. An objection to limit (b) is that almost all human actions entail consequences which are to some degree unforeseeable and irreversible.

A provocative reversal of Weizenbaum's argument is Aaron Sloman's suggestion that computers, like persons, have rights. Although the idea that limiting rights to humans (and perhaps some other animals) is a "racialist position concerning machines" (1978, p. xii) or more broadly "protein chauvinism" (a term coined by Paul Levinson) is obviously a rhetorical exaggeration, it nevertheless revives in a new context the question of the proper place of technology in a human moral framework.

Finally, there are questions to be asked regarding the "computer ethos." Is there a particular ethos of computer hackers, influenced at least in part by the computer technology itself? Does this ethos in any way exclude women, blacks, or other groups? Does it contribute (perhaps by its individualism and lack of a sense of reality) to computer crime? Is there a computer ethics that needs to be developed and taught, perhaps in the public schools, in response to the wide availability of these powerful new devices? Studies by Sherry Turkle on the ways computers enter into the self-images of those who use them provide helpful insights to all such questions.

C. Socio-Political Issues

Following a division utilized by Irene Taviss in an anthology on *The Computer Impact* (1970) – a division based on Daniel Bell's identification of three axial dimensions in society – socio-political issues can be distinguished into three broad categories: those focusing on the economic, the political, or the cultural orders.

(a) One pivotal socio-economic question concerns the impact of information technology and computers on the administration and management of economic institutions. According to the classic thesis of Max Weber, modern economic formations are characterized by centralization and rationalization (desacralization, secularization). *Prima facie*, the computer appears more compatible with these two characteristics than with archaic or premodern economic formations. Jacques Ellul (1980), for instance, explicitly argues that information technology arose at a particular point in time in order to meet demands for the greater rationalization of modern economic formations. Yet other social theorists argue that computers actually encourage individualism and make possible a new decentralization, pointing to the individualist strains information technologies place on group cohesion in both underdeveloped and communist countries as well as the "irrational" adaptations of computers in video games and astrology.

A closely related socio-economic thesis is Daniel Bell's argument concerning the changing character of work in a post-industrial, information economy. Bell (1973) argues that a dramatic shift is taking place from production to a service-oriented economy in which information (education and knowledge) and computers play the major role. Robert Lilienfeld (1978) has replied that the vogue of information theory, cybernetics, and systems theory are all forms of an ideological rationalization of quite traditional class interests.

(b) Socio-political issues come in two forms, international and domestic. With regard to international affairs, the most dramatic questions concern war and peace. Does information technology increase the likelihood of war by increasing war-making capacities and national means of surveillance without in any way moderating nationalism? Or does increased national means of surveillance, together with the increased fragility of an information network infrastructure (and nuclear weapons), make war less likely?

The positive use of computers to promote "global" analysis by means of large-scale, multi-factor modelling of population trends, the weather, pollution, resource depletion, crop production, and epidemiological phenomena should not be over-looked. It is ironic that in negotiations leading up to the new international "Law of the Sea" treaty in the late 1970s, representatives from some third world countries would not trust analyses by first world political economists until such analyses were modelled by interactive computer programs. Side by side with such an attitude is a felt need by many developing countries for a "new world information order" which would limit what they perceive as information and information technology dominance over their societies.

With regard to domestic affairs: Do information technologies promote democracy or totalitarianism? It is clear that information technologies make available enhanced powers for active propaganda and passive citizen surveillance. Political implications related to private and governmental forms of the latter (credit bureaus, the IRS, etc.) have recently been surveyed in David Burnham's *The Rise of The Computer State* (1983). An equally serious problem is that of "information overload" – too much data undermining attention, wisdom, creativity, and decision making. Jerry Mander has addressed subtle forms of this problem in relation to a specific case in his provocative *Four Arguments for the Elimination of Television* (1978).

On a more obvious level, computerized mailings and electronic polls have altered the powers of special interest groups. Polls and predictions likewise can alter the very behavior they are supposed to describe. At the same time, there are observers who present a strong case for the salutary political impact of advanced information technologies and maintain they are the means to an electronic participatory democracy.

One undiscussed danger to the socio-political consequences of information technology in both international and domestic frameworks is the potential for what might be called "information terrorism." In contrast to propaganda, in which established elites utilize information technologies to promote and defend their interests, information terrorism would be the attempt by powerless subgroups to appropriate electronic means of communication to impose political demands or to sabotage those same means on the basis of radical political theories or what might appear to be sociopathologies.

There are, as well, socio-psychological problems concerning the impact of computers on the individual citizen which are related to the ethical issues already mentioned, particularly concerning questions of human dignity and the computer ethos. But one undeveloped subject which cuts across distinctions between economics and politics – and arises from an interest in the engagement with electronic technology at the individual level – is the philosophy of work. The need to re-examine social attitudes toward work, which often gets associated with this term, is ultimately dependent on a re-thinking of the inherent nature of human work itself in the light of a cogent phenomenology of the creation and utilization of computers and information technologies. For instance, the acquisition of the skill of using a typewriter is quite different from that of learning to use a word processor, and this very difference may have important implications for our understanding of the work relationship to the world.

(c) With regard to the socio-cultural order, the key philosophical questions concern the ways computers may influence or transform the traditional branches of the liberal arts, and the cultural attitudes necessary to or brought about by high technology. Is computer music really music? Are computer-generated poems true literature? What do computer-assisted analyses of literary works (word frequency distributions, etc.) really tell us about those works? Does the automation of libraries, and an increasing reliance on computer-analyzed statistics in the social sciences (cliometrics, etc.) portend undesirable losses?

Finally, the interaction between electronic technology and cultural attitudes is a historico-philosophical question of immense proportions, with implications not only for public policy regarding technology transfer but also for the self-understanding of Western civilization. Communications studies, media theory, and the history of ideas all have contributions to make to the assessment of our problematic social history. Furthermore, if it is possible that the undifferentiated impact of information technology is not equally beneficial to each of the axial dimensions of society – i.e., that a computerized economy is good while a computerized policy is not – there may be long-term contradictions which it is not easy to assess or to meliorate.

D. Metaphysical-Epistemological Issues

Despite the large amount of energy devoted to this subject, and the provisional negative philosophical consensus of the early 1970s, the key metaphysical issue remains the ontological status of complex electronic artifacts – i.e., most succinctly, Can computers think? Are they in any meaningful sense human? Is artificial intelligence really possible? Computer scientists (such as Marvin Minsky, John McCarthy, and Roger Schank) have continued to reject the sceptical response to these questions, and in the late 1970s a few computer science philosophers, especially in Britain (Margaret Boden and Aaron Sloman), restated the argument for an affirmative response, thus reviving and deepening the philosophical discussion.

On the epistemological side, attention focuses more on the question of how computers and information technologies might alter our cognitive abilities. Here again the arguments of Patrick Heelan (1983) that all technologies affect our perception of the world would imply that all technology is implicitly information technology of some kind.

E. Religious Issues

The central religious issues concern how computers and information technology affect (to employ Christian terms) faith or the doctrine of creation and the God-human relationship, hope or eschatology and the vision of the future, and charity or daily life in the ecclesial community.

On the doctrine of creation, Wiener's *God and Golem, Inc.* (1964) remains a classic discussion, although today it seems much more likely that "self-reproducing machines" – and thus human creation rivaling God – will be achieved by means of biotechnology rather than cybernetics. On eschatology, the speculations of Walter Ong that information technology is spiritualizing the world must be moderated by an appreciation of the moral weaknesses of the computer culture and its apocalyptic prospects. On daily life in the ecclesia, there has been considerable discussion, both pro and con, regarding the meaning of the electronic church.

III

The present collection of papers, originally contributed to a week-long conference, make an important contribution both to the development of a broad-based and pluralistic philosophy of technology, and to the crucial coming-to-terms with information technology and computers which is integral to contemporary Western culture. Furthermore, the conference itself and the resultant proceedings advance the professional and institutional development of the philosophy of technology as an interdisciplinary and international community of discourse.

At the beginning of the conference Paul Durbin, the single most influential person in nursing the Society for Philosophy and Technology into existence, in an opening presentation which has not otherwise been included, made the following related remarks:

What are the arguments for the centrality of computers, automation, and information theory as a focus for philosophy of technology?

Ellul and his followers would say it is the computer that has added the ultimate technique to "the technological phenomenon" – creating an all-enveloping, inescapable technological system which dominates contemporary reality.

At the opposite extreme (from one perspective), some AI enthusiasts are convinced that their efforts will, once and for all, establish that there is no clear dividing line between artificial (man-made) and natural (God-made?) intelligence – that the old philosophical problem of mind and matter will (eventually) be laid permanently to rest.

Members of the scientific and technical community – from design engineers to computer diagnosis medical practitioners, from instrument makers to satellite builders, launchers, trackers, and users – as well as "pure" scientists in all fields – find computers and automation devices indispensable tools of their professional work. Those of us who take that community as our focus for reflection – whether engineering or computer ethicists, STR (scientific-technological revolution) theorists in Eastern Europe, "systems" philosophers, or social pragmatists like myself – must treat the computer/information revolution as seriously as do members of the scientific/technical community.

Finally, even philosophers of technology who come from an Anglo-American analytical background (especially in the philosophy of science) – whether their focus is technology assessment, decision theory, risk-cost-benefit analysis – will find that the actual techniques and methods they want to analyze are almost always embodied these days in computer programs. If they find any grist for their analytic mills in the actual work of practitioners – or, even better, if they work with the latter as participant observers – they will be forced to become computer literate (and critical) themselves.

What, then, can we expect from this conference? It is not likely to be significant advances in information theory – though the understanding of that theory might be clarified. It is not likely to be a significant breakthrough for or against artifical intelligence – though again that concept and its implications may be sharpened. Nor is it very likely – though I wish it were otherwise! – that any of us will come up with dramatic new ways of dealing, ethically or politically, with the social problems attendant on the computer/information revolution, whether in high-technology societies or in the Third World.

What we can expect, however, is to participate in the most concentrated effort to date of philosophers focusing at one time from many different perspectives on this central feature of the latest stage in the development of a worldwide technological culture (or anticulture, if one is on the critical side).

The realization of Durbin's expectations, it seems reasonable to argue, are adequately reflected in this volume of proceedings. To this end, however, it is not necessary to give summaries of individual contributions, particularly since these are otherwise provided by the "Analytic Table of Contents" and the abstracts which precede each paper. Instead, what is more appropriate would be some brief observations about the collection as a whole.

In part one, the most tightly focused and technical section, there is strong agreement about the limitations of information theory and the philosophical inadequacy of its key technical term. "Information" in the technical sense is just not the same thing as "information" in any humanly meaningful sense, stress Kramer-Friedrich, Ropohl, and Strombach. As a result, various authors make suggestions for reform. Kramer-Friedrich points out the inherently active, historical character of information; Rapp extends a now well-established (if still arguable) thesis about the theory-ladenness of scientific "facts" to apply to information; and Strombach appeals to the principles of Platonic and Aristotelian metaphysics for an ontological understanding of information. Perhaps Levinson, however, makes the most far reaching and provocative suggestion when he argues for integrating the concept of information into a biological and evolutionary context.

Part two contains the most lengthy and substantial papers and, as might be expected, is thus the core and philosophical center of gravity of the proceedings. Whereas the first five authors point out certain failures of technological theory to come to terms with reality as a whole, the contributors to the second section – especially Dretske, the Dreyfuses, Heelan, and MacCormac – build on and deepen this contention by detailed analysis of the human-computer interaction. Beck (a senior European philosopher of science) serves as a kind of transition, by presenting an anthropology of human freedom which transcends cybernetic determinism, while Margolis (a senior American analytic philosopher) makes a compatible defense of what he terms a "top-down" strategy for artificial intelligence.

At the heart of part two, however, are four versions of the argument that computers do not think. Dretske maintains that they cannot even add, in the true sense. The brothers Dreyfus (one of whom is a philosopher, the other a computer scientist) spell out the limits of calculative rationality when compared to fully embodied human thought and action. Heelan rejects machine perception as equivalent to human perception. And MacCormac, a philosopher and engineer, provides historical background and critical analysis of the brain-a-computer metaphor.

Part three contains the largest number of essays, and deals with a wide range of dangers and challenges inherent in the information-computer society. It thus picks up on concerns implicit in, say Ropohl and MacCormac, and addresses a spectrum of ethical-political concerns. Borgmann and Winner, in complementary ways, both question the revolutionary rhetoric surrounding computer development. Byrne and Laor-Agassi deal with quite specific applications of computer technologies in the work place and in medicine, and the resultant ethical-political issues that are raised. Cérézuelle (a student of Jacques Ellul) and Schirmacher (who has studied Heidegger closely) agree about the historical drift of computerization, but offer quite opposing interpretations of the human meaning of its "inevitability" – although Cérézuelle is the more circumspect of the two. Mitcham attempts to raise questions about the ability to control information technologies by appeal to some classic versions of incontinence. Zimmerli examines the same issue in considerably more detail by spelling out the hiatus between information technology and moral responsibility within all major ethical frameworks. The consensus (with Schirmacher dissenting) is certainly that ethical and political problems suggest the need for caution in the exercise of those technological powers made possible by information technology and computers. At the same time, Borgmann, Byrne, Laor-Agassi, and Lenk, hold out specific reforms and optimistic visions which should not be overlooked.

The papers included here thus broach all the major issues outlined in section two above, except for religious ones. Conceptual issues take priority in part one. The metaphysical and epistemological dimensions of a philosophical anthropology are the focus of part two. And ethical-political issues dominate in part three. Taken as a whole, and often singly, these essays clearly bridge the theory-practice gap. Individually and in concert, they maintain the pivotal position in the philosophy of technology of discussion concerning information technology and computers, as these fields continue to undergo scientific and technological change. In so doing they confirm the health and vigor of philosophy itself in a time of increasingly abstract specialization, and they give promise of further efforts in relation to aspects of our society which can be expected to play crucial roles for the foreseeable future.

Polytechnic Institute of New York

N.B. References in this introduction are keyed to the Select Bibliography found at the end of the volume.

PART I

THE METAPHYSICAL AND EPISTEMOLOGICAL CHARACTER OF INFORMATION

SYBILLE KRAMER-FRIEDRICH

INFORMATION MEASUREMENT AND INFORMATION TECHNOLOGY: A MYTH OF THE TWENTIETH CENTURY

ABSTRACT. This paper makes four arguments. First, classical information theory is not really about information. Second, electronic calculators do not in truth process information. Third, "information" is not so much a scientific concept as it is a mythical one. Finally, there is an attempt to outline the preconditions for a fully scientific theory of information.

No word sounds more exciting or propitious than "information." Though given a technical use for only several decades, "information" as a technical term loaded with connotations has already had an enormous influence on scientific modelling and even everyday language. Indeed, it has become a standard category of modern scientific, technological, and social analysis.

The technical usage suggests that "information" is precisely and scientifically defined. But seldom before has scientific theory so creatively manipulated language. Mathematical information theory, developed in its classical form toward the end of the 1940s, formulated the concept. Behind this formulation stands the protective godfather of a rich and powerful electronics information technology. And from here the term has spread to influence virtually all the sciences.

The scientific claims of information theory rest on two theses:

- Information theory provides a way to measure information content.
- The electronic calculator is an information processing technics.

The following paper will contest both theses. This will be done by means of a critical investigation of the use of the language of information measurement and information technology. As a result, the concept of information will lose its presumed scientific character and be exposed by its own inappropriate use of language for what it really is – a modern myth, a myth that scientific and technical progress has not only failed to overcome, but has actually fostered. In conclusion, a preliminary outline for an alternative theory of information will be sketched.

Carl Mitcham and Alois Huning (eds.), Philosophy and Technology II, 17–28.

SYBILLE KRAMER-FRIEDRICH

I

My initial thesis is as follows: *Classical information theory is not a theory about information at all. The formula by which the theory proposes to measure information content does not measure information but is a rule for the efficient reproduction of isomorphic signal structures.*

Mathematical information theory, as formulated by Claude Shannon in the late 1940s, claims to define the term "information." This classical information theory rapidly made its influence felt not only in the natural sciences, but in the behavioral sciences and the humanities as well. Its fundamental conceptual categories – message source, channel, receiver, with corresponding coding (transmitting) and decoding (receiving) devices – are thus postulated to be the structures common to all communication.

Information theory also claims to be able to measure the content or amount of information in a message. It does this by abstracting from all meaning or effect and focusing only on the remaining substrate.

Frequently the *messages* have meaning; that is they refer to or are correlated according to some system with certain physical or conceptual entities. These semantic aspects of communication are irrelevant to the engineering problem (Shannon, in Shannon and Weaver, 1949, p. 31).

According to the theory itself, information measurement can produce the following results: A hundred letters from a poem by Goethe and a series of a hundred random letters would have the same information content, provided both have the same statistical dependencies between the letters. When a mathematical analysis brings about results which, from the standpoint of common sense, are completely paradoxical, it naturally suggests that information theory might not really be talking about information.

The point can be made more clearly by the following situation. Consider some closed field of alternative possible choices. Take, for example, two people playing a game. Each player has a set of eight different playing pieces. The rules are that player X chooses one piece from his set, and player Y must then guess the choice. Y may ask X only questions that can be answered "yes" or "no." There are different strategies for asking questions. One strategy would be for Y simply to ask the names of all the playing pieces until he guesses correctly. If the playing pieces were, for instance, distinguished by the initial letters of the alphabet, Y could ask: "Is your piece an A?" "Is it a B?" Etc. At worst such a strategy would require eight questions.

There exists, however, a much more efficient strategy. In this other strategy, player Y in some manner divides the pieces into two equal groups and begins by inquiring whether X's piece is in the first group, say A through D. Once one has identified the half to which X's choice belongs, this half can again be subdivided, and a second question asked regarding the subgroup in which X's choice will be found. With such a strategy, player Y needs (in a game with eight pieces) exactly three questions in order to be able to identify X's choice. In general, the number of questions or information content I_c is

$$I_c = \log n,$$

where log is the logarithm to the base 2 and n is the number of possible alternative choices.

This example shows that information content is not really what is being measured. Instead, a formula has been constructed which specifies parameters for the most efficient performance of a certain operation. It tells how, given a sentence composed of certain alternative signs, the set of signs can be most easily identified, under the condition that the basic step of the identification process consists of a choice between two mutually exclusive alternatives. The formula, the explanatory value of which has just been examined, nevertheless has marked practical significance. This consists not in measuring information, but in representing a given sequence of signs with another set of signs, so that the structure of the original sequence is maintained in the medium of the new. The operation involved is called "coding." Coding is not necessary, however, for the passing of information in a living context, but only where such a context has been disturbed or broken, so that it becomes necessary to transmit messages over large spatial or temporal distances.

That the optimal question strategy in our imaginary example and the construction of an optimal code coincide can be demonstrated in the following way. Suppose player X wanted secretly to communicate to player Z which playing piece he had chosen, using a coding system whose elements range over only two symbols, say the number 0 and 1. He notices that he needs a combination of exactly three signs to cover each of the eight playing pieces, for instance

A	111	E	011
B	110	F	010
C	101	G	001
D	100	H	000

The formula for measuring information content provides directions for the optimal construction of such a code.

Such considerations lead to the following conclusion: The purported measurement of information content in classical information theory is exclusively a guide for the isomorphic transmission of structures. That this should be the case is not surprising in light of the technical problems from which information theory arose. "The fundamental problem of communication," according to Shannon, "is that of reproducing at one point either exactly or approximately a message selected at another point" (in Shannon and Weaver, 1949, p. 31). This theory was then extended, as already mentioned, into a general model of communication.

But the linear transmission from message sender to message receiver through some channel, while accurately modelling a technical system, does not apply to the passing along of information or the informing process in a living situation. The channel – through which there flows not information but only physical energy impulses – must be so structured that through the contact between a sender and the encoding or transmitting device at one end of the channel there is a production of signals which at the other end of the channel can be decoded by some receiver. This is not a communication situation but simply the mechanical preservation of signal structures. The concept of information fundamental to classical information theory reduces the act of informing to the technical transmission of isomorphic structures. But the informing processes which are fundamental to human perception – especially cognitive and communicative actions – are not able to be fully described by this theory. To try to do so is to accept a mechanical operation as a paradigm for the analysis of human interactions.

II

Let me now turn to the second thesis associated with the functioning of electronic calculators. My counter-thesis is as follows: *Electronic data processing devices are not information processing technics; they simply provide for the mechanical rearrangement of some signal series according to rules which have no relation to the meaning of the signals. Information processing is able to be simulated mechanically only if it is possible to abstract from the fact that the passing on of information is taking place.*

It is common to describe the computer as a device that processes information. To do so makes use of the following distinction. The function of the classical machine, as widely adopted during the first Industrial

Revolution, consists in the production or transformation of material or energy. The subsequent development of the calculator, by contrast, is supposed to create a machine which transforms or processes information. As an information technics, the calculator becomes – after machines which transform materials and those which transform energy – a third species of machine. And so, just as the material and energy transformers replace human physical labor, the information transformer is thought of as some kind of replacement for human mental activity.

This train of thought is superficially plausible – i.e., insofar as one considers a machine from the point of view of that human activity which is supported and/or made superfluous by the machine. But such an anthropomorphic perspective overlooks one important fact. Although mechanical processes model activities formerly carried out by human beings, they do so in ways completely different from the ways human beings do them. More precisely, in order that human activities can be delegated to machines, these activities must first be re-structured in ways that can fully alter their character. Every *action* which is to be turned over to a machine must first be transformed into an *operation*, i.e., the action must be so completely schematized that the methodical ordering of an operational process is freed from all subjective inclinations of a merely individual sort. To convert an action into an operation entails, therefore, establishing a rule for how an initial state of affairs can be transformed into an intended final state, by means of an operation that can be repeated by any person at any time in the same manner and with the same result, as long as the rule is properly followed. This element of intersubjective repeatability is perfected by a technical operation in which the operation is repeatedly executed independently of any immediate human interference.

What does this methodological reification or objectification of an action into an operation mean for activities in which human beings "process information"? The transformation of an informing process activity into an information processing operation has as a precondition that its specifically human character of passing on information be dispensed with. How does this take place? In mental activity the demand for schematizing takes the form of a demand for formalization. The formalization of an informing process has two parts: First, information must be reduced to data, because only as a meaning-free bits of data can information be designated for mechanical processing. Second, the processing of data can occur only on by means of algorithms – i.e., mathematical-logical procedures for the rearrangement of data in a finite number of steps.

It is not just a matter of finding a rule for determining some results by operational steps in every single case. The objects with which one operates must undergo an alteration as well. In consequence, the formalization of informing processes leads to an elimination of the dimension of meaning. What remains is a series of signals with rules for their rearrangement, so that the result of the rearrangement is independent of any meaning of the series. The computer has the capacity to simulate through its own physical and technical structures only such a meaning-free chain of symbols and the algorithms of their manipulation.

This production and rearrangement of a signal series according to rules which refer only to its structure and not to its meaning, can be called a calculus. That the validity and correctness of a calculus is independent of the interpretation of its signs is illustrated by a simple algebraic calculus. If one rearranges the series of signs

$$a + b = c$$

to

$$a = c - b$$

then the two remain equivalent no matter what the letters operated on might represent – be it playing pieces, number values, or terms.

In order to mechanize informing processes, we must be able to describe them as calculi in the sense of rearrangements of meaningless signs. But signs whose meaning can be dispensed with in the process of their rearrangement no longer possess the quality of informing. For a sign to have the ability "to inform" is necessarily tied to its ability "to mean." Signs become informative when they have a meaning for the actions of someone using or encountering them. Although reference is made only to the syntactic structures in a calculation operation, the real informing processes are always associated with the semantic and pragmatic dimension of the signs: the pragmatics of passing on information is its semantics.

Such considerations make it clear that the only informing process which can be delegated to the computer is one abstracted from the pragmatic and semantic dimension. Certainly human beings try to augment their information processing with the computer. Yet a computer delivers only data. Only human beings are capable of turning data into meaningful signs, i.e., of transforming data into information in the true sense. Human beings inform themselves; computers re-order meaningless strings of symbols according to rules. This being the case, a data processing system fails to be a technics for informing, but is instead no more than an instrument for performing mechanical operations on symbols. Only so far as mental activity is reducible to a pure, schematic, thoughtless rearrangement of strings of symbols, does mental work prove able to be mechanized.

The only connection between the operations of the computer and the action of human beings informing themselves is the use of symbols. Indeed, the use of symbols is *sine qua non* for all thinking. But – and this is a point which can only be stated, not developed – thinking and informing are forms of symbolic *action*, whereas all the computer does is simulate a symbolic *operation* describable as a calculus. The difference between the classical machine and modern calculating technics is not simply that today information as well as matter and energy can be processed. More important is the fact that matter and energy transformation belongs to the category of material production, whereas re-ordering a series of signals belongs to the category of symbolic production. This is not the place to develop the relationship between material and symbolic activities, the latter of which can but do not have to represent material activities. The present argument is enough to justify discarding the unreflective use of language by information theory in favor of the more accurate description of the computer as an instrument for performing mechanical operations on symbols.

To summarize at this point, the term "information" exhibits two peculiarities in classical information theory: (a) It is overly technological in character. Technical results retain a paradigmatic influence on the construction of concepts. In information theory "information" is associated with a general model for communication which is in fact a model for mechanically transferring signal structures. In information technology itself, furthermore, information takes on the form of a technical operation on symbols that can be turned over to a machine. (b) It is overly formalized in character. Processes which are unquestionably preconditions for any act of informing are taken into consideration only insofar as they can be formally reconstructed – that is, only insofar as rules of operation can be abstracted from the meaning of symbols so that they can be manipulated independently of the distinct content to which they refer.

III

With such a concept of information, atrophied by its overly technical and formalized character, the informing process cannot suitably be analyzed.

But before outlining an alternative conception of information, it is possible to follow up on another aspect of the technical concept of information. *The term "information" as it occurs in the language of information measurement and information technology is not a scientific concept, but belongs instead to the sphere of mythical consciousness.*

What is "myth"? The issue here does not require a general analysis of mythical thinking. One aspect of mythical consciousness, as explored by Ernst Cassirer in some detail, is enough to orient the present argument. Mythical thinking differs from scientific thinking in that with myth there is no distinction between the sign and the object signified. For mythical thinking, a sign does not represent some independent content lying outside the sign itself, but rather itself participates in that content. There is an immediate identification of the physical substrate of a sign with its non-physical meaning.

An example can perhaps clarify the point. The extreme unwillingness of many non-literate peoples to allow themselves to be photographed is connected to their magical interpretation of pictures. In their mythical imagination the picture of a human being is endowed with the life of the person whose image it contains. And if one wishes to harm an individual, it is sufficient to injure the picture, and the individual will suffer sickness or death. The destruction of the picture is not considered merely a symbolic act, but rather as having the same immediate effectiveness as a direct use of personal violence. If a person allows someone to take his picture, he has thus given himself into another person's hands. Scientific thinking begins when this fusion or archetype and copy is broken, when content is separated from symbol. At the birth of philosophy and science during the transition from *mythos* to *logos* among the Greeks, Plato's doctrine of ideas served as a powerful historical dividing line, precisely because it affirmed this separation in uncompromising terms.

What follows with regard to the mythical character of information measurement and information technology? For information theory, the formal manipulation of signal structures has the reality of an informing process in the sense of either their maintenance or transformation. If we start with the idea that an act of informing is always tied to the use of meaningful signs, the designation of "information processing" as the manipulation of meaningless signs can mean only this: that their "meaninglessness" consists in the syntactical, formal structures claiming meaningfulness all by themselves. With this the distinction between sign and object signified, as something that exists independently of the sign and to which the sign refers, disappears. In its place there is a mythical lack of distinction which sets up operations with formal signs as a whole informing process. The thought of the representation of information content by a sign is given up to the thought of the presentation, in which the sign, especially its structure, "is" the information. This is exactly the mechanism which is fundamental to the construction of mythical consciousness.

The fact that the concept of information carries with it certain mythical features also implies that information theory functions as an ideology. We should, for instance, ask ourselves: What social processes does such an interpretation support? At the very least, the following analogy is suggested: Just as the value of all goods and services in a capitalist society can be measured only by means of a labor-money exchange rate which abstracts from inherent meanings and is governed by technology, so something similar happens with information. All the special meanings of signs are put aside when measured against the ability to be coded – that is, exchanged.

IV

The final topic of this paper is not a complete alternative theory, but rather some preliminary ideas to indicate a path for further thinking about information. One can distinguish three kinds of human activity, each involving a different relationship between ends and means.

(a) Personal interacting – the purpose of which is the preservation or transformation of social systems, and this means membership in some social class or a social role.

(b) Material producing – which aims at preservation or transformation of the natural and artificial environment, with the means being technology.

(c) Mental informing activities – with the purpose being the preservation or alteration of symbolic worlds as given to us in myth, art, and science, by means of the use of signs and systems signs.

As a mode of symbolic action, informing processes are fundamental for all activities in which human beings create symbolic forms and orient themselves by means of symbolic representations. To this domain belongs nothing less than thinking itself. Experience involves, first of all, sensory observation and perception and, second, communication – which in turn entails coming to terms with or really understanding both human beings and texts. The conditions for the possibility and validity of informing processes and/or the passing on of information prove themselves to be the conditions for the possibility and validity of experience and thinking. Three such conditions for the possibility of all informing processes are: the *a priori* of action, the *a priori* of a common set of signs, and the *a priori* of history.

Concerning the a priori of Action

Personal activity on the part of an information processing subject is an important precondition for his ability to inform himself. An action-oriented determination of information reverses the empiricist description of human consciousness as a *tabula rasa* on which the world is supposed to engrave its representation. The receiver of information is not a passive medium over against an active and stimulating environment, but rather receives information only as someone who pursues purposes, is intentionally directed, and tries to settle issues.

The genesis of information can typically be traced to the disruption of a practical action sequence. When an "open situation" arises, where different alternatives appear for the continuation of an action, the active subject must make a decision about which is the most favorable option. Here the orienting function of information becomes operative. For one can interpret the informing process symbolically instead of as a social interaction or technical activity – i.e., as a "test case for action" in the medium of a world of reversible symbols, a world in which the irreversibility of time, though valid for actions in the external world, is removed. The situation of the subject taking in information, becomes a filter which completely selects what will become meaningful signs. This has important implications for the theory of cognition. Every starting point in cognitive research that begins not from the world of objects, but from the subject, can be traced to Kant, and accepts his theory of the existence of *a priori* forms of cognition. This *a priori* shows itself here as a starting point of action: "The world" becomes for humanity an "active existence"; what the world consists of is discovered through action. The radius of our action determines what enters into the circle of meaning.

Concerning the a priori of a Common Set of Signs

For informing to be possible, there must be a common set of signs for the sender and receiver. In contrast to the model of classical information theory, this set of signs does not refer simply to the alphabet of conventional signs and rules for their reliable transformation. Much more is it a matter of the irreducible *a priori* of the historically unfolded horizon of a colloquial language, which unifies a group of people as participants in a linguistic practice common to their daily lives. The "methodological solipsism" which is fundamental to logical empiricism no less than to Cartesian rationalism, and from which is derived the fantasy that some rational individual acting alone could recognize a thing for what it is, is avoided and limited by the necessity of the common set of signs as a condition for the informing process.

The main problem with the successful passing on of information is not the formal transformation of signal structures, but a coming to terms with "really understanding." Making oneself understood by and coming to understand another is only possible when there exists a common set of signs, a language, while at the same time these efforts are only necessary when the signs or language can be interpreted in different ways. If signs used in the informing process are not constructed as an artificial or formal language, a plurality of possible meanings will obtain. This is rooted in the pragmatic dimension of all symbol semantics. If the meanings of the signs are given less in the representations of facts than in the possibilities for actions by people in relation to those facts, then the indeterminateness of the symbolic meanings will correspond to the principle of indeterminateness of the daily practice of a society. The meaning of a symbolic action is the ensemble of possibilities for continuing the symbolic action in non-symbolic practical actions.

Concerning the a priori of a Historical Process

The receiver who is oriented toward a specific action when processing information must have at his disposal the capacity for individual memory formation. Because of this a self-repeating act of informing or passing on information is not possible – a trivial fact which is nevertheless entirely outside the scope of classical information theory. A human being cannot have the same experience twice. There is an intrinsicly ineradicable element of singularity for the active subject in every informing process. For calculation and computer operations the opposite is fundamental: there dominates the principle of unlimited repetition. For only if an event can be repeated as often as desired according to the same schema, is it able to be mechanically simulated.

This raises the question of whether an information storage system could ever be a technical simulation of memory formation as a form of the historical process. The individual human memory is not a simple storage room for fixed information, in which the past is formed as a copy that can be called up into the light of consciousness at the moment it is needed. Human memory exists only in the capacity for being constantly reconstructed. What is reconstructed is determined less by established meanings of past information than by the present problem and action context which sheds its light on the past. As a result, the meaning of "what was" constantly changes. This re-organization of information collected in the past by the re-constructing activity of consciousness according to a current information situation is an important distinction between computer information storage and human memory. The historical process of the receiving subject means, then, that information gathered in the past is liable to a constant alteration and transformation in meaning.

Marburg University

AUTHOR: Sybille Kramer-Friedrich studied history, philosophy and political science at the University of Marburg, where she did her dissertation on the philosophical problem of information theory. She has also published *Technik, Gesellschaft und Natur: Versuch über ihren Zusammenhang* (Frankfurt: Campus Verlag, 1982).

REFERENCES

Cassirer, Ernst. *The Philosophy of Symbolic Forms*, vol. 2: *Mythical Thought*. Trans. Ralph Manheim. New Haven: Yale Univ. Press, 1955.

Cherry, Colin. *On Human Communication: A Review, a Survey, and a Criticism*. 3rd edition. Cambridge, MA: MIT Press, 1978.

Shannon, Claude E. and Warren Weaver. *The Mathematical Theory of Communication*. Urbana: Univ. of Illinois Press, 1949.

PAUL LEVINSON

INFORMATION TECHNOLOGIES AS
VEHICLES OF EVOLUTION

ABSTRACT. All evolution can be viewed as the progressive development of embodied knowledge. Such cognitive evolution takes on a new form when technology is directed toward the acquisition of knowledge. This paper examines the contributions of four kinds of information technology – perception extenders (microscopes, telescopes); cognitive processing enhancers (computers); disseminators of information via high levels of abstraction (written languages); and disseminators of information via high levels of realism (photographs and televisions.)

Evolution, or the adaption of organisms to environments, is a process of knowledge accumulation, a development of behaviors which reflect an understanding of the environment at least accurate enough for survival. Assuming that the world is not an illusion, and the life we see thriving on this planet is indeed alive and kicking, we are obliged to conclude that living organisms possess some minimal degree of knowledge of the external realities they inhabit. A mountain goat unable to distinguish air from rock would very quickly fall off the mountain.

The arrival of human beings has improved this knowledge accumulation process in two unique, related ways: Unlike other species, we are consciously aware of our knowledge, and thus able to pursue it deliberately; further, we purposely embody knowledge in material technologies that reshape the world. Of course, all organisms are in the business of remaking the world, if only in the subliminal sense of redistributing particles of living and nonliving matter from one place to another. But in human technology this organic contribution to the world has reached such proportions as to become a primary constituent of the world, and, indeed, this planet is increasingly becoming a product of human knowledge rather than natural selection.

Since human understanding and its material expression in technology are consequences of the natural selection process that resulted in humans, this humanization of our part of the universe through technology can be seen as the process of evolution transcending itself. Like the goat, humans also know the difference between air and rock; but, unlike the goat, we build bridges between mountains so as not to fall off, or leap off the mountains completely in jet planes and space ships.

Of course, we must have knowledge of the external world in order to

29

Carl Mitcham and Alois Huning (eds.), Philosophy and Technology II, 29–47.

construct such technologies, and technologies are essential to this pursuit of knowledge also. Some information technologies, such as (1) telescopes and microscopes, seek to improve human knowledge by extending our perception of external reality; others, such as (2) computers, speed up our cognitive faculties themselves, allowing for rapid digestion of increased quantities of data; still others, such as (3) typewriters, books, and again computers, contribute to the growth of knowledge by widely disseminating abstract ideas and thus exposing them to multiple sources of criticism and testing; while still others, such as (4) photographs and electronic recordings, contribute by disseminating highly accurate replications of the images and sounds of life and reality.

This essay will examine some of the consequences, intended or unintended, of such information technologies for the evolution of knowledge and, by extension, the world and the universe in which human knowers reside. Human knowledge embodied in technology is the spearhead in a profound alteration of a hitherto blindly evolving, unforeseeing universe, a self-transcendence in which the undirected cosmos is beginning to turn itself inside out into a planned and deliberate existence. From this perspective, the radio telescope and electron microscope, the typewriter, photo copier, and of course, the computer – devices which help generate and disseminate knowledge, which when embodied in other devices begin to reshape the universe – are the cutting edges of cosmic evolution.

I. TECHNOLOGY AND EVOLUTIONARY EPISTEMOLOGY

Before considering the contribution of information technologies to the evolution of human knowledge, let us examine in a bit more detail the relationship between knowledge and evolution. The assessment of knowledge from an evolutionary standpoint, and of evolution from the standpoint of knowledge, has occupied such thinkers as Konrad Lorenz, Karl Popper and, most of all, American psychologist Donald T. Campbell, who has termed this approach "evolutionary epistemology," and has compiled bibliographies of more than 150 studies in this field.[1]

One of Campbell's chief contributions has been his depiction of evolution as the development of increasingly vicarious, and thus error-prone but physically safe, methods of obtaining knowledge about the environment. The primitive amoeba, for example, acquires all information about its environment by swimming through it or bumping into it – a highly accurate yet dangerous technique that obliges the amoeba to come into physical contact with everything it knows. Vision, hearing and the modes of perception of more advanced organisms function as substitutes for physical contact in the acquisition of knowledge, at once opening the door to a greater number of false reports (amoebas do not suffer optical illusions), while bestowing on their possessors a distance from the perceived object which increases maneuverability and safety. Campbell's analysis ends with the human deployment of thinking, culture and science as similarly operating proxies for perception in the attainment of knowledge, but – as we shortly shall see – technologies which extend human perception to remote parts of the universe and the subatomic realm are equivalently crowning achievements in the evolution of increasingly vicarious knowledge receptors.

In addition to such literally biological functions of knowledge, evolutionary epistemology posits various analogies between the growth of living organisms and the growth of human ideas. Again Campbell provides a compelling example, portraying the evolution of life and knowledge alike as an indirect, Darwinian, three-part process.[2]

First, both new organisms and new ideas are initially generated independently of the environment in which they will perform or be applied. This separation of creation from environmental requirements is non-Lamarckian in biological terms, and non-inductive in terms of the acquisition of knowledge.

Second, biological and cognitive creations are both subjected to a confrontation with external reality, surviving only if they evince some degree of compatibility with the outside world. This winnowing process of natural selection is carried out by criticism and testing in the realm of science and human knowledge.

Finally, the survivors of the second stage must be retained in the bodies of life and knowledge, respectively, if the processes which created and selected these survivors are to be of any value. This objective is accomplished through mechanisms of heredity in the biological realm, and by media of dissemination in the world of knowledge.

Information technologies contribute to all three stages of the knowledge process, and may be assessed as to where in the three-stage process – generation, criticism, or dissemination – they make their most fundamental contribution. Technology thus plays an important role both in the direct biological acquisition of knowledge (as an evolutionarily advanced substitute for in-person perception) and in the biological-like or evolutionary way in which knowledge itself develops.

II. TECHNOLOGY AS THE EMBODIMENT OF IDEAS

Technologies purposely applied to the pursuit of knowledge are part of a larger class, technology in general. An examination of the technological contribution to the evolution of knowledge should therefore begin with a look at how *all* technology, informationally oriented or otherwise, contributes to the growth of knowledge. The epistemic import of even a toothpick is significant, and becomes apparent when one inquires into the ontological status of technology.

What is technology? Technology is something very special in the world. It is not like the clouds or the trees (unless the trees have been informationally planted), the natural realm of non-living and living material which humans had no hand in creating. Nor is technology the same as knowledge in our heads, human creations which, though situated in the material substrates of human brains, enjoy no material realization in the world. To have knowledge of an automobile is not to have a miniature nuts-and-bolds car literally in one's head.

Without technology, all the world (and perhaps the universe) is either material and non-human, or human and unmaterialized. Human beings themselves are of course the exception, and in the marvelous stretch of synapse from the brain to the hand we begin to extend that exception, rearranging the external world bit by bit to human specification with each technology we construct, giving material expression to substanceless intention and thereby increasing the net intended content of the universe. Technology is thus material and mental, real atoms and molecules reconfigured by the magic wand of the human brain and its amanuensis, conceptions brought to material life.

This technological intermingling of mind and matter provides important lessons for philosophy and some of its oldest problems. The active intrusion of human ideas into the material world disputes one-sided empiricist and more recent behaviorist conceptions of the brain as a mere passive recipient of the world's rules, while the technological presupposition of an already existing external world to be rearranged to human specifications contradicts wholly subjectivist or idealist views of reality as mere products of the mind with no independent existence. Strengthened and indeed literally substantiated by technology are interactionist models such as Kant's which hold knowledge to be a representation of external reality shaped by the contours of our intellect. (Evolutionary theory also supports interactionist perspectives, requiring an external reality that acts upon mindless organisms and thus must be independent of mind, and accounting for innate knowledge – "mind" – as a genetically transmitted adaptation to prior environments.)[3]

But far more crucial than the technological contribution to philosophies of knowledge is the contribution technology makes to the growth of knowledge itself. By embodying one or more human ideas, every technology provides a permanent record of this knowledge, an expression of human thought far more durable and, therefore, accessible than any immaterial conception. An automobile is thus a compendium of materially embodied ideas about combustion, alloys, hydraulics and, of course, travel, a quite literal incorporation of numerous theories in glass, rubber and metal parts. Stepping back a bit, one can see that all the technologies ever produced by humans – from cave paintings to pyramids to pushcarts – constitute one huge, open-stack, unintended library of human knowledge throughout the ages – a panhistorical, transcultural Library of Congress accessible to anyone who recognizes what it is.

Despite the encyclopedic scope, however, this unintended library is quite selective in its holdings, admitting only those volumes whose embodied ideas have some degree of compatibility with external reality. Not included in this library are automotive designs whose material embodiments exploded in the laboratory, flying machines that never got off the ground or crashed shortly after take-off. In terms of the three stages in the development of knowledge mentioned earlier, any technology that ever worked is an example of an idea that has survived the strongest criticism possible: a confrontation with material reality.

In some cases, moreover, technologies not only embody knowledge in their material structures, but are deliberately designed to help generate and disseminate other knowledge by means of their operation. A telescope, for example, not only incorporates theories of optics, lens-making, etc., in its physical components, but helps increase our knowledge of the cosmos when these physical components are pointed at the stars. Insofar as such technologies give a "double assist' to human cognition, with knowledge itself rather than, say, teeth cleaning or transportation being their functional purpose, they may be considered "meta-cognitive" technologies, or artifacts of cognition whose purpose is to further cognition. And, as should be obvious, these information-intended technologies have played a much more active – indeed, explosive – role in the growth of human knowledge than the more passive, record-keeping products of technology in general.

1. Telescopes, Microscopes, and Extenders of Experience

According to Kant, knowledge results from the digestion of external experience by human cognitive capacities. The absence of either component, external or internal, renders knowledge acquisition impossible. Unconditioned experience of the external world – even were it possible – would be utterly unintelligible without a cognitive filter to sort it out; and the operation of cognitive faculties on something other than external experience becomes a self-reflexive, chimerical exercise, much like mirrors directed at mirrors, or a video camera focused on a screen displaying an image produced by the video camera. Technologies such as telescopes and microscopes help the generation of knowledge by augmenting the external experience component – by increasing our food for thought, literally, through extension of our awareness to areas which, on the basis of our naked senses, were beyond our capacity to experience and thus comprehend.

In some instances the extension is more radical than in others. The telescope brings us into better contact with the stars and galaxies which, in a limited, highly distorted way, are already perceptible to the naked eye, whereas the microscope discloses a micro-organic realm utterly invisible to unaided observation. In all cases of such technological extension, however, events whose size or distance from the human observer places them beyond the effective pale of human perception are cast in dimensions accessible to our senses and thus potentially digestible by our intellect.[4] The act of extension, then, is an act of transformation, in which aspects of the universe bordering on the infinite and the infinitesimal are transformed into human proportions. Or perhaps the transformation works the other way, perhaps it is we, who, with telescopes and microscopes in our eyes, are extended into the dimensions of the universe. Either way, the result is that aspects of existence to which we were formerly blind and deaf are opened to our scrutiny, packaged in a form that can be grasped by our intellect.[5] Telescopes and microscopes and other sense-extending technologies thus put the far and hidden reaches of the universe in our cognitive backyard. The "disproportion of man" to the cosmos which so awed Pascal is both revealed and repaired.[6]

A question inevitably arises as to what extent the observations of events through technology pertain to the events as they are, independent of the technology, and to what extent the events are artifacts or products of the technology. Aristotle's preference for the testimony of the naked senses over technology was probably warranted by the poor quality of experience-extending devices in ancient times (though Aristotle's view made life quite difficult for Galileo),[7] and, of course, all technologies, even the most advanced, necessarily function with some degree of noise or imperfection.

But the dichotomy of an artificial world served up by technologies versus a natural, real world provided by biological senses disappears when one recalls Kant's insistence that the world perceived through the senses is not and cannot be the world "as it is." The colors we see playing on a clear lake, for example, are by no means simply the colors of the lake: They are rather but a fraction of the lake, part of a much larger band of radiation filtered through the structures of our vision. Thus, our naked senses give us edited transcripts of reality, and in this light the naturally perceived world has no more ontological or epistemological legitimacy than the world revealed by technology. Both are representations or reconstructions of reality, and the fact that one is accomplished through living tissue and the other through glass and plastic confers no necessary advantage or proximity to reality on either.

Of course, our biological structures have been tested by interaction with external reality for millions of years, and thus one must credit them with at least the minimal degree of accuracy or correspondence to external reality necessary for survival. But there is no reason to suppose that, given the understanding we already have of our biological senses and the principles of perception, we are unable to build technologies that operate with at least equivalent accuracy in domains beyond the reach of our unaided perception.[8]

Granting the acceptability of our technologically engendered perceptions, however, we encounter yet another possible hazard arising from the technological augmentation of experience. Even here on Planet Earth, in everyday naked perception, we find many phenomena which elude our satisfactory understanding (e.g., magnetism) – we regularly experience far more than our cognition possibly can handle. Now when we add to this already crowded table the bounty of technological extension, do we not run the risk of overwhelming the capacities of our intellect, of placing so much food on the cognitive table that the whole structure collapses? This sensory overload would, indeed, be a serious problem, were it not for technologies specifically designed to boost our native powers of cognitive digestion, and turn the glut of experience into an advantage.

2. Computers and the Enhancement of Cognitive Processing

As just suggested, even at the dawn of humanity we probably experienced far more than our cognitive processes could handle. Numbers were an early technique for dealing with this problem (according to Alexander Marshack, a very early technique, perhaps 300 000 years old[9]). Counting the numbers of animals obtained in hunts allowed hunters to compare the success of their forays, *i.e.*, to process the raw results of the hunts into a pattern which might serve as a basis for improvement of hunting efficiency. In general, the abstraction of real events into numbers facilitates their classification and sorting, which, in turn, permits the cognitive manipulation of these events into generalities and theories. The invention of the place-holder, zero, was especially significant in this regard, permitting the development of increasingly complex computations that manipulated more and more real experiences in less and less time through multiplication, algebra and, eventually, calculus, the requisite of Newton's cosmology. (Indeed, the failure of the Roman Empire to achieve an industrial culture, despite its propensity for engineering and technological gadgetry, may have been due to the clumsiness of its number system, which, lacking the place-holder, was incapable even of multiplication.)

Computers do for numbers what numbers do for real experiences: They greatly speed the sorting and classification of numerical information. The result of this speeding-up of the speeding-up of experience-processing is that human cognitive capacities are given tractable summaries of huge quantities of experience, enabling us to get a cognitive fix on much broader and more detailed aspects of existence. Science has long sought unifying theories that tie together diverse facets of reality, but the limited quantities of experience capable of being processed by our unaided cognitive faculties have hampered this quest, and rendered such grand theories as have been produced more the result of bald speculation than empirical evidence. Computation of data at the speed of light changes this imbalance, and makes possible a big picture grounded in a big piece of external reality.

In this sense, then, the computer not only complements the telescope, but is analogous to it in its performance: The telescope reduces the dimensions of the universe to human proportions; the computer reduces the numerosity of the universe, the sheer quantity of vast expanse and myriad detail, to humanly graspable units. In predigesting vast amounts of experience, computers allow our cognitive faculties to bite off much larger portions of external reality than they could chew on their own.

The benefits of such expansion have already been felt in most areas of science.[10] But perhaps the most dramatic indication of the possible effect of computers on the growth of knowledge is suggested by the recollection that Isaac Newton spent the vast majority of his working time not in the generation of new ideas, but in the laborious, handwritten computation of his equations.[11] How might the course of science and history have run had Newton the services of even a small pocket calculator? What will result from geniuses of Newton's calibre free to devote the lion's share of their intellect to creation rather than computation?

The mention of pocket calculators, however, brings to the fore a fear, common among many educators, that computers may be *counter*-evolutionary or destructive of the human intellect, underminers of our native faculties of memory and arithmetic ability.[12] This concern, in one form or another, is actually very old, and was raised by Socrates, who in the *Phaedrus* warns that writing will result in the atrophy of memory and the end of learning through dialogue. (Fortunately or unfortunately for Socrates, his pupil, Plato, troubled to write this warning down.)

Such misgivings have about as much validity for computers as they have had for writing: Socrates was not incorrect, in the immediate sense, in that access to written or any recorded material may, indeed, dull the rote part of human memory, and one certainly cannot engage a written page in a two-way discussion. But of course he was wildly off the mark in the longer run, for writing gives us transpersonal, even transcultural memories, which have resulted in exponentially greater exchanges of information than ever could have occurred, or do occur, in purely oral cultures. In the case of computers, is not the loss of the ability to find square roots, or even to perform complex multiplications by hand, well worth the enormous gains in mathematical and communication power at the cutting edge of science and knowledge (gains which, incidentally, include the capacity for immediate written dialogue through print on computer screens)? And, even in the absurdly unlikely event that worse came to worse, with (a) no human able to do any mathematical task at all without the aid of computers, and (b) all computers and all knowledge of how to build computers somehow disappearing, is there any compelling reason to expect that humans, having developed our number system (and, indeed, many others) in far more primitive surroundings, would be unable to develop and learn the operations of mathematics again?

No evolutionary structure. whether technological or biological, is ever an unmixed blessing; rather, structures qualify as beneficial if their performances result in a net gain of some sort for the organisms they serve. The technological extension of sensory experience by telescopes and microscopes and of cognitive processing by computers performs such a service for human knowledge, covering both of the bases, external and internal, of knowledge generation. But the growth of knowledge, as has been seen, entails not only the initial generation of ideas or representations of reality, but their subsequent criticism (winnowing out) and dissemination.

Moreover, whereas the initial generation of an idea may be, and indeed at ultimate origin *must* be, a solitary activity, criticism and dissemination are social or communicative – dissemination by definition, criticism in practice – since the criticism of one's own ideas is at very least psychologically suspect. The technological contribution to the criticism and dissemination of knowledge thus takes a form different from the initial technological detection and processing of experience, the goal of technological communication of knowledge being not the production of representations of reality, but the transmission of such representations, regardless of their truthfulness (correspondence to reality) to as many people as possible as accurately as possible. These two objectives – plurality and accuracy – are not always compatible, and have been sought through two techniques, abstraction and replication.

3. Speech, Writing, and Communication Through Abstraction

Which medium offers more accurate transmission of information: speech or photography? This question cannot be answered without reference to the nature of the information or representation being transmitted. If we wished to to convey the size and ferocity of a lion seen lurking in the savannah yesterday, a photograph would obviously be more specifically accurate. On the other hand, if we wished to communicate an abstract concept, such as evolutionary epistemology, or the notion of "concept" itself, a photograph would not only be less preferable, but impossible – speech is the only one of the two media capable of transmitting information about such intangibles.

We thus get an inkling of the four-edged complexity of the communications environment: two classes of representations, of external reality and of abstractions, communicated by two techniques, replication and abstraction. Moreover, media vary not only in the accuracy of communication, in large part a function of the match between representation and technique, but in how far and wide from the source they are able to transmit their information.

How did such a multi-faceted communications system come to be? There are two things we can know with some assurance: (a) that external reality existed before abstract concepts as a possible object of representation (animals perceive external reality, but presumably do not create abstract concepts); and (b) that the highly abstracting medium of spoken language was the first major mode of human communication. Media thus seem to have arisen in a mismatch of abstract technique (speech) applied to transmission of representations of tangible, external reality, which suggests the following scenario of the origins and subsequent development of communications systems:

The initial function of speech was the transmission of information about external reality. In a technologically primitive world, lacking such niceties as tape recorders, cameras and even paper, the only way of transmitting representations of external reality, or transporting them from their source, was by translating them into abstract symbols (words) which were conveniently portable, but bore no natural connection whatsoever to the realities they ultimately described (the word *lion* bears no resemblance to any aspect of a real lion; onomatopoeia would be a trivial exception). In this early context, the distortion of abstraction, or sacrifice of accuracy, was the price paid for the transmission across space and time of any information at all about the tangible world.

Eventually, however, the abstract vehicle of speech encouraged the development of a whole new class of objects of communications: representations of abstractions themselves, entities entirely of human creation with no tangible existence in the external world to begin with. (Abstractions, of course, could have existed before language, but they would have served little purpose, being virtually incommunicable in a non-linguistic environment.) Although the word *epistemology* bears no closer resemblance to the concept of epistemology than the word *lion* does to the animal, there is no conceivably less distortive or more accurate way of communicating concepts (short of mental telepathy), and thus abstract media and abstract concepts have flourished in a mutually reinforcing relationship that has made abstraction both a central vehicle and object of human cognition. Indeed, the situation of abstract media – first speech and then speech and writing – being the only communication game in town, the only way of transmitting information abut external reality as well as abstractions, persisted until the invention of replicative photography in the 19th century.

If the above scenario is correct, then speech has had a doubly enabling role in the evolution of knowledge, and thus of humanity, initially providing the only means of communication about external reality, and eliciting a new humanly created realm of abstract concepts. This dual impact has been heightened by writing, especially the alphabet, which, as Harold Innis and Marshall McLuhan have pointed out, made possible many aspects of classic civilization, including the early rise of science and monotheism[13]; and later by the printing press, which as, again, Innis and McLuhan have stressed, transformed the medieval into the modern world by encouraging such developments as the Scientific Revolution, the Protestant Reformation, and the rise of national states.[14]

Writing improved upon speaking not only by providing permanency, which greatly increases the range of dissemination, but by eliminating the distortion that occurs every time a message is respoken: Written information, after all, need be transmitted but once to be received throughout all eternity. That writing is the last word in the communication of much of human knowledge is demonstrated in the growing presence, as indicated above, of printed exchanges on computer screens. Indeed, print-outs have always been an integral part of computers, as writing has been a part of the electronic revolution since its inception in the words on telegrams and the reading of books encouraged by electric lighting.[15]

But, as also suggested above, the abstract hegemony of speech, writing and print leaves the communication of much of the basis of human knowledge out in the cold; indeed, these media make the whole of external reality a second class citizen in communication. For the external world to receive its due, knowledge had to progress to the point where humans could invent a whole new genre of media, beginning with the development of photography 150 years ago.

4. Photography, Electricity, and the Replication of External Reality

Complete replication of an original is, in principle, impossible and self-defeating, since the duplicate obviously cannot capture, and, indeed, serves to destroy, the uniqueness of the original. The goal of replication is thus always partial, and usually amounts to the reproduction of salient aspects of the original as accurately as possible.

Pre-alphabetic pictographic systems made some attempt at replication of visual reality, but ultimately failed to survive, due to poor performance, both as replicators *and* disseminators. (Pictographic systems, which require a different symbol for each word or object recorded, are much more cumbersome to learn and use than phonetic alphabets such as English, which operate with 26 symbols. The survival of Chinese ideograms, which are part pictographic and part phonetic, is a partial exception.) Paintings, of course, also attempt to replicate the visual world, but the critical breakthrough came with the photograph and its offspring.

Photography is special because it is, to paraphrase André Bazin, free of the original sin of subjectivity[16] – that is, it captures the world without the inevitably abstracting intervention of human mentality present in all painting, and in all verbal and written descriptions of external reality. The aspects of the world conveyed in the immaculate conception of the photograph are thus as close to external reality as what can be seen of those aspects with our naked eyes or through telescopes or microscopes. Moreover, although the first photographs accurately captured only one feature of external reality, visual shape, subsequent developments in photography restored motion, synchronized sound, color, and, with the development of holography in 1949, the dimension of depth to the communicated representation.[17] This widening of the range of external reality whose representations are transmittable without abstraction has assisted not only the criticism and dissemination of these representations, but their very generation, for the life-like photographic image of external reality may serve as a substitute for the human perception of external reality in the initial encounter of the external world and human cognition that generates representations of knowledge of reality (the photographs produced by space probes would be one such case).

At the same time, the infusion of electricity (and electro-magnetic waves) into communication has resulted not only in the interactive, instantaneous, simultaneous transmission of printed information, but the immediate and simultaneous transmission to millions of people of nonabstract images and sounds. Indeed, with the hook-up of increasingly portable computer and video systems the time is rapidly approaching when anyone, at any hour, from any place in the world, will have instant access to all the knowledge possessed by humanity. This would obviously constitute a maximally optimal environment for criticism and dissemination of knowledge, and its generation as well.

Thus, in the past 150 years, the electro-chemical revolution has restored an empirical presence lacking in the communication of representations since the origin of communications and humanity in speech.

III. CONCLUSION: COGNITIVE EVOLUTION AS COSMIC EVOLUTION

In sum, then, technology contributes to the growth of human knowledge
in five distinct – though overlapping – ways:

- All technologies are material embodiments, and thus more or less
  durable records of ideas that have survived some test with external
  reality;
- Telescopes, microscopes, and similar technologies extend external
  experience, and thus the generation or knowledge in areas inaccessible to the naked senses;
- Computers enhance the operation of our internal cognitive faculties themselves, fostering the generation of knowledge from huge
  quantities of experience which might otherwise have overwhelmed
  our unaided mental capacities;
- Speech, writing and similar media facilitate the criticism and dissemination of knowledge by transmitting information or representations via abstraction, a process especially hospitable to the
  communication of abstract concepts, and perhaps even responsible for their existence; and
- Photographic and image-and-sound producing media transmit
  representations of external reality with little or no abstraction,
  enabling widespread criticism and dissemination of representations with very high degrees of correspondence to aspects of external reality as directly perceived by the senses.

The technological contribution of knowledge thus covers all necessary
cognitive bases, external and internal, abstract and sensory.

As suggested earlier, the technological contribution to, and to some
extent, constitution of, human knowledge also has consequences that go
well behind human knowledge *per se*, because human knowledge is
surely not a closed or self-contained system. To the contrary, human
knowledge or understanding and organization of the external environment appears to stand both at the summit of prior cosmic evolution and
the doorway to the future of the universe. Whether we think (with, for
example, Norbert Wiener) that human knowledge is ultimately but a
hopeless finger in the dyke against entropy,[18] or (with, for example, Ilya
Prigogine), that even non-living matter displays some of the order-
through-fluctuation characteristics of the knowledge process,[19] we can
see that knowledge has emerged from and transcended a much dumber,

undirected universe. When we further consider that the implementation of knowledge in technology quite literally and materially alters the external world, we can begin to see how existence may be increasingly shaped by the expression of knowledge in technology. Such cosmic implications of technology have been appreciated by several major philosophers, such as Marx and Heidegger, and by several minor figures, including Friedrich Dessauer,[20] but have generally been unaddressed by the intellectual currents of our time.

The decisive effect of human knowledge and technology is that, although they themselves are surely not free of unintended consequences and encouters with unforeseen elements of the environment, they inject an element of direction, deliberation, planning into a naturally selective universe which presumably previously had none. Human control, of course, does not mean that natural selection or evolution will stop or end, but rather that it will be humanly proposed and derived.[21]

The situation seems much like a freeze frame in a motion picture, where, although the film keeps moving through the projector and the motion picture process continues, one image appears in all the moving frames and thus looks frozen on the screen. In the technological projection on the cosmic screen, that image will be human. We might say that, through the expression of human knowledge in technology, the future of the universe lies in the human mind, and thus in the information technologies that assist it.

Fairleigh Dickinson University

AUTHOR: Paul Levinson was born in New York City in 1947 and received his Ph.D. from New York University in media theory in 1979. He is associate professor of communications at Fairleigh Dickinson University, adjunct professor in media studies at the New School of Social Research, and a member of the computer conferencing faculty of the Western Behavioral Sciences Institute. He has published widely in journals devoted to media studies, edited *In Pursuit of Truth: Essays on the Philosophy of Karl Popper* (Atlantic Highlands, NJ: Humanities Press, 1982), and his *Mind at Large: Knowing in the Technological Age* is forthcoming.

NOTES

1   Campbell, "Evolutionary Epistemology" in P. A. Schilpp (ed.), *The Philosophy of Karl Popper* (La Salle, IL: Open Court, 1974), pp. 413–463; "Evolutionary Epistemology: Partial Supplementary Bibliography," distributed at the First International

Convocation of the Open Society and Its Friends, New York City, November 1982 (Chair: P. Levinson, Fairleigh Dickinson University, Teaneck, NJ); representative works by Lorenz and Popper: Lorenz, "Kant's Doctrine of the *a priori* in the Light of Contemporary Biology," D. T. Campbell, trans., in R. I. Evans (ed.), *Konrad Lorenz: The Man and His Ideas* (New York: Harcourt Brace Jovanovich, 1975), pp. 181–217, and Popper, *Objective Knowledge* (New York: Oxford, 1972).

2   "Unjustified Variation and Selective Rentention in Scientific Discovery" in F. J. Ayala and T. Dobzhansky (eds.), *Studies in the Philosophy of Biology* (Berkeley, CA: University of California, 1974), pp. 179–186.

3   The relevance of technology to traditional problems in philosophy is further explored in "What Technology Can Teach Philosophy" in P. Levinson (ed.), *In Pursuit of Truth: Essays on the Philosophy of Karl Popper* (Atlantic Highlands, NJ: Humanities, 1982), pp. 157–175. These and most of the other themes discussed in the present essay are more fully developed in P. Levinson, *Mind at Large: Knowing in the Technological Age* (forthcoming).

4   Such transformation or scaling of experience to intellect is an epistemological scaling, *i.e.*, an adjustment of the relationship between external reality and an individual's cognitive structures, and thus is distinct from questions of technology and *social* scale, *e.g.*, what types of technology are appropriate to various sized groups and levels of community, as discussed, for example, by Robert McGinn, "The Problem of Scale in Human Life: A Framework for Analysis" in *Research in Philosophy and Technology*, vol. 1 (1978), pp. 39–52; and by Stan Carpenter, "Scale in Technology: A Critique of Design Assumptions," *Research in Philosophy and Technology*, vol. 8 (1985), forthcoming. The two types of scale are not totally unrelated in that group dynamics influence what and how we know, and what we know affects the formation and operation of groups. For a discussion of the impact of pooled or group mentality on evolutionarily derived cognition, see Aharon Kantorovich, "The Collective *a priori* in Science," *Nature and System* (1983), 77–96; for a discussion of the impact that the technological construction of new types of groups and social complexes has on the knowledge process, see Chapter 8 of Levinson, *Mind at Large*.

5   The benefit is enjoyed by all aspects of our cognition, not only the purely intellectual. E. H. Gombrich, for example, in *The Sense of Order* (Ithaca, NY: Cornell University Press, 1979, p. 9) reminds us that single snowflakes are nondescript dabs of white until magnification calls forth their intricate latticework and beauty, *i.e.*, transforms them into proportions that can engage our aesthetic faculties.

6   Blaise Pascal, *Pensées*, H. F. Stewart, trans. (New York: Pantheon, 1950), pp. 18–31.

7   Aristotle held that, although natural and artificial processes were capable of error (*Physica*, II:8), one natural process, perception of primary qualities (*e.g.*, color) through the senses, "is never in error or admits the least possible amount of falsehood" (*de Anima* III:3). See also Paul Feyerabend, *Against Method* (London: New Left, 1975), Chapter 10.

8   Our ability to check the accuracy of technologically rendered perceptions is, however, limited by a sort of variant of the Meno paradox (we can discover only that which

we already know. *i.e.*, which we are able to recognize in an encounter; but if we already know what we are searching for, why search?) which can be put as follows: We construct technologies because our naked senses are insufficient or incapable of performing in certain areas; indeed, if our naked senses performed well in these areas, we would not invent the technologies in the first place; hence, our unaided faculties are incapable of checking the performance of technologies in just those areas in which the technologies were designed to perform, and such technological performance can be tested only through other technologies, *i.e.*, by recourse to the very activity or class of object whose epistemological status is at issue. (Of course, we can gauge the performance of a technology in an area accessible to our naked senses – *i.e.*, training a telescope on an object on the planet Earth – and if our senses corroborate the technological performance, we could suppose an equivalently accurate performance of this technology in a remote area.) For an argument on the unity of natural and technological ways of knowing, see Aharon Kantorovich, "Quarks: An Active Look at Matter," *Fundamenta Scientiae* III (1982), 297–319; Kantorovich defines knowing as a stripping away or surface of phenomenological aspect of experience to reveal what at the time is presumed to be an unchanging or conserved set of elements, whether one is attempting to separate falsity from truth via logic, peer past the haze at the heavens through a telescope, or discover a symmetry underlying the matter by causing particles to collide in quantum physics; the technological product in these cases is not the presumed truth or reality that we uncover – which is thought to exist prior to the technological intervention – but is rather the altered environment or technologically induced circumstances that permit access to the heretofore hidden reality.

9   Discussed in B. Rensberger, "The World's Oldest Works of Art," *The New York Times Magazine* (May 21, 1978), pp. 26–29ff. See Marshack, *The Roots of Civilization* (New York: McGraw-Hill, 1972), for examination of the development of mathematics in more recent periods of the Pleistocene.

10   See W. O. Baker *et al.*, "Computers and Research", in P. H. Abelson and A. L. Hammond (eds.), *Electronics: The Continuing Revolution* (Washington, DC: American Association for the Advancement of Science, 1977), pp. 56–61, for a preliminary survey.

11   Indeed, Newton was disdainful of a rival's claim to have discovered the law of gravity on the grounds that "Dr. [Robert] Hooke could not perform that which he pretended to: let him give demonstrations of it. I know he hath not geometry enough to do it," Frank E. Manuel, *Portrait of Isaac Newton* (Cambridge, MA: Belknap, 1968), pp. 152–422. For an indication of the importance of computer programs in the calculation of equations in current theoretical physics, and a discussion of the furor that erupted at the California Institute of Technology over the possible commercial exploitation of such a program, see G. Kolata, "Caltech Torn by Dispute Over Software," *Science* (May 27, 1983), pp. 932–34. See also Heinz Pagels, "Fires in Space" (a review of Kippenhahn's *100 Billion Stars* and Friedman's *Foundations of Space-Time Theories*), *The New York Times Book Review* (August 21, 1983), pp. 9, 18, who explains that, although the laws governing physical processes inside stars were

known in the late 1920s, "astrophysicists were hampered by the sheer complexity of
the mathematical equations that describe nuclear and thermodynamic interactions.
Then after World War II, high speed computers were built. Using computers, astrophysicists could manage the mathematics involved and make detailed models simu-
lating the interiors of stars."

12   J. W. Wyatt *et al.* report in "The Status of Hand-Held Calculator Use in School," *Phi
     Delta Kappan* (November 1979), pp. 217–18, that 43% of the teachers they surveyed
     thought that pocket calculators would diminish the memory and mathematical ability
     of children. The authors also report that more than 100 studies on the effects of pock-
     et calculators show no immediate adverse results (the technology was too recent for
     longitudinal studies).

13   Representative works: Innis, *Empire and Communications* (Toronto: University of
     Toronto, 1950/1972); *The Bias of Communication* (Toronto: University of Toronto,
     1951/1964); McLuhan, *The Gutenberg Galaxy* (New York: Mentor, 1962); *Under-
     standing Media* (New York: Mentor, 1964). See also Eric Havelock, *Preface to Plato*
     (Cambridge, MA: Harvard University Press, 1963), and *The Literate Revolution in
     Greece and Its Cultural Consequences* (Princeton: Princeton University Press, 1982).

14   See also E. Eisenstein, *The Printing Press As an Agent of Change* (New York: Cam-
     bridge University Press, 1979).

15   David de Hahn suggests that "electric lighting did more to improve the habit of read-
     ing books than anything before it," *Antique Household Gadgets and Appliances*
     (Woodbury, NY: Barron's, 1977), p. 121.

16   H. Gray (trans. and ed.), *What is Cinema?* (Berkeley, CA: University of California,
     1971), pp. 12–14.

17   The evolution of communications media toward fuller and more accurate replication
     of "pre-technological" reality is traced in detail in Levinson, "Human Replay: A
     Theory of the Evolution of Media," Ph.D. dissertation, New York University, 1979.

18   *E.g.*, Norbert Wiener, *The Human Use of Human Beings* (New York: Avon, 1950/
     57), p. 66.

19   *E.g.*, R.B. Tucker, "Interview with Ilya Prigogine," *Omni* (May 1983), pp. 84ff. See
     Erich Jantsch (ed.), *The Evolutionary Vision* (Boulder, CO: Westview Press, 1981),
     for a fuller discussion of these and related themes.

20   See Daniel Bell, "Technology, Nature, and Society" in *The Frontiers of Knowledge*
     (Garden City, NY: Doubleday, 1975), pp. 28–78, for a discussion of Marx in this
     regard; Michael Zimmerman, "Technological Culture and the End of Philosophy,"
     Research in Philosophy and Technology vol. 5 (1979), pp. 137–145, for an explana-
     tion of Heidegger on technology; and Friedrich Dessauer, *Philosophie der Technik:
     Das Problem der Realisierung* (Bonn: Cohen, 1927), Part II translated as "Technolo-
     gy in Its Proper Sphere," in C. Mitcham and R. Mackey (eds.), *Philosophy and Tech-
     nology* (New York: Free Press, 1972), pp. 317–334, 375–377. Marx Wartofsky argues
     in his "Critique of Impure Reason II," *Science, Technology and Human Values* **6,**
     whole no. 33 (Fall 1980), 5–23, that human reason is best understood in its "living"
     implementation in social and technological structures. See also Levinson, "What
     Technology Can Teach Philosophy," and *Mind at Large*.

21 My point here is that we have no way of conclusively predicting, let alone guaran-
teeing, that our rationally directed technological actions will produce the intended
results. Some observers – most spectacularly Jacques Ellul in *The Technological Society* (New York: Vintage, 1964) – have gone so far as to suggest that the very implementation of rational direction in technology is self-defeating in that technology entails structures of its own which undo the rational intention. My response would be
that the very fact that we are able, in this technological age, to consider such a possibility shows that it is not so – *i.e.*, rationality conveyed through typewriters, computers, air cargo, electric lights, and a myriad of other technologies is alive and well,
amply co-existing with machines, albeit as imperfect as always. I thus agree with Stan
Carpenter, *e.g.*, "The Cognitive Dimension of Technological Change", *Research in
Philosophy and Technology*, vol. 1 (1978), pp. 213–228, that, although we must be
ever vigilant for technologically engendered damage, our rationality is eminently up
to the task of attempting to build a better world through technology.

FRIEDRICH RAPP

THE THEORY-LADENNESS OF INFORMATION

ABSTRACT: The theory of information is subject not only to scientific and technical use, but also to an epistemological analysis of the logical relations obtaining within its tacitly presupposed conceptual background. The argument here attempts to elucidate the structure of this conceptual or theoretical background, especially as it applies to information technologies and computers.

> No man sees what things are,
> that knows not what they ought to be.[1]

From the host of philosophical questions raised by the progress of hardware and software in computer science I will only consider a small selection. My concern is with the *epistemological* status of information. The following points will be discussed in sequence: (1) The relation between philosophy and computer science, (2) specification of the subject, (3) artificial and natural languages, (4) the broader conceptual framework, (5) three examples, (6) description and value-judgments, (7) a systematic scheme, (8) access to the background knowledge.

I

A strict *computer scientist* might claim that a *philosopher* has no right to deal with the theoretical status of information. The scientist could argue that only he himself is in command of the knowledge and the methods for solving any theoretical and practical problems arising in this context. Clearly this is just another case of the age-old quarrel about the relation between science and philosophy. What in fact happens is that the expert in the field concentrates on solving the problems at hand. In doing this he relies on the tacit division of labor that obtains between the different scientific disciplines on the one hand and philosophy on the other. As a rule, the specialist is not prepared either by professional training or personal interest, to deal with metatheoretical normative issues and the epistemological presuppositions behind his work, i.e. with the *a priori* elements involved. Usually he becomes aware of philosophical questions only when confronted with obstacles in his work. Examples of such cases are the discussions about quantum theory, about the theory of relativity, and about artificial intelligence. Once the obstacles have been removed or

49

Carl Mitcham and Alois Huning (eds.), Philosophy and Technology II, 49–62.

circumvented, the specialist will go on with his scientific work and the philosophical questions disappear from the stage.

One of the reasons for the success of science consists in the strategy of concentrating only on solvable problems. From the whole range of questions it is possible to ask in a certain field, only those are qualified as being "scientific" which in principle allow of a solution within the accepted, categorial framework. Others are just dismissed. This is a clever and effective procedure. But philosophers insist on dealing with these neglected questions. They deliverately enter into the intricate and controversial questions left aside by the scientific disciplines. Questions of this type are the problem of induction, the ontological status of the laws of nature, or the (tacit) *a priori* assumptions which are involved in scientific theories.

In dealing with issues of this type, a philosopher has to be careful to avoid *a posteriori* statements that could be superseded by scientific findings. Following this advice, I shall deal with epistemological status of information only in general, abstract, and *a priori* terms. so that an empirical refutation is excluded *ex hypothesi*. But this focus does not imply that the issues to be dealt with here are just empty or irrelevant. In specifying the epistemological status of information I shall concentrate on conceptual analysis and show the theoretical elements involved. In usual scientific discourse these elements are not considered but just taken for granted. Since only *a priori* issues are concerned, my claims cannot be judged by reference to factual matters, as established by sensory perception, but must be evaluated in terms of their capacity to explain a conceptual issue, namely the meaning of information.

II

It may be worthwhile to delimit my topic *ex negativo*. I will not concern myself with the borderline between man and machines or attempt to specify which human traits (always) escape simulation by computers. Nor will I enter into the issues of problem-solving, heuristics, and the programming and simulation of creative activity. Surely, any *a priori* statement about the possibilities of future software or future technology must be futile. My topic is a more specific one. I shall discuss the theory-ladenness involved in any collection and processing of information.

There are basically two ways of doing this. One might concentrate on a *genetic* analysis and investigate the chronological sequences obtaining. Such a procedure would result in something like A. Koestler's popular book *The Act of Creation*[2] or J. Piaget's "epistémologie génétique."[3] I will not adopt this approach, which would necessarily lead into the borderland between philosophy and psychology. Instead I shall turn to *structural analysis*, as applied in the philosophy of science.[4] My concern is with the logical relations that obtain within the usually tacitly presupposed conceptual background, relevant for all types of information.

The aim is to draw attention to the usually unnoticed features relevant for any *meaningful* information. In his classical paper "The Mathematical Theory of Communication," C. Shannon dealt exclusively with the problem of reproducing at one point a message selected at another. Shannon said: "Frequently the messages have meaning; that is they refer to or are correlated according to some system with certain physical or conceptual entities. These semantic aspects of communication are irrelevant to the engineering problem."[5] Since he defines the message to be transmitted as a sequence of symbols selected from a set, it is absolutely irrelevant whether this message is loaded with meaning or whether it has no meaning at all. For the study of the physical aspect of communication processes in engineering and in the mathematical theory of communication it makes no difference whether a meaning is transported or not.

It is a contingent feature of history of science that problems of information are today raised and dealt with in terms of communication theory and data processing. The very progress of computer hardware and software makes it necessary to reflect upon the problem of information, since our capacity to handle information processes effectively has outpaced our theoretical understanding of the problems of meaning involved. After all, one can play the game of *physically* transmitting a piece of information without any knowledge of the *meaning* which is possibly transported. Two points, in particular, are of relevance here. Firstly, one must account for the fact that physical messages do not *eo ipso* and by themselves have meaning. It is only by some interpreter that they are taken to convey a certain meaning. Thus the receiver must be in possession of an appropriate code in order to translate the physical message into meaningful information. Furthermore, the decomposition of a message into isolated elements (bits), which has proved so effective for coping with the engineering problems of communication, suggests that with respect to meaning, also, information can be analysed into isolable, atomic, alternative choices. At first glance the notion of logical atomism, clearly refuted and outdated in epistemological terms, [6] would seem to offer a handy theory of information; but on second look this proves not entirely adequate. Put in speculative terms, and applying the terminology of Descartes, [7] what turns out successful with respect to the *res extensa* (matter) is tacitly transferred to the *res cognitans* (mind). Yet, what can be separated in the realm of inorganic (not in the organic!) matter cannot be separated in the realm of mind; there are elementary particles and separate states in physics, but there are no complete isolated elements of meaning.

III

Consider the mode of operation of systems transmitting and processing information. The manipulative capacity of hardware consists in bringing about specific states and processes in the physical world in terms of a given program. And these physical entities are taken to represent specific conceptual entities. The correlation becomes most evident when one considers the input and output of the system in question. In order not to consist of mere noise, the input and output must have a definite meaning; one can profit from the physical processes involved only when one can correlate them to a clear–cut conceptual pattern. Digital bits, analog processes, and the physical elements of calculation processes can only be identified and understood when they are interpreted in terms of a clearly specified, definite *artificial language*. This language is so constructed as to yield to a clear-cut and unique translation of the physical structure into conceptual patterns. The starting-point for constructing this language consists in the physical processes obtaining in the hardware in question. The conceptual meaning is deliberately shaped so as to cover what the hardware yield. I.e., to put this into speculative terms again, the structure of the mind is adapted to the structure of matter. In this way operativeness, uniqueness, and clarity are obtained. The price paid for this consists in quantitative and qualitative impoverishment. The artificial language introduced has only a very restricted number of terms, each of which has a specific, well-defined meaning.

With *natural languages* things are quite different. In everyday language, too, a physical representation of meaning obtains, for example by means of sound, or letters printed on paper. But the meaning of the words is not specified in a definite way and is not independent of the context. It is the world as described by ordinary language in which we are living and experiencing, and it is always a natural language which provides the basic understanding and the conceptual framework from which all kinds of more sophisticated, scientific investigations take their departure. And everything which is formulated in scientific terms must in the last analysis be retranslated in one way or another into this world of ordinary language. Hence the question arises of the relation between the artificial, computer language and the natural, ordinary, everyday language. Clearly, computer languages can fulfill their purpose only when their elements have a definite meaning, established once and for all and independent of the context under discussion. But at the same time this language is highly impoverished as compared with the richness – and also, of course, the ambiguity – of ordinary language. This is why highly specific operations within a clear–cut logical system can be done very effectively by means of the artificial computer languages. But these languages prove inadequate as soon as the complex wealth of meaning present in ordinary language comes into play. Thus it is not surprising that simulation of creative thinking and computer translation involve serious difficulties.

In the philosophical interpretation of these issues two opposing trends can be observed. With regard to scientific theories, logical positivism holds that any vagueness or ambiguity in meaning is a shortcoming. The idea is that such ambiguities have to be eliminated by means of a *logical reconstruction* so as to arrive at a precise, well-defined artificial language. In opposition to this, ordinary language philosophy regards the *use of language* as the ultimate point of reference. The claim is that there is no way of transcending the conceptual structures embedded in ordinary language which unavoidably define the cognitive range of our world. In the analysis put forward here, a middle road between these two approaches is followed. The method of structural analysis and of logical reconstruction is borrowed from logical positivism. And the basic idea of *the broader conceptual background* which supplies the meaning of the linguistic elements involved is taken from ordinary language philosophy. (Incidentally, the theory–ladenness of information discussed here suggests that by lowering the standards for an unequivocal correlation between physical and conceptual entities (biosimulation) computer techniques could be brought closer to the working capacity of ordinary language.)

IV

The problem of correlating physical and conceptual entities can be dealt with correctly only while considering the *broader conceptual framework* involved. What comes into play here is the tacitly presupposed and usually not explicitly specified theoretical background of knowledge and understanding which is present both in ordinary language and in the language of a specific scientific discipline. It is this background which specifies the meaning of the linguistic units in question. The notion of isolated and context-free elements of language holds good only with respect to artificial languages. It is distorting and misleading when applied to the conceptual elements of non-artificial languages. At this point the teachings of ordinary language philosophy and of the philosophy of science coincide with the epistemological understanding of the coherence theory of truth. In all of these approaches one arrives at the result that there is an ultimate frame of reference which makes up an holistic web of meaning that cannot reasonably be analyzed in terms of isolated elements. The frame of reference may consist in the ordinary use of linguistic terms, in the complete conceptual framework of a scientific theory (Duhem-Quine thesis), or in the coherent system of a body of consistent propositions.[8]

In fact, the meaning ascribed to any piece of information cannot be understood in isolation. It is always part of a larger system. There are merging borderlines between the systems in question, which are made up of different theories, various scientific disciplines, and, in the last analysis, of the entire web of knowledge and understanding exhibited in ordinary language. It is against a background knowledge of this type that the meaning of the terms of artificial language can be interpreted. Since every normal, informed person is in command of this knowledge, there is usually no need to be explicit about it. But as soon as one is confronted with the task of determining the meaning of a message which is only given in the form of a series of physical patterns and about which one has no other knowledge, the relevance of the conceptual background becomes evident. It would be a mistake to regard as non-existent what in fact is being tacitly presupposed and made use of.

V

This can be demonstrated by *three examples*.

(a) Let us consider the information gathered in some *physical experiment*. The real meaning of this information can be correctly understood only by an educated and trained expert in the field. This expert must know about the theories of the physical processes investigated and about the working principles of the measuring instruments and of the transmission and amplifier systems. Only in this case will he be able to give a correct interpretation when confronted with an unexpected reading. How complicated the matter is can be learned from premature news about experimental breakthroughs on the research front and from diverging interpretations of one and the same experiment. In the broader context of the history of science the theory-bound character of physical research has been discussed in great detail under the key concepts of "models" and "paradigm cases."[9] Indeed, it is only within the theoretical context thus established that a meaning can be ascribed to a transmitted message or that mere noise can be distinguished from relevant meaningful information.

The fact that in the natural sciences one is dealing with well-defined variables seems to contradict this analysis and to support the idea of isolated data. But in fact the variables of the natural sciences represent certain states and processes which are not isolated *per se*. The states and processes considered are always part of a larger physical environment. And it is only by means of highly sophisticated experimental procedures that they can be, as it were, physically dissected and thus investigated in their pure, idealized, isolated form. This pure form is not just prescribed by nature but defined by the very theory under consideration. Each physical experiment consists in applying technological means in order to translate the theoretically preconceived conceptual isolation into a real physical one.

Similar features obtain in other fields of scientific research. In any scientific discipline there exists a specific conceptual network which includes the relevant key concepts, the basic assumptions, the standard questions, and a range of admissable answers. This theoretical machinery is constructed in such a way as to allow certain segments of the universe to be singled out and then studied in detail. And it is by investigating such segments – and not just the inexhaustible totality – that scientific findings are made. In short, all scientific knowledge deals with specific, conceptually isolated aspects of the world. For example, the sickness of one and the same person can be analyzed in terms of sociology (unfavorable social conditions), psychology (disturbed mental equilibrium), physiology (malfunction), toxicology (noxious influences), etc.

(b) In the humanities the dependence of information on a specific conceptual framework is even more obvious. In order to be able to judge the information contained in a piece of *historical evidence*, say in a treaty, correctly, one needs full knowledge of the meaning and the technical functions of the words used, of the notions and principles of international law, of the (usually hidden) aims of the agents involved, of the image they wanted to establish in public, of the antecedent state, of the consequences ensuing, etc. There is no pure information about historical facts. What we have is an abundance of heterogeneous sources, i.e. relics from the past which acquire a meaning for us only when interpretated in terms of a certain heuristic theoretical framework, which provides criteria for singling out the relevant material and for distinguishing the important sources from the less important and irrelevant ones. This is why it is possible for the events of the past to receive a new interpretation in every age; each epoch views itself differently and ascribes a new meaning to occurrences that were formerly regarded otherwise.

(c) Similar features can be detected in the more ordinary processes of *promulgating news by the media*. The allegedly pure information about simple, isolated factual matters is in fact highly theory-laden. For this reason a correct understanding which is not liable to be manipulated by falsification is possible only when the receiver knows the theoretical background which is implicitly present in the wording applied. This becomes most evident when we consider the technology in use. Whereas a Western news agency may speak of "terrorists," an Eastern one will label the same people "freedom fighters" – or vice versa, if the country in question belongs to the other block. The verbs applied ("kill," "execute," "murder") are evidence of the theoretical approach from which the event is judged. In order to understand news appropriately it is not only necessary to decipher the terminology applied. One must also know the conceptual background, for instance the Western or the Eastern interpretation of freedom, civil rights, and democracy. It is this background which decides whether the actions in question are accepted (and hence described in positive terms) or rejected (with the use of pejorative words). In addition to this, the agency is free to produce completely different messages by deliberately adding or omitting information about the context involved. Only a sufficiently broad and balanced background knowledge makes it possible to decipher the real meaning of the news published.

VI

How does it come about that phenomena of nature can be interpreted in different ways, that history can be rewritten, and that one and the same piece of news can be taken as having different meanings? The answer is that there are different scientific paradigms, that every age has its own spirit, and that there are many conceptual backgrounds for interpreting political events. Since only one of these approaches can be adopted at a time, we have to ask on what terms a choice is made. What is decisive is the interest of the scientist, the scheme of values of the historian, or the opinion of the political reviewer. As a result, propositions which at first sight seem to consist of an impartial *description* of matters of fact, turn out to be dependent on the *value-judgments* of the speaker in question.

This raises an intricate question of the value-ladenness implied in any kind of knowledge or information. By means of a simple model it can be shown that in every case some sort of interest or scheme of values is necessarily involved. Consider a universe of discourse consisting of certain elements, their properties, and the relations obtaining between them. The interest taken or the scheme of values applied are indispensable means for singling out the relevant and important objects or features to be mentioned or investigated. A certain scheme for identifying the things to be dealt with is needed, since otherwise one would have a completely homogeneous and unspecified universe of objects that could be investigated, all of them having the same *a priori* relevance.[10] The features which are not picked out in positive terms are taken to be of minor importance or even irrelevant and for that reason they are not taken into consideration. Without such a filter or scheme for selection no statement could be made and no process of research could ever start, nor could it be followed up. In actual fact, in any stage of research in which a new and hitherto unexperienced situation arises, such a discrimination scheme is needed. As everyone knows, it is characteristic of the clever scientist that he has a good nose for the really important things.

What has been described is a model of rational choice. Usually the process of selection takes place tacitly and without conscious consideration. The person selecting may not even be aware of what he is doing. But as soon as one sets out to give a consistent analysis of the steps and the elements involved one will necessarily arrive at some reconstruction of the type outlined here.

Without going into the details of the value-ladenness involved in the process of attaining knowledge and communicating information,[11] two remarks may be appropriate:

(a) Evidently a scholar is *not completely free* to choose his subject of research and to apply criteria of relevance. He can pick up a certain discipline, a specific approach, a certain paradigm case, and a specific theme. But in any age, as regards scientific disciplines and the general spirit of the time, history has, as it were, already made a global choice for posterity. Neither the scholar nor the journalist or the man on the street is free to make a choice on the basis of caprice. There is always a given over-all frame of reference to which he must adjust. But within the frame thus established there is space for individual interest, values and preference, which makes up the scientific profile of a scholar or the personal attitude of a journalist. And of course, for the genius there is always the chance of breaking the existing paradigms by a fresh approach – and thus perhaps establishing a new one.

(b) One might be inclined to interpret the close connection that obtains between evaluation and scientific theories of information statements as a *holistic interaction* which precludes all analytical distinctions. This inseparability argument is often put forward with respect to allegedly unavoidable political bias, ideology, and prejudice. But if a completely holistic interaction were taken for granted it would be impossible to distinguish between the evaluative elements of a theory and the assumptions made concerning matters of fact. Clearly there is an interrelation, but not an inseparable one. In fact it is much more helpful to stick to the idea that not only on the level of statements, but also on the metatheoretical level of theory-formation and of phrasing a piece of information a *distinction between normative and descriptive components* can be made. True, in any specific case it will be difficult to identify precisely and to demarcate the normative and the descriptive elements involved. Within a certain range their exact borderline may be open to discussion, but on the whole the distinction is indispensable. [12] Only in this way will it be possible to bring out the otherwise tacit and perhaps even hidden evaluative presuppositions present in any type of theory or information. The lesson is that by its very nature human understanding involves evaluative elements. And in order to make these evident and open to criticism, they must be made explicit by being distinguished from statements about matters of fact which are open to empirical investigation.

VII

The relevant elements having been brought together, it may be appropriate to combine them into a *systematic scheme* as depicted in Figure 1. Put in general terms, the conveying of information involves at least three elements: a physical series of states p transmitting the message; the meaning m conveyed by the message; and the state of affairs s to which the meaning refers. These states of affairs may be facts in space and/or time, or conceptual relations; for reasons of simplicity only the former case will be discussed here. In terms of the usual distinction between matter and mind, only *m* belongs to the realm of the mind, whereas *p* and *s* are part of the realm of matter.

Yet, as pointed out above, this is only the tip of the iceberg, the larger part of which is, as it were, hidden in the sea of what is simply taken for granted. It is the very abundance of information customarily available today which demands consideration of the further elements involved. These additional elements constitute the *superstructure* relevant for any kind of meaning. The meaning *m* is in turn dependent on a specific theoretical context *c*, the general background of knowledge and understanding *b*, and the interest or scheme of values *i* behind them.

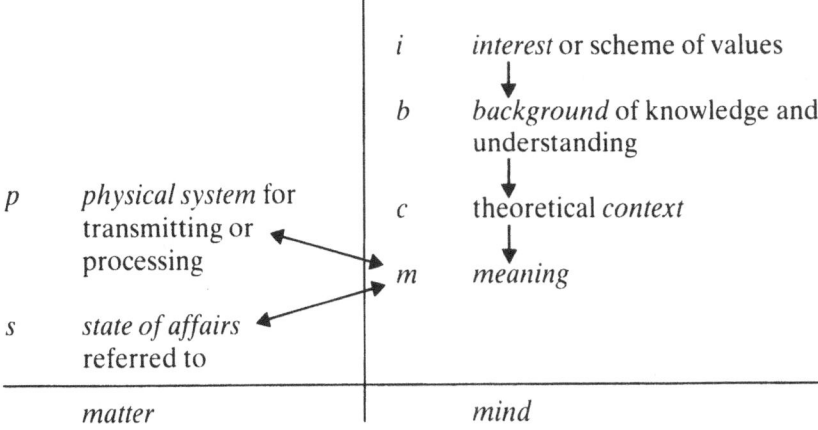

Fig. 1. Elements involved in information processes.

It is only in terms of the specific interest *i* combined with a certain background *b* that the attention of the knowing subject is directed toward a certain context *c;* and it is only in terms of this context that a meaning *m* can be ascribed to a certain state of affairs *s* which makes up the object investigated. The important point is not the specific method of classification suggested here, but rather the circumstance that in one way or another attention is drawn to the elements involved. If somebody does not accept the arguments of Section 6 above, he may apply another line of analysis and treat the background *b* and the interest *i* as a complex whole which defies further subdivision; in this case he would be led to distinguish between *m, c,* and ($b + i$).

Is the *background* of knowledge and understanding and the related pattern of *interest* and values *accessible to computer simulation*? This is a crucial question. As shown here, the usually implicit elements of the cognitive superstructure are indispensable for specifying the meaning attributed to a specific piece of information. If the artifical language to be established is to simulate a certain part of the knowledge contained in ordinary language, in one way or another the cognitive background must be represented in the computer.

In the introduction to the revised edition of *What Computers Can't Do*, H. Dreyfus has recently argued again against the chances for such a simulation.[13] Concerning the theory-ladenness of information and the decisive role of the background involved, I completely agree with Dreyfus. But the conclusions at which he arrives are exaggerated. It is true that our mode of being in the world is determined by a basic background of knowledge and understanding. It is also true that this comprehensive cognitive and cultural background of our life cannot be dealt with in the same way as other, more specific objects of knowledge. The reason is that in dealing with a specific piece of knowledge we necessarily make use of the ultimate cognitive and evaluative background, since it is only in terms of the background that relevance and meaning can be established. These connections have been analyzed by Heidegger in terms of *Sein* (Being) and *Seiendes* (being), and by Wittgenstein in terms of the transcendental function of the rules of the language game.[14] The result is that we cannot escape from our ultimate knowledge and understanding which yield the means for the way in which we cognitively deal with the world.

So far I agree with Dreyfus, and his great merit is to have drawn attention to this point. But I disagree with his sceptical and over-cautious conclusion. To use a metaphor, the ultimate background is like a pair of spectacles through which we see the world, and we are free to regard any object in the world without being troubled by these spectacles. So with respect to the simulation of any concrete and specific piece of cognition, in principle no problems should arise. We cannot reasonably speak about what is actually unknown. But this does not preclude us from turning to new horizons. Dreyfus' analysis does not demonstrate that there is a limit to the capacity of the human mind to take any subject-matter whatsoever as an object of investigation. We are even free to investigate our mode of being in the world. Otherwise investigations of this problem, including a large part of Dreyfus' book, could not have been written.

The point is that due to our finite capacities (no human being is an *intellectus infinitus*) we will never arrive at definite and final results concerning the ultimate foundations of our existence. We are always cast back into the circularity that obtains between the presuppositions made and the conclusions arrived at. But this restriction applies only to the final, uttermost elements. As soon as a certain starting-point for philosophical analysis is accepted and a concrete approach is chosen, the problem is already to some degree conceptualized and hence open to further research. It is in this way that, for example, Heidegger's *Time and Being*, Kant's *Critique of Pure Reason*, Spinoza's *Ethics*, Aristotle's *Metaphysics*, and Plato's dialogues were written. There is no unique way of conceptualizing ultimate philosophical foundations, but there are many ways of approaching this ideal.

Dortmund University

AUTHOR: Friedrich Rapp (born in 1932) studied physics and mathematics at the Technische Hochschule Darmstadt and philosophy at the University of Fribourg. From 1976 to 1985 he was Professor of Philosophy of Science at the Technical University of Berlin. His publications include *Contributions to a Philosophy of Technology: Studies in the Structure of Thinking in the Technological Sciences* (Boston: D. Reidel, 1974); *Analytische Technikphilosophie* (Freiburg: Alber, 1978), which was translated in English as *Analytical Philosophy of Technology* (Boston: D. Reidel, 1980) and into Spanish (1982); *Naturverständnis und Naturbeherrschung* (Munich: W. Fink, 1981), He also co-edited (with Paul Durbin) *Philosophy and Technology* (Boston: D. Reidel, 1983).

NOTES

1   J. Richardson, *Essay on the Theory of Painting* (1715); quoted from E. H. Gombrich, *Art and Illusion* (London: Phaidon, 1972), p. 10.

2   Arthur Koestler, *The Act of Creation* (London: Hutchinson, 1964); cf. B. Ghiselin (ed.), *The Creative Process* (Berkeley: Univ. of California Press, 1952).

3   Jean Piaget, *Principles of Genetic Epistemology* (London: Routledge, 1972); cf. H. Gruber (ed.), *The Essential Piaget – An Interpretative Reference and Guide* (London: Routledge, 1982).

4   Recent examples can be found in F. Suppe (ed.), *The Structure of Scientific Theories* (Urbana: Univ. of Illinois Press, 1974).

5   Claude E. Shannon, "The Mathematical Theory of Communication," *Bell System Technical Journal* **27** (1948), 379. This corresponds to the definition given in the *Glossary of Computing Technology* (New York: CCM, 1972), p. 109: "information: A set of symbols or an arrangement of hardware that designates one out of a finite number of alternatives, an aggregation of data which may or may not be organized."

6   Ludwig Wittgenstein, who had supported logical atomism in his *Tractatus Logico-Philosophicus* (London: Routledge, 1922), himself later criticized this notion most severely in *Philosophical Investigations* (Oxford: Blackwell, 1953).

7   René Descartes, *Principles of Philosophy* I, 48–52.

8   Wittgenstein, *Investigations* (as note 6 above); Willard van Orman Quine, *From a Logical Point of View* (Cambridge, MA: Harvard University Press, 1953); and Nicholas Rescher, *The Coherence Theory of Truth* (Oxford: Oxford Univ. Press, 1973).

9   Cf. note 4 above.

10 This notion is mentioned by Thomas S. Kuhn, *The Structure of Scientific Revolutions* (Chicago: University of Chicago Press, 1970), p. 15.

11 For a survey of the German discussion see H. Albert and E. Topitsch (eds.), *Werturteilsstreit* (Darmstadt: Wissenschaftliche Buchgesellschaft, 1979).

12 The crucial role of values and interest for human understanding has been discussed with quite different accentuation by Friedrich Nietzsche, *On the Genealogy of Morals*, trans. W. Kaufmann (New York: Vintage, 1973); Max Scheler, *Problems of a Sociology of Knowledge*, trans. K. W. Stikkers (London: Routledge, 1980); Jürgen Habermas, *Knowledge and Human Interests* (London: Heinemann, 1978). When taken in its broader meaning, H.-G. Gadamer, *Philosophical Hermeneutics* (Berkeley: University of California Press, 1978), can also be mentioned here.

13 Hubert L. Dreyfus, *What Computers Can't Do* (New York: Harper & Row, 1979), pp. 55–66.

14 M. Heidegger, *Being and Time*, trans. J. Macquarrie and E. Robinson (Oxford: Blackwell, 1967), and Wittgenstein, *Investigations* (note 6 above).

GÜNTER ROPOHL

INFORMATION DOES NOT MAKE SENSE
OR: THE RELEVANCE GAP IN INFORMATION
TECHNOLOGY AND ITS SOCIAL DANGERS

ABSTRACT. Argues that individual pieces of information cannot be mechanically synthesized into structures that make cognitive sense. Making sense out of masses of information is a human activity that is becoming progressively more difficult in an information technology world.

I

What is the human being? This is a perennial and fundamental issue of philosophy. Immanuel Kant considers the question to be the essence of three others: What can I know? What shall I do? What may I hope? Diverse answers have been given to the question of human nature, most of which stress some single feature and pretend it grasps the whole human being. There are classical "definitions" like *zoon logon echon* (the animal endowed with reason), *zoon politikon* (the animal depending on society), *homo faber* (the tool making being) and *homo ludens* (the being which plays). Modern social science have added *homo oeconomicus* (the being rationally maximizing benefits) and *homo sociologicus* (the being totally defined by social roles). Finally, in our own time, philosophical anthropology uses the definition "the acting being" (e.g. A. Gehlen, 1961).

None of these delimitations is wrong, of course, because each does identify a characteristic feature of the human being. Yet although each attribute contains some information about human nature, no single piece of information is sufficient to grasp the full sense [Sinn]* of what the human being essentially is. In fact, a human being cannot be reduced to one single trait, but must be understood as a synthesis of all the above, and possibly several other, characteristics. Not isolated information, but a significant combination of separate data is required to understand a complex phenomenon like the human being. Obviously, a single piece of information does not make sense. Sense only arises from a significant ordering of pieces of information.

Today the rise of information technology impels us to ask anew, What is the human being? And the treatment of information will turn out to be a crucial element in the development of "technological man." It will be

63

argued that the human mind needs "sense" not just "information" – the same way that human self-understanding, as indicated above, requires more than isolated data.

II

A recent model conceives of the human being in terms of a sociotechnical action system (Ropohl, 1979 and 1982). This model connects the main human characteristics inventoried in Section I above, especially the human capacities for reasoning, acting, and making tools for active use. The model grows out of philosophical anthropology as well as the ideas of cybernetics and general systems theory, and is further influenced by certain features of systems thinking in sociology.

But in contrast to the sociological use of systems theory, the action system is understood realistically – not just psychological roles, but corresponding to some empirical phenomenon. Such an action system functions at the individual (micro) institutional (meso), or social (macro) levels. These three levels of action systems form a hierarchy in the formal sense, each higher level being made up from systems of a lower level.

Acting is regarded as a function of the action system. A function may entail the transformation of matter, energy, and/or information in space and time according to a predetermined goal. Each general function may be further analysed into several subfunctions – executing (i.e. matter and energy transforming), information processing, and goal setting.

Corresponding to these subfunctions, one can assume that the action system is divided into subsystems which constitute its structure. It is important to understand that the subsystems, under conditions of advancing technicization, may be either human or technical. In the action of shaping material artifacts, the subfunctions of informing the material, tool guidance, and energy supply have, since the Industrial Revolution, been increasingly turned over to technical subsystems (tool, slide rest, power drive). For the subfunctions of materials handling, feeding, measuring, and controlling, however, the classical machine tool requires human subsystems. But through the application of information technology to automation and robotics, these subfunctions, too, can be performed by technical subsystems, and the human contribution to the manufacturing process is reduced to preparation, maintenance, and supervision. Thus the human action system is transformed into a man-machine system. Since the technical subsystems result from social labor and incorporate social skills and social knowledge, such an action system turns our to be a sociotechnical system even on the micro-level, not to speak of meso- and macro-levels.

As this discussion shows, the present stage of technical evolution is characterized by an introduction into the action system of technical subsystems which substitute, perfect, and amplify human information processing capacities. This is true not only for production and business administration, but to an increasing extend for every-day life as well. Every human action includes informational subfunctions. Now even those activities which had been supposed to be reserved to the human mind are taken over, at least in part, by technical devices like pocket calculators, personal computers, word processors, information retrieval systems, etc. the *zoon logon echon* is being transformed into an animal endowed with technical information systems.

This model of the sociotechnical action system, with one characteristic subfunction being the processing of information – and this subfunction being increasingly technicized – raises the question of how information should be conceived and whether information processing as performed by computers really is the same as the related human activity.

III

The concept of information is extremely ambiguous. Difficulties result from confounding its scientific with its common language meaning. But even when the common language meaning is put aside, there remain certain philosophical problems. Before dealing with one such problem, the scientific understanding of information needs to be made clear.

According to information theory, "information" is a strictly formal concept (Shannon and Weaver, 1949). For a given set of elements, each of them associated with a certain probability of occurrence, the information of any element is a mathematical function of that probability; the lower the probability, the greater the amount of information. Of course, this formal concept is usually interpreted by means of such empirical phenomena as signals. Signals are physical events, that is material and/or energy elements in time and space which are supposed to point beyond themselves. Thus, with information technology, information is a measure of the probability of the occurrence of signals such as electromagnetic waves, electric impulses, etc. Information in this narrow understanding, indeed, is devoid of content.

But "information," even in a scientific context, need not be restricted to this formal meaning. The key to an extended concept can be found in the character of signals not to exist for themselves, but to point toward something else. This is precisely the essence of signs in general. According to the theory of signs (Morris, 1938), every sign has three dimensions: (1) The syntactic dimension concerns the physical nature of the sign and its formal relations to other signs; obviously, it is this syntactic dimension which is covered by information theory in the narrow sense. The syntactic dimension, however, is nothing but the carrier of (b) semantic and (c) pragmatic dimensions. The semantic dimension refers to the denotation of the sign. The denotation is attributed to the sign by individual or social convention which has to be actualized each time the sign is used by a human being. Therefore, in addition to the relation between the sign and its denotation, a relation between the sign and its user must be taken into account, and this makes up the pragmatic dimension of the sign.

An extended concept of information may therefore be defined as follows: Information is a sign which (a) occurs with a certain probability (or frequency) within a sequence or arrangement of physical events, to which (b) a certain denotation may be attached, and which (c) may engage the behavior of its users in a particular way. Only the first determination (a) has as yet been formalized and quantified; theories of semantic and pragmatic information have so far escaped formalization. But from this, it cannot be concluded that information should be restricted to the syntactic dimension. It is not reasonable to accept as scientific only those ideas which have been completely quantified and formalized. And even in the special case, by definition the syntactic dimension of a sign cannot exist for itself. Insofar as any phenomenon is regarded as a sign, it is obviously implied that there exists a denotation and a user's engagement, even when, in the interests of formalization or technical realization, a certain project abstracts from the semantic and the pragmatic dimensions.

For instance, the central processing unit of a computer or pocket calculator really does not work on anything but signals. Nothing more is happening than a transformation of sequences which consist of trivial physical events like "current on" and "current off." From this, some scholars conclude that computers do not process information at all, and, therefore, that there is no functional equivalence between human and technical information processing. But this conclusion neglects both the essence of signs and the actual organization of computers. Signals are processed only because and insofar as they serve as potential carries of denotation. The computer would be incomplete if there were not appropriate mechanisms for assigning denotation to the signals, as is actually done not only through the so-called interpreters, but also through fixed storage devices and peripheral units. In fact, the user of a pocket calculator – to use a simple illustration – when pressing buttons, does deliver information to the machine, because each button has a definite denotation (a number, arithmetic operation, etc.); and when reading the result from the display, the user again receives information from the machine. The manipulation of mere signals applies to the central processing unit, it is true, but input and output units, in connection with the fixed interpreting program, allow for the semantic dimension as well.

The distinction between signal and information processing, although useful, is not a convincing reason to reject any man-machine equivalence as far as the semantic dimension is concerned. Real problems do not arise before turning to the pragmatic dimension, which covers the relations between the sign and its user. Even in this respect, the equivalence applies to a certain set of information processes. Whenever two or more technical information systems are linked, one system must function as the user of the signs delivered by the other and, in fact, the second will change its behavior according to certain commands from the first system. But whenever the user is a human being, the technical objectification of information comes up against fundamental limits, because what is defined as a relation between object and subject cannot abstract from that subject. Above all it is the intentionality of the human mind which is dependent on sense to deal with information.

IV

The word "sense" is somewhat ambiguous. "Sense," in the present context, should not be interpreted as a faculty of perception, "feeling," "practical reason," or "apprehension" – just to cite some common definitions. And above all, "sense" [*Sinn*] should be distinguished from "denotation" [*Bedeutung*].

It was Gottlob Frege (1892) who introduced the distinction between sense (in German *Sinn*) and reference or nominatum (*Bedeutung*) into the philosophy of logic. According to Frege, *Bedeutung* refers to an object denoted by a name; therefore, in English texts this term is usually translated as "reference," "nomination," or "denotation," *Sinn*, on the contrary, characterizes the specific manner in which that object is given. A famous example for this distinction is the pair of names "morning star" and "evening star." Both names have the same reference, since they both denote the planet Venus. But the sense of the one is distinct from the sense of the other, because the time of day when Venus is observed distinguishes them. Frege, to be sure, used such a distinction to deal with some subtle problems of logic; and among logicians it is controversial whether Frege's approach is convincing. But for present purposes Frege's distinction can nevertheless serve as a heuristic suggestion that the sense of complex designations is constituted by the very combination of independent pieces of information. In Frege's example, sense combines the pieces of information, one, about a heavenly body at a certain place in the sky and, two, about the time of day when this heavenly body is observed.

Turn to another example from the field of information technology. Imagine a data bank of the secret service of a fictitious government designed to store all available personal characteristics about certain "interesting" people. Assume that the system, being queried about a certain person, responds with the following data:

1. male
2. aged 28
3. unmarried
4. studies sociology
5. does not watch TV commercials
6. hobby photographer
7. frequently dines in a vegetarian restaurant
8. member of the local golf club
9. does not own a car, travels by bicycle
10. conscientious objector
11. purchases a number of Mozart records
12. does not subscribe to any newspaper
13. etc.

Now suppose the analyst, or an analyzing computer program, is focussed on the combination of the items 2 (under thirty), 4, 5, 6, 7, 9 and 10, insinuating that every person subsumable under this personality profile will most probably be a member of the ecological movement. Then the investigative system will deduce that said person is very likely to be a "green" as well, and therefore expected to join a planned demonstration against a nuclear power plant.

"Green" is nothing else than the sense of the collected pieces of information. Each isolated piece of information has its own meaning and reflects a specific fact attributed to that person, but does not make sense when regarded in isolation from other attributes. Some of the attributes are neutral or even opposite to the presumed sense, but anyone who wants to identify a "green" does not bother about the golf club membership. Sense is not a property of reality, but is constituted by subjective construction. Sense is a cognitive pattern which covers the actual facts in order to better organize understanding. And, of course, this understanding may sometimes be misleading, when the need for "making sense" is stronger than the ability to recognize embarrassing facts.

Sense is a necessary scheme for perceiving and understanding the world. The manifold of perceptable data cannot be managed by human consciousness unless it is able to restrict that variety and reduce it to a tolerable level (Ashby, 1956; Luhmann, 1971). This is the essence of the concept of understanding in the humanities as opposed to the concept of explanation in science. As the German historian Th. Schieder has pointed out, "understanding" involves "singling out from thousands of phenomena a certain set and synthetically combining them into a unity" (1968, p. 38). Max Weber, too, said something of the same thing when he introduced into his program for an "understanding sociology" the "ideal type" as a fundamental cognitive tool. The ideal type "combines certain relations and events of historical life to a consistent contextual universe" (1973, p. 234). Although Weber's concept of sociological meaning is not completely clear, the ideal type is certainly an attempt to make sense of the social world.

In general, the typological approach, even if it does not follow all of Weber's methodological rigor, is a means for constituting sense. The typological method has also recently been recommended as the way to deal with technological forecasting and technology assessment (Ropohl, 1973), and can serve as an extensional approach to describe and understand complex objects.

Following this line of argument, a tentative formal clarification of the difference between information and sense can be established. For this we return to the definition of information developed in the previous section.

Let Z be a set of signs

$$Z = \{z_1, z_2, \ldots, z_i, \ldots, z_n\}$$

where the following is true for every z_i:

(a) z_i is a physical (material or energy) event, which occurs with the frequency of probability p_i;
(b) z_i has a certain denotation as agreed upon by convention;
(c) z_i has a certain engagement with the behavior of its users.

Then, any element z_i which is a member of the set Z may be defined as *information*, with (a), (b), and (c) being its respective syntactic, semantic, and pragmatic dimensions. The amount of information, according to mathematical information theory, may be calculated only with regard to the syntactic dimension, which provides for an important aspect, but not for a complete understanding of information.

Now form the set E of all the subsets of Z, that is the power set of Z:

$$E = \mathscr{P}(Z)$$

Next, define a partial set F which is a subset of E as consisting of the empty set φ, the one-element-sets $\{z_i\}$ and the full set Z:

$$F = \{\{\varphi\}, \{z_1\}, \ldots, \{z_i\}, \ldots, \{z_n\}, \{Z\}\}.$$

These special subsets of the power set must be excluded, because they contain either no information or only individual pieces of information. The remaining and larger portion of the power set, however, indicated all the possible combinations of data, and therefore we call it the sense set S^*

$$S = E / F$$

Sense can thus be defined as an element s_i of the set S. And we may define the *relevance* of information as the quality or the disposition of the respective sign z_i to be included or to be apt to be included in that specific subset of Z which describes a certain sense s_i.

Reconsidering the introductory example, we see that "green" is a subset of the personality cross-section comprising the element $\{2, 4, 5, 7, 9, 10\}$. Hence, these pieces of information are relevant with regard to the supposed sense, whereas other features of the personality profile are not. Moreover, the example shows that the extensional definition of sense may only serve as a preliminary means for increasing the precision of that concept. The specific arrangement of pieces of information which constitutes this very sense is by no means a mere coincidence. Rather, it results from certain assumptions concerning intrinsic relations between the said features, such as the empirical generalization that young people more frequently support the ecological movement than older people, or that giving up a private car is related to ignoring advertisements. So it can be argued that sense not only implies the mere aggregation of signs, but additionally a set of relations constituted within the subset of signs, so that sense could be conceived in terms of a system (Lenk and Ropohl, 1978).

The principles which determine the selection of one sense from among the multitude of possible senses must also be examined. These principles cannot be found in any single sign nor in any combination of signs. Until this point it was possible to treat the problem in an objective way, but now the subjectivity of the human mind must be taken into account. Obviously the principles for selecting and thus constituting sense are rooted in the history and present state of consciousness of a subject (Weizenbaum, 1976, chapters 7 and 8), his internal model of the world (Sachsse, 1974, p. 213), his background knowledge and his values (Rapp), and his active intentions (Kramer-Friedrich).

Sense is not caused by mere facts, but grows out of the total situation in which the individual mind finds itself. Sense mediates between objectivity and subjectivity. Sense is the mode in which the world is *given* to the human mind, *and* it is the scheme which the mind has to *construct* in order to provide orientation and identity within the absurd variety of the world. As the constitution of sense is bound to human subjectivity, it is sense, not information, which demarcates the limits of technical objectification. Computers do handle information, indeed, but that information will not make sense unless the individual human mind is involved in applying its own sense to the given pieces of information.

V

The constitution of sense, however, is not just an individual performance. On the contrary, it may be that the main traits of personal sense patterns originate in societal mechanisms, and that personal sense is rather a variation of "the social construction of reality" (Berger and Luckmann, 1966). Nevertheless, individual variations do exist, especially in modern societies which tend to become open and pluralistic, so that the individual person experiences his or her personal sense as an expression of freedom.

The interplay between societal and individual processes of sense constitution is manifold and highly complex. It becomes more complex with the introduction of information technology. In terms of the sociotechnical system, as described in the second section, human abilities, knowledge, and attitudes are increasingly being transferred into technical subsystems. Extensive aspects of institutionalization and socialization are thus going to be technicized, because what was once individual or social knowledge is being objectified in technical devices and transmitted to the personal system by technical mechanisms.

But there remains a difference between knowledge and sense. The question is whether information technology influences the quality of sense constitution, and if so, in what ways. Mere information does not make sense, but there is no sense without information. Hence, changes in the quantity and quality of information necessarily affects the character of sense frameworks and their constitution.

One problem arises from the ever growing quantity of information. Through data banks, information retrieval systems, expert systems, and their spreading accessability through telecommunication networks, the amount of available information is increasing dramatically. This has in fact been a trend since the invention of printing, the development of modern science, and the diffusion of mass media. But never before has there been the kind of omnipresent pressure from information accumulation as can be expected in the next few years. Before the end of this century, to mention only one factor, a high percentage of private households will be connected to telecommunication networks and subjected to incredible amounts of commercial and non-commercial data.

The more information available, however, the more difficult it will be to integrate such information into comprehensive sense frameworks. Human consciousness will be confronted with an abundance of information, the relevance of which will be unrecognizable. The individual capacity of integrating new information into existing sense frameworks is limited. When information overwhelms the mind, cognitive dissonance results and, to avoid complete confusion, there will be irrational reactions. This is what can be called a "relevance gap" in information technology, a gap which may well become a universal crisis of sense.

Obviously the relevance gap is the main difficulty in the field of artificial intelligence. Pattern recognition, language translation, information retrieval, and problem solving depend at least as much on understanding of sense as on the technical representation of mere knowledge. For instance, when files concerning a certain issue are requested from an information retrieval system, the system will include only a certain fraction of all relevant files and only a certain fraction of the included files will really be relevant to the stipulated issue. Neither "recall" nor "precision" can be one hundred percent accurate. Usually both will be very inexact, because the restricted set of descriptors used to characterize the issue in question is insufficient to indicate the real sense of the problem. Nevertheless, the point is not to add to the popular "computers-cannot" literature, since artificial intelligence may be expected to succeed in a lot of practical tasks during the next decades.

But on principle, artificial intelligence will never be fully equal to the individual user's engagements. Since it relies, like all information technology, on the objectification of information processing, it will be unable by definition to provide for personal sense. Of course, it is possible to write computer programs which will be able to handle superstructures of information which comply with the formal definition of sense given above. But – and this alludes to one of Weizenbaum's arguments (1976, Chapter 7) – the computerized copy of sense means nothing but the deindividualization of sense. Computer programs for expert systems, for instance, must be provided with some kind of professional sense. But that will not be my sense nor your sense. Instead it will be a socially generalized sense, at best bare of any individual character, at worst deformed by the onesided perspectives of illegitimate minorities. The historiography of Orwell's ministry of truth in *1984* provides an illustration of this danger. In consequence, the relevance gap may be replaced by a universal domination of sense.

VI

Returning to the model of the sociotechnical system, it has to be recognized that the human ability to process information is shifting to technical subsystems. But the human being is a sense constituting being more than just an information processing being. An individual sense constitution, on principle cannot be objectified. When information technology multiplies information, either it leads to a wholesale confusion of sense, or it brings about a computerized sense domination.

To guard against such dangers is a techno-political problem of organizing information technology, and increasing awareness among computer scientists of their social responsibility. But there is no domination without slaves: Sense domination will not work insofar as the individual crisis of sense can be overcome. This is the very point where the philosophical task appears. In the face of the multiplication of information through science and technology, philosophy again has to assist in constituting sense. It has to *assist* – which means that philosophy is not entitled to constitute sense on behalf of individuals. Technological sense domination must not be overcome through philosophical sense domination. In an open society, philosophy has to confine itself to elaborating the categories of sense constitution and to instructing people about the possibilities for constituting their own personal sense.

Frankfurt University

74 GÜNTER ROPOHL

AUTHOR: Günter Ropohl, born 1939, received a Ph.D in engineering with a dissertation on flexible manufacturing systems, then turned to the sociology and philosophy of technology. He has taught at the University of Karlsruhe, and was from 1979 to 1981 Director of the Institute for General Studies. Since 1981 he has been Professor of General Technology at the Johann Wolfgang Goethe University in Frankfurt. He has written widely on systems theory and the philosophy of technology.

* The German *Sinn* is regularly rendered as "sense" and used as a technical term. Although understanding of the term is indicated by the idiom in the title of this paper; a more detailed analysis is given in Section IV below.

REFERENCES

Ashby, W. Ross. *An Introduction to Cybernetics*. New York: Wiley, 1956.

Berger, Peter and Luckmann, Thomas. *The Social Construction of Reality*. Garden City, NY: Doubleday, 1966.

Frege, Gottlob. "On Sense and Nomination" (1892), trans. H. Feigl, in Herbert Feigl and Wilfrid Sellars (eds.), *Readings in Philosophical Analysis* (New York: Appleton-Century-Crofts, 1949).

Gehlen, Arnold. *Anthropologische Forschung*. Hamburg: Rowohlt, 1961.

Lenk, Hans and Ropohl, Günter (eds.), *Systemtheorie als Wissenschaftsprogramm*. Königstein: Athenäum, 1978.

Luhmann, N. "Sinn als Grundbegriff der Soziologie," in J. Habermas and N. Luhmann (eds.), *Theorie der Gesellschaft oder Sozialtechnologie* (Frankfurt: Suhrkamp, 1971).

Morris, Charles W. *Foundations of the Theory of Signs*. Chicago: University of Chicago Press, 1938.

Ropohl, Günter. "The Use of Systems Technology and Morphological Method in Technological Forecasting," in H. Blohm and K. Steinbuch (ed.), *Technological Forecasting in Practice* (Lexington, MA: Lexington Books, 1973), pp. 29–38.

Ropohl, Günter. *Eine Systemtheorie der Technik*. Munich and Vienna: Hanser, 1979.

Ropohl, Günter. "Some Methodical Aspects of Modelling Sociotechnical Systems," in R. Trappl et al. (eds.), *Progress in Cybernetics and Systems Research*, vol. X (Washington, DC: 1982), pp. 525–536.

Sachsse, Hans. *Einhfürung in die Kybernetik*. Braunschweig: Vieweg, 1974.

Shannon, Claude E. and Weaver, Warren. *The Mathematical Theory of Communication*. Urbana: Univ. of Illinois Press, 1949.

Schieder, Theodor. *Geschichte als Wissenschaft*. Munich and Vienna: 1968.

Weber, Max. "Die Objektivität sozialwissenschaftlicher und sozialpolitischer Erkenntnis," *Archiv für Sozialwissenschaft und Sozialpolitik* **19** (1904), 22–87. Quoted and trans. from the reprint in: Max Weber, *Soziologie, Universalgeschichtliche Analysen, Politik*, J. Winckelmann (ed.) (Stuttgart: A. Kröner, 1973).

Weizenbaum, Joseph. *Computer Power and Human Reason*. San Francisco: Freeman, 1976.

WERNER STROMBACH

"INFORMATION" IN EPISTEMOLOGICAL AND ONTOLOGICAL PERSPECTIVE

ABSTRACT. As a human phenomenon, information is involved with communication, whereas logically it is a kind of relation. Reviews the definitions of G. Klaus, N. Wiener, and G. Günther. Argues that information has the same ontological status as K. Popper's "world 3," and that C.F. von Weizsäcker's suggestion of a relation between Aristotelian *forma* and information is essentially correct.

The problem to be discussed concerns the meaning of "information." Since C. Shannon developed a way to measure information (although in my opinion it is not really information that is measured), many authors distinguish between a scientific and a non-scientific use of the concept, identifying the scientific with the metrical. Y. Bar-Hillel remarks, however, that while measurement is an interesting aspect of information, it cannot be the last word concerning this subject.[1] Our question is: What can be said from a philosophical perspective?

We first encounter the phenomenon of information in the processes of communication: between an informed sender and a partially uninformed receiver there exists a differential in information which is to be equalized. To inform means to reduce what is unknown, surprising, new.

Information can be produced by a person or can be discovered through observation, conversation, experiments, books, etc.; such discovery may even be reluctant or involuntary. The information transmitted is packaged in a message; it is the content of a message or a material signal, which may also be called the material carrier of a sign. In the material state information is only *in potentia*, due to become actual in a receiving and decoding consciousness or (to put it in general terms) in a decoding system.

Having described information as it occurs in a process of communication, we should try to explicate its logical use and status. In this we can turn for help to the work of the Marxist philosopher, G. Klaus. According to Klaus, "when information is understood as the sign of a class of equivalent signals (representing physical facts), it is logically neither an object nor an attribute of an object, but an attribute of attributes (a predicate of predicates) like, for example, the natural numbers, which can be defined as abstracted classes of equivalent sets."[2] E. Oeser, a Vien-

Carl Mitcham and Alois Huning (eds.), Philosophy and Technology II, 75–81.

nese philosopher, extends Klaus by adding that information, as an attribute of attributes of material systems, must not itself be material.[3]

This explication is, however, inadequate because it does not take into account the relational character of information. Let me explain. Certain concepts involve what are called correlated ideas. One of them, according to the German philosopher, B. von Freytag-Löringhoff, is the concept of freedom, which from a logical point of view, refers neither to a substance nor to a simple attribute but involves at least a triadic relation.[4] Similar notions are movement, action, and likewise information, because information involves a sender, a receiver, a carrier, and a content.

To capture this, it is not enough to describe "information" as a predicate of predicates (meta-predicate). Consider an example: rocket flares. This material can have the attribute of being red. Furthermore, red phosphorescent rocket flares can have the character (meaning) of a distress signal (as in the mountains). No doubt, this is an attribute of attributes, but I do not think it is yet information in a human sense. To become information there must also be a receiver to understand the signal as a sign, to decode its content, and to react to it. This factor of bringing about human activity is completely neglected when one speaks of "information" as only a meta-predicate. Logical analysis must take into account the latter operation, which makes information into a relation, more precisely a tetradic relation. This example also demonstrates the need to distinguish between information and message or between sign and signal, as will become even more evident when we consider the ontological status of information.

From an ontological perspective we must consider the place of information in the over-all order of things, its essence or "quidity" as traditional philosophers said. N. Wiener has pointed out only what information is not: that it is neither matter nor energy.[5] Klaus defines the term more explicitly by calling information a third essential aspect of matter, but for G. Günther it is a third aspect of the world in a metaphysical sense.[6]

Günther calls information a third proto-metaphysical component of the world, and is correct in arguing that information cannot be placed exclusively either on the physical-material side of being or on the subjective-spiritual one. But what is this third reality? If we were to agree with Günther to take it as a fundamental metaphysical component, would we not have to interpret other phenomena which lie between subjectivity and material being in a proto-metaphysical sense as well?

In recent discussions we find two other important ideas concerning the ontological status of information: K.R. Popper's "world 3" and C.F. von

Weizsäcker's reference to Aristotle's concept of *forma*. According to Popper, there are fundamental weaknesses to both monistic and dualistic views of the world. This is why we should turn our attention to philosophers who assert the existence of a third world – for example Plato, the Stoics, Leibniz, Bolzano, and Frege.[7] Popper does not wish to identify his position with any of these philosophers, but he accepts the idea of a pluralistic world-interpretation and distinguishes as follows: world 1 is the physical world; world 2 is mind-immanent (subjective); world 3 is the world of objective intelligibilia, of ideas (in an objective sense), of theories and their logical relations, the world of arguments, the products of the human mind as recorded in languages, the arts, sciences, and technologies. Unlike Plato's, this world is not a world of eternal ideas; although it is autonomous, it is produced by human beings, which means that it would not exist if human beings had not created the objects or entities in it. It thus contains not only truth but error, and it can give rise to new objects and problems. The rise of world 3 out of worlds 1 and 2 is brought about by humanity.

Moreover, human beings realize ideas in an active process, in a kind of re-creation or actualization of possibility. When distinguishing between actual information, which influences systems and brings about action, and potential information, which is only stored as in an undecoded message, we can say with Popper: this potential information, these signals and languages, are nevertheless objects of world 3.

"Information in language" is also central to the work of C.F. von Weizsäcker.[8] For Weizsäcker too information has an objective character and, like Popper, he thinks there is a relation to the Platonic *eidos*. More important for him, however, is the Aristotelian concept of form, which is in fact an etymological component of the word "information." Initially (1959) Weizsäcker used *forma* synonymously with "Gestalt" or "structure" and neglected the dynamic aspect of the form concept in favor of a static interpretation. This changes ten years later (1969) when, in accordance with the classic conception, "form" becomes what we can know of an object. Since information now measures the growth of knowledge produced by an event, it also measures an "amount of form." This connection between information and "being able to know" leads to the corollary that information is present only when something is being understood. Furthermore, since the receiver of information reacts (which means that he becomes a sender) Weizsäcker puts forth the thesis that "information is only what produces information." He thus reclaims that dynamic element in the concept of information which has always been immanent in the notion of form.

For an ontological definition of information let us rely then on the two elements of "activity" (*energeia*) and "form" (order, structure). Because a formal order in being entails the unification of a manifold, which is thus informed with meaning or by some law, the concept of order is, on the one hand, closely related to the concepts of wholeness and system[9] and, on the other, to that of the regularity of events. But not every order is information. There must also be

(a)   a differential in information similar to the energy differential which causes motion or work;
(b)   the intelligibility of the message to a receiving system; and
(c)   the generation of activity in the receiving system.

All three of these supplementary characteristics are attributable to human beings. Human beings investigate the unknown, decode the code of nature, and act in accordance with received information. Information consists in the representation of the order of reality which becomes manifest in *human* beings – i.e., in logical thinking, conceptual understanding, ethical evaluation, and significant or useful human action.

But such a definition remains inadequate, because it fails to consider those cases in which the receiver is not a person but some other system, i.e., a computer (in relation to genetic information). In search of a more comprehensive definition, we can make more use of the notions of structure and system and the distinction between acting and causing. A message is anything that acts on a system, maintains the German philosopher of technology and science, H. Sachsse.[10] Nevertheless, not all messages cause activity on the part of the receiver, either because the receiver already knows the content of the message or because it is not understood. Messages can likewise act on a receiving system without being information for it. As H. Dolch, a German theologian at the University of Bonn, recently wrote to me, a nutcracker, in cracking a nut, does not inform it. The nutcracking is not information for the nut, although it is a message in Sachsse's sense.

In view of all this, one can formulate the following definition: *Information is a structure which causes activity in a receiving system simply by virtue of its form.*[11] Structure is the form or order of a system by virtue of which it becomes an information-carrier.

But to what extent is the Aristotelian concept of *forma* to be found in our concept of information? Of course, since matter and form in the Aristotelian sense are metaphysical conceptions, our use of form can only be an analogous one.[12] Nevertheless, this analogy can be used to argue that the features of actuality and activity involved in the traditional concept of *forma* are illustrated by the ways that

(a) information makes a sign-carrier actual, not in its physical being but in its being a sign-carrier; and

(b) a message causes activities on the part of the receiver, which makes the potential information actual, although in its own way and in conformity with its own possibilities.

This touches on a use of the notion of form in the theory of cognition. According to the classical view, a cognizable object is *in* the cogniscient being in the way of the cogniscient being (*modo cognoscentis*), and a sentient impression is a cause of actions. Hence a distinction was drawn between receiving an impression (receptive, passive) and comprehending an object (spontaneous, active). Accordingly, a receiving system stands in a certain relation to the potential information of a signal or a series of signals: passive and active, accepting and making actual – decoding, as Popper says.

From an evolutionary standpoint, the necessity of processing information is based on the fact that living beings must adapt to their environments. Weizsäcker therefore defines organic evolution as an adaptation of the genetic code and of resulting behavior to the facts of the natural world. In this sense, evolution consists in the obtaining of information, or in the words of the physicist A. Unsöld: obtaining, utilizing, and passing on information.[13]

At lower levels of being, environmental influences cause spontaneous reactions. But at higher levels the activity of the receiver leads to the construction of an inner model. According to Sachsse, this is indeed grounded in perceptions of the environment, but the model is no simple copy of the received information: it is derived from the processing of the information. Finally, the individual also directs his activities toward the environment, changes it and makes it a source of new information. It seems important to take into consideration the fact that in our consciousness these actions can be modelled without any risk to oneself, and that this modelling capacity is even augmented by the modern possibilities of computer-simulation.

Let me conclude by observing that a theory of cognition based on the ideas of information and evolution is argued against by many philosophers, especially by proponents of a transcendental philosophy.[14] According to the objection, this kind of theory may have a certain empirical value, but it does not deal with fundamental questions such as the very possibility of cognition.

This is not the place to reply to such arguments. Nevertheless, it should also be noted that there are many authors, especially scientists, who consider the concept of information to be fundamental for our understanding of the world and of human nature. Whereas hitherto an increasing complexity of structures was characteristics of cosmic and biological evolution, a further evolution in human beings and societies may be distinguished by an increasing density of information. In spite of the human problems involved with computerization, a higher rationality obtained by information-processing may well diminish the threat of irrational actions in our highly sensitive society. Nonetheless, rationality must also preserve a sphere of freedom, not only for rational action but also for personal development. That is, the irrational element in human existence must not necessarily be extinguished in the process of securing more rational conditions for human life.

Dortmund University

AUTHOR: Werner Strombach (born 1923) is Professor of Philosophy at Dortmund University. Since 1964 he has been actively in engineering education at both the high school and college level, and in 1981 he founded the colloquium on Philosophy and Information Theory at Dortmund. His works include *Natur und Ordnung* (Munich: Beck, 1968), *Die Gesetze unseres Denkens* (Munich: Beck, 1970; 3rd edition, 1975), and *Mathematische Logik* (Munich: Beck, 1972).

NOTES

1   Yehoshua Bar-Hillel, "Wesen und Bedeutung der Informationstheorie," in H. von Ditfurth (ed.), *Informationen über Information* (Hamburg: Hoffmann and Campe, 1969).
2   Georg Klaus, *Rationalität-Integration-Information* (Munich: Fink, 1974); and "Information," in Georg Klaus and M. Buhr (eds.), *Philosophisches Wörterbuch* (Berlin: Enzyklopädie Leipzig, 1972).
3   Erhard Oeser, *Wissenschaft und Information*, vol. 2 (Vienna and Munich: R. Oldenbourg, 1976).
4   Baron von Freytag-Löringhoff, "Die logische Struktur des Begriffs Freiheit," in J. Simon (ed.), *Freiheit* (Munich: K. Alber, 1977).
5   Norbert Wiener, *Cybernetics* (Cambridge: MIT Press, 1948).
6   Gotthard Günther, *Das Bewusstsein der Maschinen* (Krefeld: Agis, 1953).
7   Karl R. Popper, *Objective Knowledge, An Evolutionary Approach* (New York: Oxford University Press, 1972); and Karl R. Popper and John C. Eccles, *The Self and Its Brain* (New York: Springer, 1977).

8 Carl Friedrich von Weizsäcker, *The Unity of Nature*, trans. Francis J. Zucker (New York: Farrar, Straus, Giroux, 1980); and *Der Garten des Menschlichen – Beiträge zur geschichtlichen Anthropologie* (Munich and Vienna: C. Hanser, 1977). *The Unity of Nature* collects a number of papers, among them "Language as Information" (from 1959) and "Matter, Energy, Information" (from 1969).

9 Werner Strombach, "Wholeness, Gestalt, System: On the Meaning of these Concepts in the German Language," *International Journal of General Systems* **9** (1983), 65–72.

10 Hans Sachsse, *Einführung in die Kybernetik* (Braunschweig: Vieweg, 1971).

11 This *objective* (actual or potential) information should not be confused with its *subjective* varient, in which human beings must be respected with regard to their knowledge, expectations, interests, feelings, curiosity, etc.

12 Werner Strombach, "Der forma-Begriff in der neuscholastischen Naturphilosophie und sein Beitrag zum Informationsverständnis – eine Erinnerung an Joseph Kleutgen," in U. Neemann and E. Walther-Klaus (eds.), *Logisches Philosophieren: Festschrift für A. Menne zum 60. Geburtstag* (Hildesheim, Zurich, New York: G. Olms, 1983).

13 Albrecht Unsöld, *Evolution kosmischer, biologischer und geistiger Strukturen*, 2nd edition (Stuttgart: Wissenschaftliche Verlagsgesellschaft, 1983).

14 Friedrich Kaulbach, "Philosophische und informationstheoretische Erkenntnistheorie," in E. Börger, D. Barnocchi, and F. Kaulbach (eds.), *Zur Philosophie der mathematischen Erkenntnis* (Würzburg: Königshausen and Neumann, 1981).

PART II

PHILOSOPHICAL ANALYSES OF THE INTERACTIONS BETWEEN HUMAN BEINGS AND COMPUTERS

HEINRICH BECK

BIO-SOCIAL CYBERNETIC DETERMINATION
— OR RESPONSIBLE FREEDOM?

ABSTRACT. Part one of this paper describes in some detail how the principles of cybernetics are manifested at all levels of reality – from the physical to the organic and even the social. Part two, however, argues that human freedom, grounded in self-consciousness, transcends cybernetic determinism, and makes possible moral responsibility.

Today we face a dilemma: On the one hand, we can progressively understand reality as a closed system of causal relations or feedback control systems which can be mechanically modelled. With science and technology it is even possible to describe the living organism, the human being, and society as self-regulating functional units which adapt to various environmental conditions. This transforms reality into a huge automaton, completely integrating all human beings. One the other hand, this world, which is wholly possessed by technology and science, imposes on us the exercise of responsibility. The need for responsibility exists at a more radical and universal level than ever before.

Responsibility, however, requires and presupposes freedom. But how can the human being, as a feedback control system, be free? Thus arises the philosophical task of conceiving man, on the scientific-technical model as a feedback control system, in such a way that freedom and responsibility remain possible for him.

This is our subject, and we will proceed according to the following outline. As a *first step*, the world and man are presented from the perspective of science and technology as a closed causal system, constituted by a number of feedback control loops. This provides the framework for the above-mentioned philosophical task. But the existence of responsible freedom requires that we take a *second step* towards identifying the special cybernetic structure of mind and freedom. In this step we must transform the scientific-technical framework into a philosophical one.

I. THE PHILOSOPHICAL PROBLEM: THE WORLD AS A CLOSED
CAUSAL SYSTEM OF FEEDBACK CONTROL LOOPS

The term "cybernetics" comes from the Greek κυβερνητική and originally meant "art of steering" or "art of directing." As is well known, the

Carl Mitcham and Alois Huning (eds.), Philosophy and Technology II, 85–95.

term was adapted by Norbert Wiener to refer to the study of all self-regulating processes, and since then has been expanded to include the basic
principles concerning developments in nearly every field of science and
technology. A cybernetic system exists wherever causal factors are
arranged in such a way that by themselves they create and maintain a
structure. In that case we can also speak of "automatic structure" or of an
"automaton."[1]

Take, for example, a system in the inorganic field or engineering technology. Consider an automatic heating plant: Rising room temperature
causes (by means of a thermostat) a decrease in heat production and
therefore a lowering of the room temperature. The room temperature
regulates itself. The principle of self-regulation entails that the effect
(room temperature) is fed back to one of its causes (heat generation) so
as to regulate it.

At the level of organic life, the regulation of sugar content H_2 ions, pH,
and body temperature are brought about in a similar manner. Here the
transmission of messages is carried out by nerves and hormones. We can
even say that all reactions in an organism regulate themselves by means
of a fluctuating balance between disorders. All important life organs,
such as the brain, heart, etc., as well as each important life process, depend on self-regulation. Individual growth and self-preservation, along
with preservation of the species and the transmission of characteristics,
are only possible as self-regulating processes involving an automatic
adaption to changing environmental conditions.

The psycho-mental interactions of man and society, *mutatis mutandis*,
can also be understood as cybernetic processes. Every efficient human
activity appears as some kind of "operation arc of experience and action"
consisting of different components. Some are inorganic-physical, some
organic-physiological, others psycho-mental. Action is continuously
modified by experience, which means action regulates its own development by feeding information back to its cause.

The cybernetic structure thus is analogously present at the inorganic,
organic, and psycho-mental levels of reality. In consequence, this opens
up certain special ways for understanding the developmental relationships between life, science, and technology. But recognizing structural similarities between psychic, organic, and mechanical automatic
processes, it becomes possible to construct models of life and mental operations which contribute to a more exact study of those biological and
mental functions. Such progressive scientific observations further enable
us to construct more and more perfect automatons. Therefore we can

speak of a kind of cybernetic interaction between science and technology, within which they mutually stimulate one another.

With the mechanical modelling of the natural functions of organs such as the kidneys, heart, and even certain parts of our brain, more and more possibilities are also offered to medical science and technology. When, because of accident or disease, it becomes necessary surgically to repair or remove natural organs, it is now sometimes possible to augment or replace them with technological artifacts. Often survival is only possible through such means, which are likewise capable of improving the quality of life.

Furthermore, the reproduction, growth, and behavior of species and individuals depend on inherited dispositions which find their physical expression in the structures of genes and chromosomes. Hereditary dispositions are in some way fixed patterns or programs of behavior. This is the basis for experiments to alter the individual programs of behavior by subjecting the gene structures to certain types of mutagens. Scientists thus try to create new species by technological means. In the same way, since human mental behavior has a biological basis, some scientists hope to create new mental behavior programs, thereby bringing to birth a new and higher human race, perhaps one better adapted to abstract mathematical thinking, certain social interactions, or even living in space colonies. This attempt by human beings to take technological control of their own evolution and to reach some kind of perfection reminds us of Nietzsche's philosophy of the "superman" as the end of history. But one must ask: Will this new man be so superior that he will obtain technological control of the whole future of humanity? Will he be able to control even the creation of human beings? At this point, we become aware of the extent of the responsibility which falls to us as a result of the rapid increase of the cybernetic possibilities of power.

But the power to control technologically the evolution of society is not restricted to the bio-technological alteration of human nature. Social processes also exhibit in themselves the structure of feedback control systems, the primary causal factors of which can be manipulated to a certain extent. Contemporary social science describes society as a closed system of automatic rationalization, of adaption between supply and demand, price and advertisement. The demand orients and fixes itself according to the supply, and in the same way the supply adjusts to bring about a certain demand through pricing policies and advertisement. Each factor enters into a dynamic cybernetic dialectic of act and potency which governs the rhythm of social change.

Culture and education within society further exhibit these same feedback processes. Society seeks to form a young generation according to its
needs, by state-supported educational programs and institutions. At the
same time society can only achieve its goals by adaptation to the conditions of the younger generation, if it lets itself be changed, for example,
by certain educational reforms. Neither will a younger generation
achieve its goals without accepting the justified limitations of society.

Finally, let us look at world politics. Ultimately, no state can pursue its
self-interests without recognizing the interests of other states to a high
degree. Only by this means, can one state persuade others to go along
with its own aims and ideas. This is one of the principles of political prudence. Humankind thus moves toward a consistent feedback control system in the area of international affairs. Within this feedback control system all states are causal members, regulated by the necessity to adapt to
each other. Advances in information technology and computers, i.e.,
technologies explicitly designed with the aid of cybernetics, contribute
especially to developments in these last three areas of sociology, education, and politics.

Just as in biology, such examples from the field of sociology show that
along with positive benefits from scientific and technological progress in
cybernetics there are increasing negative dangers. While science and
technology make available more and more technical means, we must
concern ourselves more and more with how we should use them and the
goals they should serve. If today, and even more so in the future, human
nature can be altered and provided with new biological and mental abilities, to what end should it be altered and what abilities are really desirable? If some day it should prove possible to increase abstract mathematical abilities, this does not guarantee a better way of life for all humanity.
Although it might be possible to control economic, educational, and
world-political processes in a more perfect manner by cybernetic direction, such control could lead to a more complete destruction of the real
and necessary dimension of human value by making man more and more
an object of anonymous powers.

In the automated society regulated with the help of information technologies and computers the question of responsibility becomes more
acute than ever before. Today we feel ourselves provoked to counteract
the temptation and the danger of abusing technical possibilities in a destructive way. But such feeling seems like a delusion, and the rhetoric of
responsibility sounds like meaningless phrases, if the world functions
purely as a closed feedback control system and man operates as part of an

automatic system of causal factors, physically, mentally, and socially. If every action is completely determined by the feedback principle, where is there a place for something like responsibility, which depends on the possibility of free engagement even when it goes against the rules of adaptation and opportunity? Truly, if one accepts the scientific method, objectifies the world as a closed deterministic system,[2] responsibility would be meaningless. Therefore the difficult philosophical task arises, to open a space for a human subjectivity which is free and responsible for itself, a space which integrates as well as transcends the previously outlined scientific view of the world, and thus makes possible technological action with responsible freedom.

II. TOWARD A SOLUTION: THE SPECIAL CYBERNETIC STRUCTURE OF MIND AND FREEDOM

One initial solution to the problem presents itself as typical of modern philosophy. This goes back to Descartes and Kant, and distinguishes two levels of human existence. On the one hand, there is the material-biological level; on the other, the psycho-mental one. At the first level there is strict causal determination, which is objectified by natural science; in the second, there is freedom and responsibility, characterized by subjective experience, which cannot be objectively analyzed.

But such a dualistic perspective on human nature is undermined by the following experiences: *First*, the psycho-mental phenomena of consciousness is largely dependent on material dispositions, especially in the brain. *Second*, subjective freedom and responsibility have an influence on the objective field and the material dimension of our existence. We cannot, therefore, understand the essential unity of the objective physical and the subjective psychic aspects of human existence by splitting that existence up into two different levels.

A philosophical conception of freedom and responsibility must be developed by means of a comprehensive analysis of human experience. For example, we need to take into account the experience that we are influenced by social institutions – political parties, professional organizations, the Christian churches, and our own family, all of which seek in their own way to determine and direct our behavior. But we must further recognize that attempts to justify our behavior by appeal to the influence of such institutions is a rationalization, because we are able to criticize those institutions, even to keep our distance from them, by asking them to justify themselves. Sometimes we are even forced to choose between opposing institutional programs with different origins, or to criticize traditional conventions of behavior. Although exposure to the influence of social groups often has more unconscious influence than we consciously realize, we are never completely in their power. If we do not exercise our ability critically and consciously to analyze such influences, this is already a decision for which we are responsible.

This is equally true with regard to our biologically determined hereditary dispositions – whether we work against our own negative dispositions, whether we develop or turn away from our positive abilities – lies for the most part in our power. Neither sociological influences nor biologically determined dispositions and inclinations are decisive in our behavior, but it is we who decide in confrontation with those determining forces. We ourselves decide whether and how far we want to follow them.[3]

Here values like honesty and justice, conceived by moral cognition, serve as criteria for judging our behavior and themselves become "programs" for our lives. Such values are recognized as having an absolute claim which we can fulfill by acting morally. The structure of responsibility thus reveals its dialogic character: Responsibility refers to experiencing the claim and the call of a value which is to be answered by acting. The one who acts responsibly considers the consequences of his behavior under the guiding principle of an experienced value claim which entails the respect and appreciation of our fellow human beings. But this kind of action, experienced in our conscience, would be impossible if we were totally determined by a bio-social cybernetics. To be *obliged* to behave in a certain way supposes that we are not forced by nature to behave that way, but that we can also behave differently; a moral claim is not to be confounded with biologically, psychologically, or sociologically conditioned constraint but requires, applies, and even provokes personal freedom.[4]

By such analyses of phenomena in the area of freedom and responsibility, the foundation is laid for a better understanding of the structure of freedom and intellectual consciousness. What happens when I define myself by the act of deciding? By doing so I obviously become my own vis-à-vis and from that position, I face and conceive myself with my various possibilities, and I direct my determinative force toward myself to choose one of these possibilities and to determine myself to this or that way of acting. Becoming one's own actually happens whenever I say "I," which further means the same as to say "you" to oneself. Whenever I "consult myself" about something, ask myself about its character and try to give an answer to myself, I talk to myself like I talk to my own partner, I appeal to and realize myself. By praising or reproaching myself, by judging my behavior positively or negatively, by realizing my abilities and my limits, or by trying to form and determine my existence differently and anew, I face myself like an object by knowledge of myself, and I get closer to myself, with one of my thus recognized abilities, by self-decision. I express myself by the act of recognizing myself, I go out of myself and put me in front of myself; and by the act of deciding who I am I enter into my abilities more deeply and realize myself. So my intellectual existence describes a circle – yes, a full circle – not a closed but an open circle, since I develop by self-determination. Thus my self-determination is cyclic causality, once again a pure cybernetic structure.[5]

Although we find a similarity in the structure of individual conscience and the cybernetic structures of the material and social world, there is an important difference. For it is a characteristic of the material as well as of the social feedback control systems that they are always a combination of elements – of people who determine each other. Input and output, conductors, transmission factors, and programs are spatially separate and it is possible to isolate them from each other physically. What is more, the respective determining – i.e., conditioning, controlling, and regulating power on the one hand, and the determined, conditioned, controlled, and regulated on the other – lie opposite each other like different parts of the system.[6]

An essentially different situation exists, however, when I determine myself by an act of self-consciousness and free decision. Then it is not the case that one part of myself recognizes and determines another one; I as the one who recognizes and determines and I as the one who is recognized and determined by myself are not two parts within me, but identical. I as the "subject" of self-recognition and self-determination, and I as the "object" of self-recognition and self-determination, are one and the same. In this center of myself, where I am with myself, there I do not at all consist of different parts, there I am undividedly and indivisibly myself.

The common aspect of the *sameness* between a material (or sociological) feedback control system and mental self-consciousness is based upon the fact that in every case the effect is turned back to the cause and thus determines itself.[7] The *difference*, however, is to be seen in the fact that in every case both are different members of a system, in the other they are not.

Strictly speaking, only where human mind is concerned can we speak of self-determination, self-control, and self-regulation, whereas in all the remaining systems one member determines and regulates another. If we define an automaton as a self-regulating structure, mind is the most perfect automaton and *only mind* is an automaton in the proper sense. Ordinary language usage, however, does not allow analogies between the terms "automaton" as a determinative structure without any physical parts in free self-determination. To some extent a human being could be called a "free automaton," i.e., an automaton which does not function according to principles that are totally conditioned from outside. As a "free automaton" human beings cannot be totally determined and controlled from outside but, because of their mental abilities, have the possibility and the task of giving and forming their own patterns of behavior, in an act of free and responsible decision. This is why modern man in a highly technicized industrial society, who is to a high degree controlled from outside, feels so self-alienated. Thus there is a pedagogical need and necessity to help people discover their identity in critical self-responsibility and self-determination, by showing them the important values which can challenge and encourage them to respond freely and to accept responsibility. Indeed, precisely this need must be integrated into the information technologies of a computer culture.

To sum up, we find the following cybernetic model of the human, according to which human behavior stems from three factors: (a) a material and biological constitution, inherited as inborn dispositions or "basic programs"; (b) an external influence, above all the programs and conventions coming from education and social groups; and (c) the personal decision of human beings themselves – their ways of dealing with the first two groups of factors, of answering them and more or less assimilating them according to experience and an understanding of values. In any case, the human is not exclusively and totally determined by them, but puts their determination to some use or end. Man is not absolutely fixed by nature or society, but is relatively open – given to himself, and therefore given up to himself, as his own task.

We may suppose that the biological and intellectual feedback control system are connected to each other and take part of each other, because of the unity of mind and body which is basic to human nature. This means that the biological phenomena of the human body do not function exclusively according to biological laws of causality, but that they are also codetermined by the intellectual self-determination of man; to this corresponds the well-known phenomenon, that whether a person recovers from a serious illness depends largely on his psychic attitude toward the meaning of his life and on his will to live. For the same reason the free intellectual self-determination of man is not absolute but only limited "because it participates in the biological and sociological feedback control system and is also determined by it."

With this, we recognize that the circle of self-determination takes place at the same time in intellectual freedom and in organic constraint, and thus shows itself on two different levels of its own being. By an act of intellectual awareness and freedom it is with itself; and by an act of organic life it is spread into the organs which are spatially separate, and which are thus transformed into the living unity of a human individual. By doing so the intellectual form "in-forms" the living material of the body and disposes (or indisposes) accordingly physical events; and by intellectual awareness the self-determining and controlling life of the organism recognizes itself and attains free self-disposal.

From this results the human task of existence: *First*, to internalize his proper physical life and the social data – i.e., all external feedback control systems – and to accept them with intellectual self-determination and responsibility; and *second*, to express the latter as perfectly as possible, which means making the subjective objective and the objective subjective. Outside movement by self-expression, and inside movement by world impression, are the two complimentary directions of living act in the human being.

As a consequence of his simultaneous external and internal existence, of his being subject and object at the same time, it is impossible completely to condition or determine the human being, either from outside or from inside. It is not possible, either for others to objectify him totally, or for him fully to subjectify himself. That is why a human being is responsible for himself at the same time that he depends on the personal assistance from his fellow human beings. Future projects that are purely abstract and anonymous, or the attempt absolutely to determine the future by information technologies and statistical predictions, must be criticized as ideologies which do not recognize the reality of the self and its free initiatives. Conditioned progress – yes, but only on the condition of social responsibility, which takes care of the *individuum* and his personal needs.[8]

University of Bamberg .

AUTHOR: Heinrich Beck was born in 1929 and studied philosophy, psychology, and theology. He has lectured widely at universities in Europe and Latin America, and is currently Professor of Philosophy at the University of Bamberg. He has written extensively of the philosophy of science and technology, with *Philosophie der Technik* (Trier: Spee-Verlag, 1969) and *Kulturphilosophie der Technik: Perspektiven zu Technik – Menschheit – Zukunft* (Trier: Spee-Verlag, 1979) being two of his major works.

NOTES

1   Cybernetics is "the entire field of control and communication theory, whether in the machine or in the animal," i.e., in fully closed functional aggregates (as N. Wiener puts it in *Cybernetics* [Cambridge, MA: MIT Press, 1963], p. 11). H. J. Flechtner, in *Grundbegriffe der Kybernetik* (Stuttgart: Wissenschaftliche Verlagsgesellschaft, 1968) says that cybernetics is the "general science of structure, relation, and reaction in dynamic systems" (p. 10). Cybernetics can also be described as the science of the "structures of efficiency" (Hans Sachsse, *Einführung in die Kybernetik, unter besonderer Berücksichtigung von technischen und biologischen Wirkungsgefügen* [Braunschweig: Vieweg, 1971], p. 3), or of "systems" which are open for energy, closed for information, regulation and direction" (W. Ross Ashby, *An Introduction to Cybernetics* [London: Chapman & Hall and University Paperbacks, 1973], p. 19). According to H. Walter and W. D. Keidel, cybernetics applies itself to the "mathematical-quantitative and structural consideration of reactions concerning complex systems, that is systems which work out information in the same way as machines, living creatures, or groups of living creatures; the *theory of information*, however, applies itself to a corresponding consideration of communication between human being and human being (by signs) and between human being and environment (by observation)" ("Vorwort" to Werner Meyer-Eppler's *Grundlagen und Anwendungen der Informationstheorie* [Berlin, Heidelberg, and New York: Springer Verlag, 1969]).

2   This is not altered by the fact that such objectivity is limited by Heisenberg's uncertainty principle. Because a physicist changed the condition of an elementary particle in the act of observing it, he cannot fix what that condition is like, if it is not observed. But from the fact that we are not able to comprehend the causal determination of the microcosmic elementary processes it does not follow that these processes are not completely determinated in themselves. Such determinism may, however, be formed in a different way than in the macrocosmic field, which can be comprehended by the laws of nature.

3   If such self-determination is limited or abolished by means of "forced actions," we are dealing with "borderline cases" or disease – which presupposes the existence of the opposite basic tendency and the "normal case."

4   If the values and the ways of behaving to which society conditions us by education were radically determining, we would never be able to ask critical questions about their justification. But we are able to break loose from such a conditioning, which is often felt as an unwarranted restraint on our conscience, as soon as we recognize the nonsense of its claim. The task of forming a conscience requires this critical distance

and an openness toward absolute values which are the measuring base of every criticism of senseless constraints. See my *Génération en conflict* (Quebec: 1972).

5 For a closer analysis and interpretation of the cybernetic structure typical of mind and organism, see my *Der Akt-Character des Seins: Eine spekulative Weiterführung der Seinslehre des Thomas v. Aquin aus einer Anregung durch das dialektische Prinzip Hegels* [The act-character of being: speculative continuation of Thomas Aquinas' theory of being, motivated by Hegel's dialectic principle] (Munich: Heuber, 1965); "Die rhythmische Struktur der Wirklichkeit" [The rhythmic structure of reality], *Philosophia naturalis* **9** (1965), 485–504; and "Analogia Trinitatis – ein Schlüssel zu Strukturproblemen der heutigen Welt" [Key to structural problems of today's world], *Salzburger Jahrbuch für Philosophie* **XXV** (1980), pp. 87–99.

6 Even if an automaton is determined, controlled, or produced by another one, and this one again by another, and so on, the situation does not essentially change, because all function as parts of a comprehensive system. According to Gödel's Theorem it is impossible in principle that the law of producing a series to which all its members conform and which fixes their order is produced or determined by one member of that specific series; it has to be found and reflected on a higher mathematical level. Cf. Beda Thum, "Die Selbsttechnisierung des Denkens" [The self-technification of thinking], *Naturwissenschaft und Theologie* **9** (Freiburg: 1967).

7 Here we may remember that, for example, Nicolai Hartmann sees the character of "finality," i.e., of a behavior fixed to an objective, in a structure where the cause is determined and controlled by the effect (although he wanted to ascribe finality only to a human mind). Cf. N. Hartmann, *Teleologisches Denken* [Teleological thinking] (Berlin: De Gruyter, 1951), pp. 64–88, esp. pp. 71ff.; and *Ethik* (Berlin: De Gruyter, 1962), pp. 192ff. and 669ff.

8 For a closer study: Gotthard Günther, *Das Bewußtsein der Maschinen: eine Metaphysik der Kybernetik* (Krefeld: Agis-Verlag, 1963); Georg Klaus, *Kybernetik und Gesellschaft* (Berlin: Deutscher Verlag der Wissenschaften, 1964), Helmar G. Frank (ed.), *Kybernetik: Brücke zwischen den Wissenschaften* (Frankfurt: Umschau Verlag, 1964), and *Kybernetik und Philosophie* (Berlin: Duncker und Humbolt, 1966); D. J. Bartholomew, *Stochastic Models for Social Processes* (New York: Wiley, 1967); John Cohen, *Human Robots in Myth and Science* (London: Allen & Unwin, 1966); Eberhard Lang, *Zu einer kybernetischen Staatslehre: Eine Analyse des Staates auf der Grundlage des Regelkreismodells* (Salzburg and Munich: Pustet, 1970); Karl Steinbuch, *Automat und Mensch: Auf dem Weg zu einer kybernetischen Anthropologie* (Berlin, Heidelberg, and New York: Springer Verlag, 1971), and my *Kulturphilosophie der Technik: Perspektiven zu Technik-Menschheit-Zukunft* [Philosophy of culture in the epoch of technics: perspectives of technics-mankind-future] (Trier: Spee-Verlag, 1979).

FRED DRETSKE

MINDS, MACHINES AND MEANING

ABSTRACT. Computers do not think. This thesis is defended, initially, by distinguishing between agents and the actions they perform with instruments, and those operations which instruments themselves perform. Furthermore, it is argued that computers do not even perform the arithmetic operations, because they merely manipulate electrical impulses to which human beings have assigned certain numerical meanings. Numerical information in the true sense is not what computers deal with, since in order to be information symbols must have a relevance for the entity manipulating them.

Computers are machines and there are a lot of things machines can't do. But there are a lot of things *I* can't do: speak Turkish, understand James Joyce, or recognize a hemlock when I see one. Yet, numerous as are my incapacities, they do not materially affect my status as a thinking being. I lack *specialized* skills, knowledge and understanding, but nothing that is *essential* to membership in the fraternity of rational agents. With machines, though, and this includes the most sophisticated modern computers, it is different. They *do* lack something that is essential.

Or so some say. And in saying it they are, or *should* be, prepared to tell us what *is* essential, what *are* the conditions for membership in this exclusive fraternity. If one doesn't have to understand James Joyce to gain admission, is there, then, something *else* one has to understand? If so, what? How to order a meal in a restaurant? If one doesn't have to be able to recognize hemlocks, is there something else one must be able to recognize? What? Arches? Or is it that although there is nothing, no *specific* thing, one has to be able to understand, identify, or know, there is nonetheless something *or other* toward which one must exhibit a modicum of cognitive skill? If so, it is hard to see how to deny computers admission to the club.

I happen to be one of those philosophers who, though happy to admit that minds compute, and in this sense *are* computers, have great difficulty seeing how computers could be minded. I'm not (not *now* at least) going to complain about the impoverished inner life of the computer — how it doesn't *feel* pain, fear, love or anger. Nor am I going to talk about the mysterious inner light of consciousness. For I'm interested in the *cognitive* abilities of machines, and I'm not at all sure one needs feelings or reflective consciousness to solve problems, play games, recognize things,

97

Carl Mitcham and Alois Huning (eds.), Philosophy and Technology II, 97–109.

or understand instructions. Nor am I going to talk about how *bad* computers are at doing what most children can do – e.g., speak and understand their native language, make up a story or appreciate a joke. For such comparisons make it sound like a competition, a competition in which humans, with their enormous head start, and barring dramatic breakthroughs in A1, will remain unchallenged for the foreseeable future. I don't think the comparison should be put in these terms because I don't think there is a genuine competition in this area at all. It isn't that the best machines are still at the level of two-year olds – requiring only greater storage capacity and fancier programming to grow up. Nor should we think of them as idiot savants, exhibiting a spectacular ability in a few isolated areas, but having an overall IQ too low for fraternal association. For machines, even the best of them, don't have an IQ. They don't *do* what we do – at least none of the things that, when we do them, exhibit intelligence. And its not just that they don't do them the *way* we do them or as *well* as we do them. They don't do them at all. They don't solve problems, play games, prove theorems, recognize patterns, let alone think. They don't even add and subtract.

To convince you of this, it is useful to look at our relationship to various instruments and tools. This preliminary examination will not take us far, but it will set the stage for a clearer statement of what I take to be the fundamental, and I think unbridgeable, gulf between minds and machines. In our descriptions of instruments and tools we display a tendency to assign them the capacities and powers of agents who use them. Despite the National Rifle Association's efforts to convince us that people, not guns, kill people, we all tend to think, or at least talk, of artifacts as telling us things, recognizing, sensing, remembering and, in general, doing things that, in our more serious, literal, moments, we acknowledge to be the province of rational agents. In most cases this figurative use of language does no harm. No one is really confused. Though we open doors, and keys open (locked) doors, no one seems to worry about whether keys open doors better than we do, whether we are still ahead in this competition. Why not? Since both keys and people open doors, why doesn't it make sense to ask who does it better? Because, of course, we all understand that this isn't the same sense of the verb "to open". We open locked doors *with* keys and the only sense in which the key is said to open doors is the sense in which that is the instrument typically used to perform this act. *We* are the agents. The key is the instrument. Because we sometimes speak of the instrument in terms appropriate to the agent, speak of the key as doing what the agent does with the key, we should not allow ourselves to accept the silly idea that, therefore, there are some things we do that keys can also do.

But before jumping to the hasty conclusion that the computer is, like the key, merely a fancy instrument in our cognitive tool box – and, thus, taken by itself, unable to do what we can do with it – we should look around. After all, don't amplifiers really amplify? Surely it isn't *we* who amplify with this electronic device in the way a carpenter pounds nails with a hammer. And who really picks up the dust: the maid or the vacuum cleaner? Is the vacuum cleaner merely an instrument that the maid uses to pick up dust? Well yes, but *not* quite the way one uses a key to open a door or a hammer to pound a nail. One pushes the vacuum cleaner around but *it* picks up the dust. In this case (unlike the key case) the question: "Who picks up dust better: people or vacuum cleaners?" *does* make good sense, and the answer, obviously, is the vacuum cleaner. We may never have had any real competition from keys for opening doors, but we seem to have lost the race for picking up dust to vacuum cleaners.

What such examples indicate is that the agent-instrument distinction is no certain guide to who or what is to be given credit for a performance. We do things. Instruments do things. And sometimes we do things with instruments. Who gets the credit depends on what is done and how it is done. To ask whether a simple pocket calculator can really multiply or whether, instead, it is *we* who multiply *with* the calculator is to ask, whether relative to this task, the agent-instrument relation is more like our use of a key in opening a door or our use of a vacuum cleaner in picking up dust. *Some* of the instruments we use *literally* perform the tasks for which they are used. Others are mere tools, incapable of doing what we do with them.

Well, then, are computers our computational keys? Or are they more like our vacuum cleaners? Do they literally do the computational tasks that we sometimes do without them but do it better, faster, and more reliably? This may sound like a rather simple-minded way to approach the issue of minds and machines, but unless one gets clear about the relatively simple question of *who* does the job, the person or his pocket calculator, in adding up a column of figures, one is unlikely to make much progress in penetrating the more baffling question of whether more sophisticated machines exhibit (or will some day) some of the genuine qualities of intelligence. For I assume that if a machine can really play chess, prove theorems, understand a text, and recognize an object, if these descriptions are *literally* true, then to that degree it participates in the intellectual enterprise. To that extent it is minded.

So let me begin with a naive question: can computers add? We may not feel very threatened if this is *all* they can do. Nevertheless, if they can do even this much, then the barriers separating mind and machine have been breached and there is no reason to think they won't eventually be removed.

The following argument is an attempt to show that whatever it is that computers are doing when we use them to answer our arithmetical questions, it isn't addition. Addition is an operation on numbers. We add 7 and 5 to get 12, and 7, 5 and 12 are numbers. The operations computers perform, however, are not operations on numbers. At best, they are operations on certain physical tokens that *stand for* or are interpreted as standing for, the numbers. Therefore, computers don't add.

In thinking about this argument (longer than I care to admit) I decided that there was something right about it. *And* something wrong. What is right about it is the perfectly valid (and relevant) distinction it invokes between a representation and what it represents, between a sign and what it signifies, between a symbol and its meaning. We have various ways of representing or designating the numbers. The written numeral "2" stands for the number 2. When uttered in the right context, so does the sound "tu". Unless equipped with special pattern recognition capabilities, machines are not prepared to handle these particular symbols. But they have their own system of representation: open and closed switches, the orientation of magnetic fields, the distribution of holes on a card, and so on. But whatever the form of representation, the machine is obviously restricted to operations on the symbols or representation themselves. It has no access, so to speak, to the *meaning* of these symbols, to the things the representations represent, to the numbers. When instructed to add two numbers stored in memory, the machine manipulates representations in some electro-mechanical way until it arrives at another representation – something that (if things go right) stands for the sum of what the first two representations stood for. At no point in the proceedings do numbers, in contrast to numerals, get involved. And if, in order to add two numbers, one has to perform some operation on the numbers themselves, then what the computer is doing is not addition at all.

But this argument, as I am sure you are aware, shows *too* much. It shows that *we* don't add either. For whatever operations may be performed in our nervous system when we add two numbers, it quite clearly isn't an operation on the numbers themselves. Brains have their own coding systems, their own means of representing the objects (including the numbers) about which its (or our) thoughts and calculations are directed.

In this respect a person is no different than a computer. Biological systems may have different systems for representing the objects of thought, but they, like the computer, are necessarily limited to manipulating these representations. This is merely to acknowledge the nature of thought itself. It is a *vicarious* enterprise, a *symbolic* activity. Adding two numbers is a way of thinking *about* two numbers, and thinking *about X* and *Y* is not a way of pushing *X* and *Y* around. It is a way of pushing around their symbolic representatives.

What is wrong with the argument, then, is the assumption that in order to add two numbers, a system must literally perform some operation on the numbers themselves. What the argument shows, if it shows anything, is that in order to carry out arithmetical operations, a system must have a way of representing the numbers and have the capacity for manipulating these representations in accordance with arithmetic principles. But isn't this precisely what computers do?

I have discussed this argument at some length only to make the point that all cognitive operations (whether by artifacts or natural biological systems) will (assuming the truth of materialism) inevitably be realized in some electrical, chemical or mechanical operation over physical structures. This fact alone doesn't tell us anything about the cognitive nature of the operation being performed – whether, for instance, it is an inference, a thought or the taking of a square root. For what makes these physical operations into thoughts, inferences, or arithmetical calculations is, among other things, the meaning or semantics of those structures over which they are performed. To think about the number 7, or our cousin George, you needn't do anything with the number 7 or our cousin George. But you do need the internal resources for representing 7 and George and the capacity for manipulating these representations in ways that stand for operations on, or conditions of, the things being represented.

This should be obvious enough. Opening and closing relays doesn't count as addition, or as moves in a chess game, unless the relays, or their various states, stand for numbers and chess moves. But what may not be so obvious is that these physical activities cannot acquire the relevant kind of meaning merely by *assigning* them an interpretation, by letting them mean something *for us*. Unless the symbols being manipulated mean something *for the system manipulating them*, their meaning, whatever it is, is irrelevant to evaluating what the system is doing when it manipulates them. I cannot *make* you, someone's parrot, or a machine think about my cousin George, or the number 7, just by assigning meanings in accordance with which this is what your (the parrot's, the machine's) activities stand for. Everything depends on whether this is the meaning these events have for you, the parrot, or the machine.

To illustrate, consider a simple galvanometer. We can write anything we choose on the face of this instrument, thereby investing its behavior (the position of the pointer) with any meaning we please. Call it an orgone energy detector. This cosmic energy, the *elan vital* of the universe, is measured in *orgs* so we divide the scale in *milliorgs*. If you clasp the handles of this wondrous machine, it will measure the amount of orgone energy pulsing through your body. A movement of the pointer *means* that there is a change in your orgone energy potential (a very bad sign – a sign of an unstable personality). At least it would mean this, and *has* meant this, to crackpot inventors and their devotees. But this, clearly, is something the galvanometer's fluctuations mean *to naive users* of the instrument. The instrument itself should hardly be credited (or blamed) for meaning this. It is, after all, a simple galvanometer. Movements of its pointer signify changes in the flow of electric current. If we want to assign these movements an additional meaning, if we want to interpret them as meaning something about orgone energy, personality, or the strength of a person's sex drive, that is up to us. This doesn't change what the machine is doing. It means what it always meant – something about the flow of electric current. We can't change what *it* is doing by changing what its activities mean to us.

Or consider a dog that has been trained to detect marijuana. Custom's agents can use the dog to detect concealed marijuana. When the dog barks, wags its tail, or does whatever it was trained to do when it smells marijuana, this alerts the agent to its presence. As a result of the dog's behavior, the custom's official comes to believe that there is marijuana in the suitcase. But what does the dog believe? In this case we are using the dog in the way we might use an instrument – as a device whose sensitivity to one thing (a certain smell in the case of the dog) tells *us* something about the object emitting that smell: that it is marijuana. But it doesn't tell *the dog* this. The internal states aroused *in us* as a result of the dog's sensitive discriminations have a semantics, a meaning, which completely transcends the meaning that we can plausibly assign the dog. If the dog has any beliefs, they are, presumably, beliefs concerning a certain smell. *That* is what the stimulus means *for the dog* and, hence, what must be considered in determining what the dog is doing. *He* is identifying a certain smell. *We* (are using the dog to) identify concealed marijuana. To describe the dog as recognizing marijuana (*as* marijuana) is to transfer to the dog something *we* use the dog to do. We can certainly use dogs to solve our detection problems, just as we can use machines to solve our computational problems, but we should not let this mislead us into assigning an inflated significance to what our instruments are doing.

As a final illustration of this important point, suppose we designed a machine to determine the relative lengths of objects from their photographs. What the machine does is to determine the relative lengths of the two photographic images (of, say, *A* and *B*) and it types out "*A* is longer than *B*" if and only if the photographic image of *A* is longer than the photographic image of *B*. Question: is this system comparing the lengths of *A* and *B*? Is this the correct way to describe what it is doing? We have made it say something (namely, "*A* is longer than *B*") which suggests that it is comparing the lengths of *A* and *B*, but what reason is there for thinking that this is what these symbols mean to the machine? If we are careful in the way we produce the photographs, always arranging to have *A* and *B* at equal distances from the camera and perpendicular to the line of sight, we can use the machine to tell which is longer: *A* or *B*. And since this is the question whose answer we seek, there is no reason (from the standpoint of user convenience) not to let the machine's output express exactly what we learn when we use it; *viz.*, that *A* is longer than *B*. But if we ask what the machine is doing, it seems clear that this output sentence does not describe what the machine itself is doing. It is comparing representations – photographic images. The meaning or semantic value of these representations, what they stand for, are completely irrelevant to its activities. By making the machine print out the words "*A* is longer than *B*" we create an illusion that what the machine is doing is what we are doing with the machine – comparing the lengths of *A* and *B*. But the machine no more thinks, judges, or says that *A* is longer than *B* than our dog thinks, judges, or says (by wagging his tail) that the suitcase contains marijuana. Though the machine can be made to print out the words "*A* is longer than *B*" what *it* means by these words (if it means anything) is that *this* is longer than *that* where *this* and *that* happen to be representations of *A* and *B*.

But this makes it sound as though our descriptions of machines, though perhaps a bit inflated with our own cognitive purposes, are nonetheless in the right ballpark. In comparing the machine to our marijuana sniffing dog, I encouraged this view by suggesting that the machine is in somewhat the same position as the dog. Though they don't do exactly what we do in using them, they nonetheless do *something* of cognitive interest. If the dog doesn't identify or recognize marijuana *as* marijuana, it at least recognizes it as that funny smelling stuff. And so it may be with the machine. In industrial applications of machine vision, for example, it may be an exaggeration to say that the machine recognizes short circuits on the printed circuit board it examines. After all, it merely searches for breaks or discontinuities in the metallic deposit. The machine is concerned with *spatial* discontinuities; its users are worried about *electrical* discontinuities. Under the right circumstances, we can use the former as a sign of the latter, but the two are quite different. Still, the machine does do something worth dignifying with the word "recognition" even if *what* it recognizes isn't quite what we say it is in our careless moments. Or does it?

I indicated at the outset that I don't think it is merely that machines lag behind us in the cognitive competition. Rather, they, or most of them anyway, haven't even entered the race. And the reason for this wholesale skepticism is, as I have just argued, that to understand *what* a system does when it engages in the manipulation of symbols, it is necessary to know, not just what these symbols mean, what interpretation they can be *assigned*, but what they mean for the system performing the operations. And my deflationary view of the capabilities of machines arises from my conviction that however versatile machines may be in handling symbols, the symbols they handle have no meaning for the machine itself. This should not be taken to imply that machines cannot serve as useful models for our intellectual operations. On the contrary. Their prevalent use in cognitive psychology indicates otherwise. What it does imply is that the machines do not literally do what we do when we engage in those activities for which they provide an effective model. A computer simulation of a hurricane needn't, and obviously doesn't, blow trees down. Why should we suppose that a useful computer model of problem solving must itself solve problems, reason or compute?

Why do I think that the input to, and activities of, a machine are totally devoid of meaning for the machine itself? To answer this question I need to talk a moment about information. I need to talk about information because, despite our sloppy talk in this area, machines, most of them anyway, have no need for, in no way depend on, and have no way of getting, information about the objects and conditions its symbols stand for. And without information, without some cognitive access to the world its representations represent, the machine's symbol structures, however meaningful they may be to others, are meaningless to the machine itself.

These remarks about information and the computer's lack of access to it may sound paradoxical. Aren't computers our information processors *par excellence*? Isn't this THE AGE OF INFORMATION and isn't the computer responsible for ushering in this age?

Forget about Madison Avenue hyperbole for a moment and think about why information is important, why it is such a valuable commodity, why we invest billions in its collection, storage and retrieval. If one consults a dictionary, one will find information described in terms of such notions as "intelligence," "news," "instruction," and "knowledge." These terms are suggestive. They have a common nucleus. They all point in the same direction – the direction of *truth*. Information is what is capable of yielding knowledge, and since knowledge requires truth (you can't know I have a toothache if I don't have one), information requires it also. We say that a pamphlet contains information about how to probate a will, for example, and we say this because we believe that someone could *learn* something about probating a will by consulting the pamphlet. Information booths are not merely places where clerks are paid to utter meaningful sounds. What makes them information booths is that the clerks either know, or can quickly find out, about matters of interest to the average patron. One can learn, come to know, by making inquiries at such places. Hence, *information* booths.

When scientists tell us that we can use the pupil of the eye as a source of information about another person's feelings or attitudes, that a thunder signature contains information about the lightning channel that produced it, that the dance of a honeybee contains information as to the whereabouts of the nectar, or that the light from a star carries information about the chemical constitution of that body, the scientists are clearly referring to information as something carried by reliable signs, trustworthy indicators, as something from which we can learn. And a state of affairs, condition, or signal contains information about X to just that degree to which its reliable indication of the condition of X permits one to learn how things stand with respect to X. *This* is why information is important. Despite some people's tendency to speak of anything stored on a magnetic disc as information, a random set of symbols carries no information. We sometimes speak of *mis*information, but this is not a species of information anymore than fools gold is a kind of gold or decoy ducks are a species of duck. To get information about X is to get something whose reliable connection with X tells you how things stand with respect to X. This is why the fuel gauge in your car is useful. It carries information about the amount of fuel you have left. It tells you this. It lets you know. Broken gauges don't carry information however much we may, through ignorance, depend on them.

There is, of course, a statistical theory of information which tells us that the amount of information in the encyclopedia remains the same (perhaps even increases) when the letters are scrambled so that gibberish results. According to this theory, there is more information contained in a randomly generated sequence of letters than there is intelligible prose (because less redundancy). But this theory of information, though useful for limited engineering purposes, is largely irrelevant to cognitive studies. This, surely, isn't the sense of information which explains why, in order to extract information, captured prisoners are tortured. This isn't what suspicious husbands hire detectives to provide.

Even when a string of symbols makes perfectly good sense, it may or may not carry information. It all depends on what, if anything, you can learn from it. If I tell you I have a headache, I have given you *zero* information about my head. The reason I haven't communicated any information by this form of words is that I do not, in fact, have a headache. You might come to believe I do when you hear me utter these words, but you will not have been *informed*. The words I uttered *meant* that I had a headache, but that is not the information they carried.

I digress about this topic because it is important to understand that to pick up, process and store information it isn't enough to have a symbol manipulation capacity. Unless the symbols being manipulated carry information about the mattters whereof they speak, unless these symbols stand to the world in a certain regular way, a way which indicates, reveals, or somehow tells the system how the world is, these symbol structures, however rich they may be in meaning *for us*, are devoid of the kind of meaning that plays a role in the life of the system itself.

Information is irrelevant to the operation of a machine in a way it is not irrelevant to the operation of sea snails or bacteria. If the sea snail doesn't get information about the turbulence in the water, it risks being dashed to pieces when it swims to the surface to obtain the micro-organisms on which it feeds. If a certain strain of bacteria do not get information about the direction of geomagnetic north, they cannot orient themselves so as to avoid the toxic surface water rich in oxygen. If, in other words, an animal's internal sensory states were not rich in information about the presence of prey, predators, cliffs, obstacles, water, and heat, it would not survive. It isn't enough to have the internal states of these creatures mean something *to us*, for it to have symbols it can manipulate. If these symbols don't mean something to the animal itself, if they don't somehow register the goings-on in its surroundings (in something like the way a thermometer registers conditions in its surroundings), its symbol manipulation capacity is completely worthless to the animal. Of what possible use is it to be able to operate on symbols for food, danger, and sexual mates if the occurrence of these symbols is wholly unrelated to the actual presence of food, danger and mates?

With machines it is different. It matters not at all whether the so-called "information" we supply the computer is really information or not. The machine doesn't need information. Nothing it does depends on it receiving information. *Our* purposes may be frustrated if we supply the machine with fiction rather than fact, but the machine's operation is in no way impaired by such deception. A machine is, as it were, a *preserver* or *transmitter* of information in the same way the principles of deductive logic are preservers or transmitters of truth. If you start with true premises, and reason validly (i.e., in accordance with the principles of logic) you will reach true conclusions. But the principles themselves are totally insensitive to the truth value, the informational status, of the sentences on which they operate. They are as happy in concluding that "Elephants can fly" from the premises that "Elephants are lizards" and "All lizards can fly" as they are with a corresponding argument about robins and birds. And a computer is no different. Its function is to manipulate (process, store, transform) what it is given, and if what it is given is information, it will be happy to return the favor. But nothing *it* does depends on the informational value of the symbols it manipulates.

Perception is the name we give to those processes by means of which a system obtains information about its surroundings. Just as the blinking light on your car dash carries information about your oil pressure, so the electrical pulses surging down your optic nerve carry information about the distribution of light in your surroundings. Without some kind of perceptual contact with the outside world, nothing happening in the system means anything in the relevant sense about what is happening outside the system. We can still arrange to *assign* a meaning to the system's internal states (or its output) by letting it manipulate (or print out) symbols that are meaningful to us. We can assign an interpretation that makes our use of the machine more convenient. But this surgical graft of meaning onto a system is clearly irrelevant to determining what the system itself is actually doing.

These remarks should make clear that I think work in machine perception, pattern recognition and robotics has greater relevance to the cognitive capacities of machines than the most sophisticated programming in such purely intellectual tasks as language translation, theorem proving, or game playing. For a pattern recognition device is at least a device whose internal states mean something about what is happening, or the conditions that exist, in its environment. It is (to use some of Bert Dreyfus's language) at least *in the world*.

Nevertheless, even here, there remains a gulf between what machines do and what we do when we recognize a hemlock, identify a chair, or perceive a sequence of numbers. For the machine, though it can absorb information about its immediate surroundings, has no use for this information. Nothing to *do* with it. Even the lowly thermostat can extract information (about temperature) from its surroundings. In a sense, it even has something it is supposed to do with this information: namely, regulate the furnace. But there is still no sense in which the system itself (in contrast to we who use the system) needs *this* information. Nothing *it* does depends on the information contained in the input. A badly calibrated thermostat will certainly make our life uncomfortable, but it wouldn't make the slightest difference to the thermostat itself.

Consider a compass needle. Its orientation carries information about the directon of magnetic north. Its pointing in *that* direction means that north is in *that* direction. But though we sometimes speak, figuratively, of the compass sensing or perceiving the direction of north, the compass itself does not literally perceive. It "perceives" north in the way a key "opens" doors. Suppose, however, that we surgically install this compass in the leg of a man. Assuming the compass is now part of the man, have we given *the man* a new perceptual capacity? Is he now magnetotactic? Can *he* now perceive the direction of magnetic north? Of course not. As long as the system in which the information is carried has no need for it or, though having a need for it, no way to exploit this information in the furtherance of its own ends or purposes, the information is meaningless to the system itself.

Contrast this case with the bacteria I mentioned earlier. Such bacteria have internal magnetosomes (as they are called) harnessed to a motor control system whose function it is to enable the creature to satisfy its needs (to escape oxygen-rich environments) by exploiting the information (about geomagnetic north) embodied in these internal magnets. Here we have something approximating a *real* perceptual system. What is the difference? Why are we tempted to say that the bacteria can (literally) sense magnetic north, but not the compass or the surgically altered man? The difference, of course, is that we have embedded a natural indicator, a carrier of information, into a system which has *both* a need for this information *and* the resources for exploiting it in promoting its own vital purposes. Unlike the compass or the man, the orientation of the magnets mean something *to the bacteria*.

I do not mean to be keeping machines at arm's length by denying them legitimate needs of their own. I'm not quite sure what needs are (except, perhaps, requirements for survival or well-functioning), but I am, for the sake of argument, willing to let machines have them. They don't need vitamins, fresh air, or companionship, to be sure, but they do need, let us say, electricity, oil and low humidity. But if these are genuine needs of the machine, I don't think we will get any closer to intelligent machines, machines capable of doing something of cognitive interest, until we build them to not only get the information they require to satisfy these needs, but give them the capability of *using* this information for this purpose. Only then will the symbols a computer manipulates mean something to the computer itself. And, I might add, only then will we be in a *real* competition with machines (not just their owners) for the scarce resources of this world.

University of Wisconsin, Madison

AUTHOR: Fred Dretske (born 1932) has a B.S. in electrical engineering and advanced degrees in philosophy. He is Professor of Philosophy at the University of Wisconsin, Madison. His publications include *Seeing and Knowing* (Chicago: University of Chicago Press, 1969) and *Knowledge and the Flow of Information* (Cambridge, MA: MIT Press, 1981).

HUBERT L. DREYFUS AND STUART E. DREYFUS

FROM SOCRATES TO EXPERT SYSTEMS:
THE LIMITS OF CALCULATIVE RATIONALITY

ABSTRACT. This paper examines the general epistemological assumptions of artificial intelligence technology and recent work in the development of expert systems. These systems are limited because of a failure to recognize the real character of expert understanding, which is acquired as the fifth stage of a five-step process. A review of the successes and failures of various specific expert system programs confirms this analysis.

For the past quarter of a century researchers in Artifical Intelligence (AI) have been trying without success to write programs which will enable computers to exhibit general intelligence like Hal in 2001. Now out of this work has recently emerged a new field called knowledge engineering which by limiting its goals has applied this research in ways that actually work in the real world. The result is the so-called expert system which has been the subject of recent cover stories in *Business Week* and Edward Feigenbaum's book *The Fifth Generation: Artificial Intelligence and Japan's Computer Challenge to the World.*[1] The occasion for this new interest in machine intelligence is no specific new accomplishment but rather a much publicized competition with Japan to build a new generation of computers, with built-in expertise. This is the so-called fifth generation. (The first four generations were computers whose components were vacuum tubes, transistors, chips, and large scale integrated chips.) According to a *Newsweek* headline: "Japan and the United States are rushing to produce a new generation of machines that can very nearly think."

Feigenbaum, one of the original developers of expert systems, who stands to profit greatly from this competition, spells out the goal.

In the kind of intelligent system envisioned by the designers of the Fifth Generation, speed and processing power will be increased dramatically; but more important, the machines will have reasoning power: they will automatically engineer vast amounts of knowledge to serve whatever purpose humans propose, from medical diagnosis to product design, from management decisions to education.[2]

What the knowledge engineers claim to have discovered is that in areas which are cut off from everyday common sense and social intercourse, all a machine needs in order to behave like an expert are some general rules and lots of very specific knowledge. As Feigenbaum puts it:

111

Carl Mitcham and Alois Huning (eds.), Philosophy and Technology II, 111–130.

The first group of artificial intelligence researchers. . . was persuaded that certain great, underlying principles characterized all intelligent behavior. . . .

In part, they were correct. . . . [Such strategies] include searching for a solution (and using "rules of good guessing" to cut down the search space); generating and testing (does this work? no; try something else); reasoning backward from a desired goal; and the like.

These strategies are necessary, but not sufficient, for intelligent behavior. The other ingredient is knowledge – specialized knowledge, and lots of it. . . . No matter how natively bright you are, you cannot be a credible medical diagnostician without a great deal of specific knowledge about diseases, their manifestations, and the human body.[3]

This specialized knowledge is of two types:

The first type is the facts of the domain – the widely shared knowledge. . . that is written in textbooks and journals of the field, or that forms the basis of a professor's lectures in a classroom. Equally important to the practice of the field is the second type of knowledge called *heuristic knowledge*, which is the knowledge of good practice and good judgement in a field. It is experiential knowledge, the "art of good guessing" that a human expert acquires over years of work.[4]

Using all three kinds of knowledge Feigenbaum developed a program called DENDRAL which is an expert in the isolated domain of spectrograph analysis. It takes the data generated by a mass spectrograph and deduces from this data the molecular structure of the compound being analyzed. Another program, MYCIN, takes the results of blood tests such as the number of red cells, white cells, sugar in the blood, etc. and comes up with a diagnosis of which blood disease is responsible for this condition. It even gives an estimate of the reliability of its own diagnosis. In their narrow areas, such programs are almost as good as the experts.

And is not this success just what one would expect? If we agree with Feigenbaum that: "almost all the thinking that professionals do is done by reasoning. . . ."[5] we can see that once computers are used for reasoning and not just computation they should be as good or better than we are at following rules for deducing conclusions from a host of facts. So we would expect that if the rules which an expert has acquired from years of experience could be extracted and programmed, the resulting program would exhibit expertise. Again Feigenbaum puts the point very clearly:

[T]he matters that set experts apart from beginners, are symbolic, inferential, and rooted in experiential knowledge. Human experts have acquired their expertise not only from explicit knowledge found in textbooks and lectures, but also from experience: by doing things again and again, failing, succeeding. . . getting a feel for a problem, learning when to go by the book and when to break the rules. They therefore build up a repertory of working rules of thumb, or "heuristics," that, combined with book knowledge, make them expert practitioners.[6]

Since each expert already has a repertory of rules in his mind, all the expert system builder need do is get the rules out and program them into a computer.

This view is not new. In fact, it goes back to the beginning of Western culture when the first philosopher, Socrates, stalked around Athens looking for experts in order to draw out and test their rules. In one of his earliest dialogues, *The Euthyphro*, Plato tells us of such an encounter between Socrates and Euthyphro, a religious prophet and so an expert on pious behavior. Socrates asks Euthyphro to tell him how to recognize piety: "I want to know what is characteristic of piety. . . to use as a standard whereby to judge your actions and those of other men." But instead of revealing his piety-recognizing heuristic, Euthyphro does just what every expert does when cornered by Socrates. He gives him examples from his field of expertise; in this case situations in the past in which men and gods have done things which everyone considers pious. Socrates persists throughout the dialogue in demanding that Euthyphro tell him his rules, but although Euthyphro claims he knows how to tell pious acts from impious ones, he will not state the rules which generate his judgments.

Plato admired Socrates and sympathized with his problem. So he developed an account of what caused the difficulty. Experts had once known the rules they use, Plato said, but then they had forgotten them. The role of the philosopher was to help people remember the principles on which they act. Knowledge engineers would now say that the rules the experts use have been put in a part of their mental computers where they work automatically.

When we learned how to tie our shoes, we had to think very hard about the steps involved. . . Now that we've tied many shoes over our lifetime, that knowledge is "compiled," to use the computing term for it; it no longer needs our conscious attention.[7]

On this Platonic view the rules are there functioning in the expert's mind whether he is conscious of them or not. How else could we account for the fact that he can perform the task?

Now 2000 years later, thanks to Feigenbaum and his colleagues, we have a new name for what Socrates and Plato were doing:

[W]e are able to be more precise. . . and with this increased precision has come a new term, *knowledge acquisition research.*[8]

But although philosophers and even the man in the street have become convinced that expertise consists in applying sophisticated heuristics to masses of facts, there are few available rules. As Feigenbaum explains:

[A]n expert's knowledge is often ill-specified or incomplete because the expert himself doesn't always know exactly what it is he knows about his domain.[9]

So the knowledge engineer has to help him recollect what he once knew.

[An expert's] knowledge is currently acquired in a very painstaking way; individual computer scientists work with individual experts to explicate the expert's heuristics – to mine those
jewels of knowledge out of their heads one by one. . . the problem of knowledge acquisition is the critical bottleneck in artificial intelligence.[10]

When Feigenbaum suggests to an expert the rules the expert seems to be
using he gets an Euthyphro-like response. "That's true, but if you see
enough patients/rocks/chip designs/instruments readings, you see that it
isn't true after all."[11] and Feigenbaum comments with Socratic
annoyance: "At this point, knowledge threatens to become ten thousand
special cases."[12]

There are also other hints of trouble. Ever since the inception of Artificial Intelligence, researchers have been trying to produce artificial experts by programming the computer to follow the rules used by masters in
various domains. Yet, although computers are faster and more accurate
than people in applying rules, master-level performance has remained
out of reach. Arthur Samuel's work is typical. In 1947, when electronic
computers were just being developed, Samuel, then at IBM, decided to
write a checker playing program. Samuel did not try to make a machine
play checkers by brute force calculation of all chains of moves clear to the
end. He calculated that if you tried to look to the end of the game with the
fastest computer you could possibly build, subject to the speed of light, it
would take 10 followed by 21 zeros centuries to make the first move. So he
tried to elicit heuristic rules from checker masters and program a computer to follow these rules. When the rules the experts came up with did not
produce master play, Samuel became the first and almost the only AI
researcher to make a learning program. He programmed a computer to
vary the weights used in the rules, such as the trade-off between center
control and loss of a piece, and to retain the weights that worked best.
After playing a great many games with itself the program could beat
Samuel, which shows that in some sense computers can do more than
they are programmed to do. But the program still could not beat the sort
of experts whose heuristic rules were the heart of the program.

The checkers program is not only the first and one of the best experts
ever built, but it is also a perfect example of the way fact turns into fiction
in AI. The checkers program once beat a state checkers champion. From
then on AI literature cites the checker program as a noteworthy success.
One often reads that it plays at such a high level that only the world champion can beat it. Feigenbaum, for example, reports that "by 1961
[Samuel's program] played championship checkers, and it learned and
improved with each game."[13] Even the usually reliable *The Handbook of*

Artificial Intelligence states as a fact that "today's programs play championship-level checkers."[14] In fact, Samuel said in a recent interview at Stanford University, where he is a retired professor, that the program did once defeat a state champion but the champion "turned around and defeated the program in six mail games." According to Samuel, after 35 years of effort, "the program is quite capable of beating any amateur player and can give bettter players a good contest." It is clearly no champion. Samuel is still bringing in expert players for help but he "fears he may be reaching the point of diminishing returns." This does not lead him to question the view that the masters the program cannot beat are using heuristic rules; rather, like Socrates and Feigenbaum, Samuel thinks that the experts are poor at recollecting their compiled heuristics: "the experts do no know enough about the mental processes involved in playing the game."[15]

The same story is repeated in every area of expertise, even in areas unlike checkers where expertise requires the storage of large numbers of facts, which should give an advantage to the computer. In each area where there are experts with years of experience the computer can do better than the beginner, and can even exhibit useful competence, but it cannot rival the very experts whose facts and supposed heuristics it is processing with incredible speed and unerring accuracy.

In the face of this impasse it was necessary, in spite of the authority and influence of Plato and 2000 years of philosophy, for us to take fresh look at what a skill is and what the expert acquires when he achieves expertise. One must be prepared to abandon the traditional view that a beginner starts with specific cases and, as he becomes more proficient, abstracts and interiorizes more and more sophisticated rules. It might turn out that skill acquisition moves in just the opposite direction: from abstract rules to particular cases. Since we all have many areas in which we are experts, we have the necessary data, so let's look and see how adults learn new skills.

Stage 1: Novice

Normally, the instruction process begins with the instructor decomposing the task environment into context-free features which the beginner can recognize without benefit of experience. The beginner is then given rules for determining actions on the basis of these features, like a computer following a program. The beginning student wants to do a good job, but lacking any coherent sense of the overall task, he judges his performance mainly by how well he follows his learned rules. After he has acquired more than just a few rules, so much concentration is required during the exercise of his skill that his capacity to talk or listen to advice is severely limited.

For purposes of illustration, we shall consider two variations: a bodily or motor skill and an intellectual skill. The reader wishing to see real-life examples of the process we shall outline should consult Patricia Benner's *From Novice to Expert: Excellence and Power in Clinical Nursing Practice.*[16] The student automobile driver learns to recognize such interpretation-free features as speed (indicated by his speedometer) and distance (as estimated by a previously acquired skill). Safe following distances are defined in terms of speed; conditions that allow safe entry into traffic are defined in terms of speed and distance of oncoming traffic; timing of shifts of gear is specified in terms of speed, etc. These rules ignore context. They do not refer to traffic density or anticipated stops.

The novice chess player learns a numerical value for each type of piece regardless of its position, and the rule: "always exchange if the total value of pieces captured exceeds the value of pieces lost." He also learns that when no advantageous exchanges can be found center control should be sought, and he is given a rule defining center squares and one for calculating extent of control. Most beginners are notoriously slow players, as they attempt to remember all these rules and their priorities.

Stage 2: Advanced Beginner

As the novice gains experience actually coping with real situations, he begins to note, or an instructor points out, perspicuous examples of meaningful additional components of the situation. After seeing a sufficient number of examples, the student learns to recognize them. Instructional maxims now can refer to these new *situational aspects* recognized on the basis of experience, as well as to the objectively defined *non-situational features* recognizable by the novice. The advanced beginner confronts his environment, seeks out features and aspects, and determines his actions by applying rules. He shares the novice's minimal concern with quality of performance, instead focusing on quality of rule following. The advanced beginner's performance, while improved, remains slow, uncoordinated, and laborious.

The advanced beginner driver uses (situational) engine sounds as well as (non-situational) speed in his gear-shifting rules, and observes demeanor as well as position and velocity to anticipate behavior of pedestrians or other drivers. He learns to distinguish the behavior of the distracted or drunken driver from that of the impatient but alert one. No number of words can serve the function of a few choice examples in learning this distinction. Engine sounds cannot be adequately captured by words, and no list of objective facts about a particular pedestrian enables one to predict his behavior in a crosswalk as well as can the driver who has observed many pedestrians crossing streets under a variety of conditions. Already at this level one leaves features and rules and turns to learning by prototype, now being explored by researchers such as Eleanor Rosch at Berkeley and Susan Block at M.I.T.

With experience, the chess beginner learns to recognize over-extended positions and how to avoid them. Similarly, he begins to recognize such situational aspects of positions as a weakened king's side or a strong pawn structure despite the lack of precise and universally valid definitional rules.

Stage 3: Competence

With increasing experience, the number of features and aspects to be taken account of becomes overwhelming. To cope with this information explosion, the performer learns, or is taught, to adopt a hierarchical view of decision-making. By first choosing a plan, goal or perspective which organizes the situation and by then examining only the small set of features and aspects that he has learned are the most important given that plan, the performer can simplify and improve his performance.

Choosing a plan, a goal or perspective, is no simple matter for the competent performer. It is not an objective procedure, like the feature recognition of the novice. Nor is the choice avoidable. While the advanced beginner can get along with recognizing and using a particular situational aspect until a sufficient number of examples makes identification easy and sure, to perform competently *requires* choosing an organizing goal or perspective. Furthermore, the choice of perspective crucially affects behavior in a way that one particular aspect rarely does.

This combination of necessity and uncertainty introduces an important new type of relationship between the performer and his environment. The novice and the advanced beginner applying rules and maxims feel little or no responsibility for the outcome of their acts. If they have made no mistakes, an unfortunate outcome is viewed as the result of inadequately specified elements or rules. The competent performer, on the other hand, after wrestling with the question of a choice of perspective or goal, feels responsible for, and thus emotionally involved in, the result of his choice. An outcome that is clearly successful is deeply satisfying and leaves a vivid memory of the situation encountered as seen from the goal or perspective finally chosen. Disasters, likewise, are not easily forgotten.

Remembered whole situations differ in one important respect from remembered aspects. The mental image of an aspect is flat in the sense that no parts stand out as salient. A whole situation, on the other hand, since it is the result of a chosen plan or perspective, has a "three-dimensional" quality. Certain elements stand out as more or less important with respect to the plan, while other irrelevant elements are forgotten. Moreover, the competent performer, gripped by the situation that his decision has produced, experiences and therefore remembers the situation not only in terms of foreground and background elements but also in terms of senses of opportunity, risk, expectation, threat, etc. These gripping, holistic memories cannot guide the behavior of the competent performer since he fails to make contact with them when he reflects on problematic situations as a detached observer, and holds to a view of himself as a computer following better and better rules. As we shall soon see, however, if he does let them take over, these memories become the basis of the competent performer's next advance in skill.

A competent driver beginning a trip decides, perhaps, that he is in a hurry. He then selects a route with attention to distance and time, ignores scenic beauty, and as he drives, he chooses his maneuvers with little concern for passenger comfort or for courtesy. He follows more closely than normal, enters traffic more daringly, occasionally violates a law. He feels elated when decisions work out and no police car appears, and shaken by near accidents and traffic tickets. (Beginners, on the other hand, can perpetrate chaos around them with total unconcern.)

The class A chess player, here classed as competent, may decide after studying a position that his opponent has weakened his king's defenses so that an attack against the king is a viable goal. If the attack is chosen, features involving weaknesses in his own position created by his attack are ignored as are losses of pieces inessential to the attack. Removal of pieces defending the enemy king becomes salient. Successful plans induce euphoria and mistakes are felt in the pit of the stomach.

In both of these cases, we find a common pattern: detached planning, conscious assessment of elements that are salient with respect to the plan, and analytical rule-guided choice of action, followed by an emotionally involved experience of the outcome.

Stage 4: Proficiency

Considerable experience at the level of competency sets the stage for yet further skill enhancement. Having experienced many situations, chosen plans in each, and having obtained vivid, involved demonstrations of the adequacy or inadequacy of the plan, the performer sees his current situation as similar to a previous one and so spontaneously sees an appropriate plan. Involved in the world of the skill, the performer "notices," or "is struck by" a certain plan, goal or perspective. No longer is the spell of involvement broken by detached conscious planning.

There will, of course, be breakdowns of this "seeing," when, due perhaps to insufficient experience in a certain type of situation or to more than one possible plan presenting itself, the performer will need to take a detached look at his situation. But between these breakdowns, the proficient performer will experience longer and longer intervals of continuous, intuitive understanding.

Since there are generally far fewer "ways of seeing" than "ways of acting," after understanding without conscious effort what is going on, the proficient performer will still have to think about what to do. During this thinking, elements that present themselves as salient are assessed and combined by rule to produce decisions about how best to manipulate the environment. The spell of involvement in the world of the activity will thus temporarily be broken.

On the basis of prior experience, a proficient driver approaching a curve on a rainy day may sense that he is travelling too fast. He then consciously decides whether to apply the brakes, remove his foot from the accelerator, or merely to reduce pressure on the accelerator.

The proficient chess player, who is classed a master, can recognize a large repertoire of types of positions. Recognizing almost immediately and without conscious effort the sense of a position, he sets about calculating the move that best achieves his goal. He may, for example, know that he should attack, but he must deliberate about how best to do so.

Stage 5: Expertise

The proficient performer, immersed in the world of his skillful activity, sees what needs to be done, but *decides* how to do it. For the expert, not only situational understandings spring to mind, but also associated appropriate actions. The expert performer, except of course during moments of breakdown, understands, acts, and learns from results without any conscious awareness of the process. What transparently *must* be done *is* done. We usually do not make conscious deliberative decisions when we walk, talk, ride a bicycle, drive, or carry on most social activities. An expert's skill has become so much a part of him that he need be no more aware of it than he is of his own body.

We have seen that experience-based similarity recognition produces the deep situational understanding of the proficient performer. No new insight is needed to explain the mental processes of the expert. With enough experience with a variety of situations, all seen from the same perspective or with the same goal in mind, but requiring different tactical decisions, the mind of the proficient performer seems gradually to de-compose this class of situations into subclasses, each member of which shares not only the same goal or perspective, but also the same decision, action, or tactic. At this point, a situation, when seen as similar to mem-bers of this class, is not only thereby understood but simultaneously the associated decision, action or tactic presents itself.

The number of classes of recognizable situations, built up on the basis of experience, must be immense. It has been estimated that a master chess player can distinguish roughly 50 000 types of positions. Auto-mobile driving probably involves a similar number of typical situations. We doubtless store far more typical situations in our memories than words in our vocabularies. Consequently these reference situations, un-like the situational elements learned by the advanced beginner, bear no names and, in fact defy complete verbal description.

The expert chess player, classed as an international master or grand-master, in most situations experiences a compelling sense of the issue and the best move. Excellent chess players can play at the rate of 5–10 seconds a move and even faster without any serious degradation in per-formance. At this speed they must depend almost entirely on intuition and hardly at all on analysis and comparison of alternatives. We recently performed an experiment in which an international master, Julio Kaplan, was required rapidly to add numbers presented to him audibly at the rate of about one number per second while at the same time playing five-second-a-move chess against a slightly weaker, but master level, player. Even with his analytical mind completely occupied by adding numbers, Kaplan more than held his own against the master in a series of games. Deprived of the time necessary to see problems or construct plans, Kaplan still produced fluid and coordinated play.

The expert driver, generally without any awareness, not only knows by feel and familiarity when an action such as slowing is required, but he generally knows how to perform the act without evaluating and compar-ing alternatives. He shifts gears when appropriate with no conscious awareness of his acts. Most drivers have experienced the disconcerting breakdown that occurs when suddenly one reflects on the gear shifting process and tries to decide what to do. Suddenly the smooth, almost automatic, sequence of actions that results from the performer's involved immersion in the world of his skill is disrupted, and the performer sees himself, just as does the competent performer, as the manipulator of a complex mechanism. He detachedly calculates his actions even more poorly than does the competent performer since he has forgotten many of the guiding rules that he knew and used when competent, and his performance suddenly becomes halting, uncertain, and even inappropriate.

It seems that a beginner makes inferences using rules and facts just like a heuristically programmed computer, but that with talent and a great deal of involved experience the beginner develops into an expert who intuitively sees what to do without applying rules. Of course, a description of skilled behavior can never be taken as conclusive evidence as to what is going on in the mind or in the brain. It is always possible that what is going on is some unconscious process using more and more sophisticated rules. But our description of skill acquisition counters the traditional prejudice that expertise necessarily involves inference.

Given our account of the five stages of skill acquisition, we can understand why the knowledge engineers from Socrates, to Samuel, to Feigenbaum have had such trouble getting the expert to articulate the rules he is using. The expert is simply not following any rules! He is doing just what Feigenbaum feared he might be doing – recognizing thousands of special cases. This in turn explains why expert systems are never as good as experts. If one askes the experts for rules one will, in effect, force the expert to regress to the level of a beginner and state the rules he still remembers but no longer uses. If one programs these rules on a computer one can use the speed and accuracy of the computer and its ability to store and access millions of facts to outdo a human beginner using the same rules. But no amount of rules and facts can capture the understanding an expert has when he has stored his experience of the actual outcomes of tens of thousands of situations.

The knowledge engineer might still say that in spite of appearances the mind and brain *must* be reasoning – making millions of rapid and accurate inferences like a computer. After all the brain is not "wonder tissue" and how else could it work? But there are other models for what might be going on in the hardware. The capacity of experts to store in memory tens of thousands of typical situations and rapidly and effortlessly to see the present situation as similar to one of these apparently without resorting to time-consuming feature detection and matching, suggests that the brain does not work like a heuristically programmed digital computer applying rules to bits of information. Rather it suggests, as some neurophysiologists already believe, that the brain, at times at least, works holographically, superimposing the records of whole situations and measuring their similarity. Dr. Karl Pribram, a Stanford neurophysiologist who has spent the last decade studying holographic memory, explicitly notes the implication of this sort of process for expertise. When asked in an interview whether holograms would allow a person to make decisions spontaneously in very complex environments, he replied, "Decisions fall out as the holographic correlations are performed. One doesn't have to think things through. . . a step at a time. One takes the whole constellation of a situation, correlates it, and out of that correlation emerges the correct response."[17]

We can now understand why, in a recent article in *Science*, two expert systems builders, Richard Duda and Edward Shortliffe, who assume rather cautiously but without evidence that "experts seem to employ rule-like associations to solve routine problems quickly"[18] are, nonetheless, finally forced by the phenomenon to conclude:

The identification and encoding of knowledge is one of the most complex and arduous tasks encountered in the construction of an expert system. . . Even when an adequate knowledge representation formalism has been developed, experts often have difficulty expressing their knowledge in that form.[19]

We should not be surprised that, in the area of medicine, for example, we find doctors concluding that:

The optimistic expectation of 20 years ago that computer technology would also come to play an important part in clinical decisions has not been realized, and there are few if any situations in which computers are being routinely used to assist in either medical diagnosis or the choice of therapy.[20]

In general, based on the above model, our prediction is that in any domain in which judgments improve with experience, no system based upon heuristics will consistently do as well as experienced experts, even if they were the informants who provided the heuristic rules. Since there already seem to be many exceptions to our prediction, we will now deal with each alleged exception in turn.

To begin with there is a system developed at M.I.T. called MACSYMA, for doing certain manipulations required in calculus. MACSYMA began as a *heuristic* system. It has evolved, however, into an *algorithmic* system, using procedures guaranteed to work which involve so much calculation people would never use them, so the fact that, as far as we can find out, MACSYMA now outperforms all experts in its field, does not constitute an exception to our hypothesis.

Next there are expert systems that are, indeed heuristic, and which perform better than anyone in the field. This happens when there are no experts at the particular task such systems perform. This is certainly the case with the very impressive R1 developed at Digital Equipment Corporation to decide how to combine components of VAX computers to meet customer's needs. Configuring VAXes is a new problem and the relevant facts, viz. performance characteristics of components, are rapidly changing. Thus, no one has had time to develop the repertoire of typical cases necessary for expertise. This is also the case with spectrograph analysis. Duda notes that "For the molecular families covered by [its] empirical rules, [DENDRAL] is said to surpass even expert chemists in speed and accuracy."[21] But expert chemists need not be expert spectrograph interpreters. Before DENDRAL, chemists did their own spectrograph analysis, but it was not their main work so no one chemist need have dealt with sufficient cases to become an expert. Thus it would be no surprise if DENDRAL outperforms all comers.

Chess seems an obvious exception to our prediction, since chess programs have already achieved master ratings. The chess story is complicated and stimulating. Programs that play chess are among the earliest examples of expert systems. The first such program was written in the 1950s and by the early sixties fairly sophisticated programs had been developed. The programs naturally included the facts of the chess world (i.e. the rules of the game) and also heuristics elicited from strong players.

Master players, in checking out each plausible move that springs to mind, generally consider one to three plausible opponent responses, followed by one to three moves of their own; etc. Quite frequently, only one move looks plausible at each step. After looking ahead a varying number of moves depending on the situation, the terminal position of each sequence is assessed based on its similarity to positions previously encountered. In positions where the best initial move is not obvious, about one hundred terminal positions will typically be examined. This thinking ahead generally confirms that the initial move intuitively seen as the most plausible is indeed best, although there are occasional exceptions.

To imitate players, the program designers attempted to elicit from the masters heuristic rules that could be used to generate a limited number of plausible moves at each step, and evaluation rules that could be used to assess the worth of the roughly one hundred terminal positions. Since masters are not aware of following any rules, the rules that they suggested did not work well and the programs played at a marginally competent level.

As computers grew faster in the 1970s, chess programming strategy changed. In 1973, a program was developed at Northwestern University by David Slate and Larry Atkin which rapidly searched *every* legal initial move, every legal response etc. to a depth determined by the position and the computer's speed, generally about three moves for each player. The roughly one million terminal positions in the look-ahead were still evaluated by rules. Plausible-move-generation heuristics were discarded, the program looked less like an expert system, and quality of play greatly improved. By 1983, using these largely brute-force procedures and the latest, most powerful computer (the Cray X-MP capable of examining about ten million terminal positions in choosing each move), a program called Cray-Blitz became world computer chess champion and achieved a master rating based on a tournament against other computers which already had chess ratings.

Such programs, however, have an Achilles heel. While they are perfect tacticians when there are many captures and checks and a decisive outcome can be found within the computer's foreseeable future (now about four moves ahead for each player), computers lack any sense of chess strategy. Fairly good players who understand this fact can direct the game into long-range strategic channels and can thereby defeat the computer, even though these players have a somewhat lower chess rating than the machine has achieved based on play against other machines and humans who do not know and exploit this strategic blindness. The ratings held by computers and reported in the press accurately reflect their performance against other computers and human players who do not know or exploit the computer's weakness, but greatly overstate their skill level when play is strategic.

A Scottish International Master chess player, David Levy, who is a computer enthusiast—and chairman of a company called Intelligent Software in London—who is ranked as roughly the thousandth best player in the world, bet about $ 4000 in 1968 that no computer could defeat him by 1978. He collected, by beating the best computer program at that time 3.5 games to 1.5 games in a five game match. He was, however, impressed by the machine's performance and the bet was increased and extended until 1984, with Levy quite uncertain about the outcome. When the 1984 match approached and the Cray-Blitz program had just achieved a master-level score in winning the world computer championship, Levy decided to modify his usual style of play so as maximally to exploit the computer's strategic blindness. Not only did he defeat the computer decisively, four games to zero, but, more importantly, he lost his long-held optimism about computer play. As he confessed to the *Los Angeles Times* of May 12, 1984,

During the last few years I had come to believe more and more that it was possible for programs, within a decade, to play very strong grandmaster chess. But having played the thing now, my feeling is that a human world chess champion losing to a computer program in a serious match is a lot further away than I thought. Most people working on computer chess are working on the wrong lines. If more chess programmers studied the way human chess masters think and tried to emulate that to some extent, then I think they might get further.

Levy summed up his recent match by saying "The nature of the struggle was such that the program didn't understand what was going on."[22] Clearly, when confronting a player who knows its weakness, Cray-Blitz is not a master level chess player.

We could not agree more strongly with Levy's suggestion that researchers give up current methods and attempt to imitate what people do. But since strong, experienced, chess players use the holistic similarity recognition described in the highest of our five levels of skill, imitating people would mean duplicating that pattern recognition process rather than returning to the typical expert system approach. Since similarity for a strong chess player means similar "fields of force" such as interrelated threats, hopes, fears, and strengths, not similarity of the location of pieces on the board, and since no one can describe such fields, there is little prospect of duplicating human performance in the foreseeable future.

The only remaining game program that appears to challenge our prediction is Hans Berliner's backgammon program, BKG 9.8. There is no doubt that the program used heuristic rules obtained from masters to beat the world champion in a seven-game series. But backgammon is a game involving a large element of chance, and Berliner himself is quite frank in saying that his program "did get the better of the dice roles" and could not consistently perform at championship level. He concludes:

The program did not make the best play in eight out of 73 non-forced situations. . . . An expert would not have made most of the errors the program made, but they could be exploited only a small percent of the time. . . . My program plays at the Class A, or advanced intermediate, level.[23]

The above cases are clearly not counter examples to our claim. Neither is a recent SRI contender named PROSPECTOR, a program which uses rules derived from expert geologists to locate mineral deposits. Millions of viewers heard about PROSPECTOR on the CBS Evening News in September 1983. A special Dan Rather Report called "The Computers are Coming" showed first a computer and then a mountain (Mount Tolman) as Rather authoritatively intoned "This computer digested facts and figures on mineral deposits, then predicted that the metal molybdenum would be found at this mountain in the Pacific Northwest. It was." Such a feat, if true, would indeed be impressive. Viewers must have felt that we were foolish when, later in the same program, we were shown asserting that, using current AI methods, computers would never become intelligent. (While we explained and defended this claim during an hour-long taped interview with CBS, all of this was necessarily omitted during the 5 minute segment on computers that was aired.) In reality, the PROSPECTOR program was given information concerning prior drilling on Mount Tolman where a field of molybdenum *had already been found.* The expert system then mapped out undrilled portions of that field, and subsequent drilling showed it to be basically correct about where molybdenum did and did not exist.[24] Unfortunately, economic-grade molybdenum was not found in the previously unmapped area; drilling disclosed the ore to be too deep to be worth mining. These facts do not justify the conclusion that the program can outperform experts. So far there is no further data comparing experts' predictions with those of the system.

This leaves MYCIN, mentioned earlier, INTERNIST-I, a program for diagnosis in internal medicine, and PUFF, an expert system for diagnosis of lung disorders, as the only programs that we know of which meet all the requirements for a test of our hypothesis. They are each based exclusively on heuristic rules extracted from experts, and their performance has been compared with that of experts in the field.

Let us take MYCIN first. A systematic evaluation of MYCIN was reported in *The Journal of the American Medical Association.* MYCIN was given data concerning the actual meningitis cases and asked to prescribe drug therapy. Its prescriptions were evaluated by a panel of eight infectious disease specialists who had published clinical reports dealing with the management of meningitis. These experts rated as acceptable 70% of MYCIN'S recommended therapies.[25]

The evidence concerning INTERNIST-I is even more detailed. In fact, according to *The New England Journal of Medicine,* which published an evaluation of the program, "[the] systematic evaluation of the model's performance is virtually unique in the field of medical applications of artificial intelligence."[26] INTERNIST-I is described as follows:

From its inception, INTERNIST-I has addressed the problem of diagnosis within the broad context of general internal medicine. Given a patient's initial history, results of a physical examination, or laboratory findings, INTERNIST-I was designed to aid the physician with the patient's work-up in order to make multiple and complex diagnoses. The capabilities of the system derive from its extensive knowledge base and from heuristic computer programs that can construct and resolve differential diagnoses.[27]

The program was run on 19 cases, each with several diseases, so that there were 43 correct diagnoses in all, and its diagnoses were compared with those of clinicians at Massachusetts General Hospital and with case discussants. Diagnoses were counted as correct when confirmed by pathologists. The result was:

[O]f 43 anatomically verified diagnoses, INTERNIST-I failed to make a total of 18, whereas the clinicians failed to make 15 such diagnoses and the discussants missed only eight.[28]

The evaluators found that:

The experienced clinician is vastly superior to INTERNIST-I in the ability to consider the relative severity and independence of the different manifestations of disease and to understand the temporal evolution of the disease process.[29]

Dr. G. Octo Barnett, in his editorial comment on the evaluation, wisely concludes:

Perhaps the most exciting experimental evaluation of INTERNIST-I would be the demonstration that a productive collaboration is possible between man and computer – that clinical diagnosis in real situations can be improved by combining the medical judgment of the clinician with the statistical and computational power of a computer model and a large base of stored medical information.[30]

PUFF is an excellent example of an expert system doing a useful job without being an expert. PUFF was written to perform pulmonary function test interpretations. One sample measurement is the patient's *Total Lung Capacity* (TLC), that is, the volume of air in the lungs at maximum inspiration. If the TLC for a patient is high, this indicates the presence of *Obstructive Airways Disease*. The interpretation and final diagnoses is a summary of this kind of reasoning about the combinations of measurements taken in the lung test. PUFF's principal task is to interpret such a set of pulmonary function test results, producing a set of interpretation statements and a diagnosis for each patient.

Using thirty heuristic rules extracted from an expert, Dr. Robert Fallat, PUFF agrees with Dr. Fallat in 75–85% of the cases. Why it does as well as the expert it models in only 75–85% of the cases is a mystery if one believes, as Robert MacNeil put in on the MacNeil-Lehrer television news, that researchers "discovered that Dr. Fallat used some 30 rules based on his clinical expertise to diagnose whether patients have obstructive airway disease." Of course, the machine's limited ability makes perfect sense if Dr. Fallat does not in fact follow these 30 rules or any others.

But in any case, PUFF does well enough to be a valuable aid. As Dr. Fallat puts it:

There's a lot of what we do, including our thinking and our expertise, which is routine, and which doesn't require any special human effort to do. And that kind of stuff should be taken over by computers. And to the extent that 75% of what I do is routine and which all of us would agree on, why not let the computer do it and then I can have fun working with the other 25%.[31]

Feigenbaum himself admits in one surprisingly honest passage that expert systems are very different from experts:

Part of learning to be an expert is to understand not merely the letter of the rule but its spirit. . . . [The expert] knows when to break the rules, he understands what is relevant to his task and what isn't. . . Expert systems do not yet understand these things.[32]

But because of his philosophical commitment to the rationality of expertise and thus to underlying unconscious heuristic rules, Feigenbaum does not see how devastating this admission is.

Once one gives up the assumption that experts must be making inferences and admits the role of involvement and intuition in the acquisition and application of skills, one will have no reason to cling to the heuristic program as a model of human intellectual operations. Feigenbaum's claim that "we have the opportunity at this moment to do a new version of Diderot's *Encyclopedia*, a gathering up of all knowledge – not just the academic kind, but the informal, experiential, heuristic kind"[33]; as well as his boast that thanks to Knowledge Information Processing Systems (KIPS) we will soon have "access to machine intelligence – faster, deeper, better than human intelligence"[34] can both be seen as a late stage of Socratic thinking, with no rational or empirical basis. In this light those who claim we must begin a crash program to compete with the Japanese Fifth Generation Intelligent Computers can be seen to be false prophets blinded by these Socratic assumptions and personal ambition – while Euthyphro, the expert on piety, who kept giving Socrates examples instead of rules, turns out to have been a true prophet after all.

University of California, Berkeley

AUTHORS: Hubert L. Dreyfus (born 1929) received his Ph.D. from Harvard, has taught at Brandeis University and MIT, and is currently Professor of Philosophy at the University of California, Berkeley. He has done extensive research in phenomenology and in artificial intelligence, and his well-known book *What Computers Can't Do* (New York: Harper & Row, 1972; revised edition, 1983) has been translated into five languages. His brother,

Stuart E. Dreyfus (born 1931), received his Ph.D. from Harvard, has worked at the RAND Corporation, been Visiting Professor at Harvard and MIT, and is currently Professor of Operations Research at the University of California, Berkeley. This paper is adapted from their book *Mind Over Machine: Putting Computers in Their Place* (New York: Free Press, 1986).

NOTES

1 Edward Feigenbaum and Pamela McCorduck, *The Fifth Generation: Artificial Intelligence and Japan's Computer Challenge to the World* (Reading, MA: Addison-Wesley, 1983), p. 56.

2 *Ibid.*, p. 56.

3 *Ibid.*, p. 38.

4 *Ibid.*, pp. 76–77.

5 *Ibid.*, p. 18.

6 *Ibid.*, p. 64.

7 *Ibid.*, p. 55.

8 *Ibid.*, p. 79.

9 *Ibid.*, p. 85.

10 *Ibid.*, pp. 79–80.

11 *Ibid.*, p. 82.

12 *Ibid.*

13 *Ibid.*, p. 179.

14 Avron Barr and Edward A. Feigenbaum, *The Handbook of Artificial Intelligence*, Vol. 1 (Los Altos, CA: Wm. Kaufmann, 1981), p. 7.

15 These quotations are taken from an interview with Arthur Samuel, released by the Stanford University News Office, April 28, 1983.

16 Patricia Benner, *From Novice to Expert: Excellence and Power in Clinical Nursing Practice* (Reading, MA: Addison-Wesley, 1984).

17 Daniel Coleman, "Holographic Memory: An Interview with Karl Pribram," *Psychology Today* **12**, no. 9 (Feb. 1979), p. 80.

18 Richard O. Duda and Edward H. Shortliffe, "Expert Systems Research," *Science* **220**, whole no. 4594 (April 15, 1983), p. 266.

19 *Ibid.*, p. 265.

20 G. Octo Barnett, M.D. "The Computer and Clinical Judgment," *New England Journal of Medicine* **307**, no. 8 (Aug. 19, 1982), p. 493.

21 Richard O. Duda and John G. Gashnig, "Knowledge-Based Expert Systems Come of Age," *Byte* (Sept. 1981), p. 254.

22 *Los Angeles Times* (May 12, 1984).

23 Hans Berliner, "Computer Backgammon," *Scientific American* (June 1980), 64–72.

24 See *Byte* (September 1981), p. 262, caption under figure. The CBS News report is not the only sensationalized and inaccurate report on PROSPECTOR spread by the mass media. The July 9, 1984 issue of *Business Week* reports in its cover story, "Artificial Intelligence: It's Here": "Geologists were convinced as far back as World War I that a rich deposit of molybdenum ore was buried deep under Mount Tolman in eastern Washington. But after digging dozens of small mines and drilling hundreds of test borings, they were still hunting for the elusive metal 60 years later. Then, just a couple of years ago, miners hit pay dirt. They finally found the ore because they were guided not

by a geologist wielding his rock hammer, but by a computer located hundreds of miles to the south in Menlo Park, California."

25   Victor L. Yu *et al.*, "Antimicrobial Selection by a Computer", *Journal of the American Medical Association* **242,** no. 12 (Sept. 21, 1979), 1279–1282.
26   Randolph A. Miller, M.D., Harry E. Pople Jr., Ph.D. and Jack D. Myers, M.D., "*INTERNIST-1,* an Experimental Computer-Based Diagnostic Consultant for General Internal Medicine," *New England Journal of Medicine* **307,** no. 8 (Aug. 19, 1982), p. 494.
27   *Ibid.*, p. 468.
28   *Ibid.*, p. 473.
29   *Ibid.*, p. 494.
30   *Ibid.*, p. 494.
31   The MacNeil-Lehrer Report, on Artificial Intelligence, April 22, 1983.
32   E. Feigenbaum and P. McCorduck, *The Fifth Generation*, pp. 184–185.
33   *Ibid.*, p. 229.
34   *Ibid.*, p. 236.

PATRICK A. HEELAN

MACHINE PERCEPTION

ABSTRACT. The computational theory of perception rests on three principles: first, a scientific realism which takes visual objects to be verdical only if they exhibit the structure of physical objects in Euclidean space; second, an identity of mind and body, so that to see X is equivalent to having a brain state X' which is the product of a computational process applied solely to the retinal image; and third, the machine principle, which states that to understand seeing is equivalent to knowing how to build a seeing machine. It is argued that all of these principles are fallacious.

I. THE COMPUTATIONAL PROGRAM AND MACHINE PERCEPTION

The computational theory of visual perception is a powerful, multidisciplinary experimental program proposed by the late David Marr and his associates[1], and represents perhaps in its most developed form the research program of contemporary neurophysiological psychology.[2] According to this program, visual perception is a form of machine seeing accomplished by the neurophysiological system as its machine, and machine seeing is the twofold process, first, of constructing a symbolic representation of an external object based on elementary features of a grey-level image such as the retinal image and, secondly, of correctly identifying it by comparing it with a repertory of symbolic representation types (or "descriptions") of possible (external world) objects possessed antecedently by the brain as algorithms.

The computational program rests on three philosophical principles, two of which are common to all current neurophysiological research, the third being an extension or specific application of these: firstly, a *scientific realism* which takes the goal of verdical vision to be to exhibit correctly as a visual object the gross *scientific* structure (rather than, say, the pragmatic Life-World structure) of physical objects, in particular their Euclidean structure, otherwise processing failures in the perceptual system are assumed to have taken place[3]; secondly, *and identity principle of mind and body* – to see X is "denotationally equivalent" to being in a specific coded brain state X' which is the end product of a computational process applied to the grey-level retinal image[4]; thirdly, the *machine principle of the computer* – to understand seeing is equivalent to knowing how to build a seeing machine, envisioned as a robot computer control-

Carl Mitcham and Alois Huning (eds.), Philosophy and Technology II, 131–156.

led by the same algorithms that control seeing that would act – that is, make discriminations outside of itself – in all ways like a human seer.[5]

I hold that all of these principles in so far as they make general (or philosophical) claims are fallacious. However, the basic conception that acts of perception imply language-like algorithms in the retino-cortical system is, I believe, philosophically important and in my view sound. With regard to the philosophical understanding of perception, however, the program needs to be reconceived as part of a larger story, more on the analogy of language structure, use and interpretation, and when so reconceived, the strengths of the empirical part of the program can be assessed and certain specific weaknesses discerned.[6]

II. HERMENEUTICAL PHENOMENOLOGY OF PERCEPTION

The philosophical background of my approach is that of a hermeneutical phenomenology – of the phenomenology of the late M. Merleau-Ponty, and of the hermeneutical turn characteristic of M. Heidegger, H-G. Gadamer, and P. Ricoeur. I take the view that, while it is false to identify Mind with Body in the sense assumed by the computational program, it is also false to regard Mind and Body as two separable Cartesian entities. In place of identity and dualism, I take the position that people are bodily knowers or psychic bodies. This kind of knowing with the body is called by Merleau-Ponty, the "Flesh" ("*la Chair*"), and the interpretative aspect of this knowing is signified by the term Heidegger uses for people, "Dasein".[7] Part of what this concretely implies will become clear below. I have used such an approach in my recent book *Space-Perception and Philosophy of Science*[8] from which many of the succeeding points will be taken. Against a scientific realism – that only scientific accounts are capable of truly describing reality – I defend the primacy of perception – only perception can give the contours of reality. I take the pragmatic (or practical) view that only the Life-World (or World – with a capital – for short) is the home of the real, and that scientific entities can and do enter the Life-World as genuine perceptual phenomena under conditions to be described.

The Life-World is properly studied by a phenomenological method, such as that introduced by E. Husserl and developed so remarkably by M. Merleau-Ponty and others. The ontic commitments one's Life-World makes for one are articulated in and through the natural language one speaks; the natural language, as it were, speaks out the historical and historically changing milieu in which one finds oneself, others, and the environing reality of things; this is the milieu presupposed by every act of human communication. Mediating the World to us then is language; it is in and through this language – corrected and criticized – that phenomenological studies are done; the necessary use of natural language then does not interfere with the directness with which "the things themselves" – in this case, objects in the Life-World – are given within a phenomenological study.

But although natural language emerged together with human identity and Worldly objects, the general grounds for the interpretation which language brings to the World, and the specific grounds which justify the use of specific terms and phrases in particular perceptual situations need to be further studied. Central to the position I am taking is the thesis that all perception is hermeneutical, that is, mediated by text-like structures in the environment and in the brain to which we respond existentially with an interpretative act, an act of perceiving. Among such text-like structures in the environment, some are natural, that is, independent of human action on the environment, and others artifactual, the latter I call readable technologies.

The hermeneutic or interpretative aspect of all human knowing is epitomized in the term Heidegger used for humanity, *Dasein* or *There-Being*. Dasein is the being whose essence is to understand Being, where to understand is to enter a pre-conceptual union with the horizons of the World, from which union issues articulated knowledge via the hermeneutical circle. Such a pre-conceptual union – *Vorhabe*, in Heidegger's term – has a language-like structure. I hold that, since perception is a form of hermeneutical understanding, it is based both on text-like structures in the World and brain, and *parole*:like structures (see below) in the sensory-cortical system. such text-like and *parole*-like structures could be called by the general term "representations." This analogy between, on the one hand, language or *langue*, text, and *parole*[9] and, on the other hand, environmental and neurological structures associated with perception, is central to the proper reconception of the computational program. Though the explicit attendant philosophy of the computational program is incompatible with a hermeneutical phenomenology of perception, it is interesting that the computational program nevertheless speaks of the neurophysiological processes used in perceiving as "descriptions" using a perceptual "vocabulary," as "representations" and "symbols" – all terms with linguistic analogies. For the computational program of vision to make sense to a philosopher, however, it would be necessary to work out the philosophical role of the language-like structures in the brain, and how these may be thought to function in the presence of retino-cortical input from the surrounding world.

III. PERCEPTUAL CONTENTS OF THE WORLD

To the thesis of *scientific realism*, I oppose two considerations: (1) that we naturally and primitively see our World, not in the Euclidean geometry of classical physics but in the two-parameter family of hyperbolic Riemannian geometries.[10] For short let me speak of this as hyperbolic vision. The text-like cues or 'texts' on which hyperbolic vision depends is a structure of angles in the optical field, two of orientation with respect to the viewer and one of parallax – binocular or monocular – on which the estimation of depth is based. (2) To the extent that we come to see our World in a Euclidean way, we depend for their 'text' on *artifactual clues* in the carpentered environment which speak the 'language' of scientific or physical structure.[11] Such 'texts' depend on the existence in the environment of furniture, buildings, and other features which exhibit regular, geometrical, and modular features, these embody a virtual Euclidean coordinate frame; the 'text' for Euclidean vision will then comprise the set of coincidences between features lying on the virtual coordinate frame. Hyperbolic vision gives the primitive or naive horizons of the World; Euclidean vision adds to them horizons of a scientific culture mediated by readable technologies. According to the principle of the primacy of perception, both can be judged with equal ground to be manifestations of the real. The visual includes the physical – by *the physical,* I mean, objects as described by the natural sciences – but the physical does not coincide with the visual. The real, however, is the perceptual and includes both the visual and the physical.

The phenomenology of visual experience indicates that what kind of vision one uses, Euclidean or hyperbolic, depends of the *interest* one has in looking – vision is pragmatic – and on the 'text' one chooses to use – supposing there is a choice of 'text.' In hyperbolic vision there is a two-parameter family of possible visual spaces, differentiated roughly by the location of the true point (where visual size and shape match physical size and shape – this is the center of what Arnheim calls the Newtonian oasis of vision)[12] and by the distance – always finite – to the farthest visual horizon. Which of these spaces shapes a particular visual experience will depend on the interest, purpose, and anticipations of the viewer. Evidence for this can be gleaned from everyday perception, from optical illusions, from the use of optical instruments, and from the history of pictorial representations.[13] If the viewer is primarily interested in classical objectivity the choice will be to see in a Euclidean way, but the realization of such a vision will depend on the presence in the visual field of the modular environmental structures which 'speak' the 'language' of this kind of vision. Classical or Cartesian objectivity in vision, essential to engineers and others who build and service mass-produced machine technologies, has characterized the culture of the modern world since the sixteenth century.

IV. PERCEPTION AS HERMENEUTICAL

The phenomenology of visual experience can be used as a paradigm case from which certain general philosophical conclusions can be made about perception in general. I conclude that perceiving has essentially a hermeneutical character; that is, (1) it is a function of the viewer's prior interests in the antecedent possibilities of his/her World, (2) it uses language- or text-like structures in the environment, some of which such as physical structures in the incident optical array are natural, and some such as the carpentered environment are artifactual, and (3) it responds to states of retino-cortical stimulation which play the role of *paroles*,[14] the artifactual text-like structures are readable technologies. The main differences between *parole* (spoken word), on the one hand, and *text* (written word), on the other, are the following: the spoken word, *parole*, is always addressed to a living and present listener in the midst of a Life World situation which contextualizes the discourse and permits questions to be raised by the listener and answered by the speaker; the written word, *text*, is always (or almost always) addressed to absent readers, who may inhabit a variety of historical situations and cultures, and who cannot usually enter into dialogue with the author. Consequently, while the meaning of a spoken word is usually univocal and easily grasped because speaker and listener easily come to share the same communication situation, a text can have multiple meanings; these, moreover, are defined, not by the writer as such, but by communities of readers, the writer being as it were the first reader among his contemporaries; readers, however, may have to be transformed by the hermeneutical process itself in order properly to appropriate the text. Within a hermeneutic phenomenology of perception, I take the retino-cortical input from the World to play a *parole*-like role, and the antecedent structures in brain and World to play text-like roles.

The discovery that perception is mediated by text-like structures in the environment does not of course change the phenomenology of perceptual acts, perception is still the experience of the direct giveness of horizons of the World. Hermeneutical mediation is not new knowledge mediated by old, like that produced by deductive or inductive inference. It follows the process of the hermeneutical circle which moves in a gradually elucidating spiral between the anticipation of specific structure and the success or failure experienced in finding such structure; the outcome of this process is the emergent giveness of real or possible experience itself. All perception works this way, it is then influenced by the anticipations or interests of the viewer as well as by the text-like structures in the environment and *parole*-like structures in the retino-cortical system through which the perceiver is physically united with the World. Perception then is causal, since it receives the '*parole*' to which it is responding. But this is not the full story, perception is also hermeneutical, since the context in which it responds to potentially meaningful stimulation is governed by the structure of the hermeneutical circle. The processes of the hermeneutical circle mediate between the potential text-like and *parole*-like structures of signifiers given in pre-experience and the domain of the signified. These processes themselves cannot be perceived, nor do they belong to the object of the natural sciences; consequently, they are not describable by either. They are, however, presupposed by both, and can be studied by a hermeneutic phenomenology.

According to the analysis just given, the realism associated with the primacy of perception does not distinguish between naive or primitive perception on the one hand, and a perception which depends on artifactual cues on the other. Artifactual cues, as I have said, are provided by readable technologies, among which pre-eminently stands the carpentered environment. However, the class of readable technologies includes certain kinds of scientific instruments which permit direct (though hermeneutical) access to the profiles of a scientific quantity, the way thermometers do for thermodynamic temperature. The realism of the primacy of perception does not then preclude realism for scientific entities; these latter have the opportunity of becoming horizons of the Life-World provided their existence and perceptual essence in the World can be mediated by a suitable publicly available readable technology.[15] Such a realism, I have called a *hermeneutical* or *horizonal realism*.

V. THE COMPUTATIONAL PROGRAM OF THE
NEUROPSYCHOLOGICAL SCIENCES

Let me schematize what I believe to be the total philosophical context within which the aims of the computational program can best be understood. The location in that scheme of the computational program can then be shown, and the program assessed for its strengths, weaknesses, and ambiguities.

The process of inquiring into perception contains three interrelated phases: phase 1. pre-conditions of the inquiry; phase 2. the scientific program; and phase 3. the act of vision. The computational program focusses on phase 2, and here its importance lies. Biases created by the principles of scientific realism and the identity principle lead to some careless and incorrect assumptions about phase 1, and to a confused and partial account of phase 3.

Phase 1

The pre-conditions of the inquiry include the perceiver-subject (S1), the scientific-researcher (S3), and the World of perceptual objects, the last I take to be the World of Real Objects. I take S1's relation to perceptual object to be in the first-person – he/she *sees* the object, and S3's relation to be in the third-person – he/she is focussed on the *representations* of that object in S1. Add to this list the philosopher – S2? – who takes in what S1 and S3 are doing; S2 adds his/her own philosophical reflections to the perceptual phenomenology of S1 and the scientific accounts of S3.

It is important to note that S1 cannot be existentially S3; in fact no two of the trinity of S1, S2, and S3 can be existentially the same, though they may take turns in what they existentially do. Each, however, is in possession of knowledge, though of different sorts. Speaking as S2, the object of philosophical study – Merleau-Ponty's *la Chair* (The Flesh) or Heidegger's *Dasein* - is nevertheless in some sense both S1 and S3, but also more; it is humanity in its total manifold of complimentarily embodied Worldly understanding. A philosophy of perception is a philosophy of *la Chair* or *Dasein*, it is not just a phenomenology done by S1 or the science done by S3.

Right at the very start, the program makes assumptions about the world of the real (the term "world" will be capitalized only where it has the technical signification of the Life-World); in keeping with the

assumption of scientific realism, it is assumed that all perceivable objects
are Euclidean in structure, not merely in their physical structure but also
in any phenomenological re-presentation for which truth is claimed.[16] As
I have already explained, in order for the visual structure of the phenomenological object to be the same as its physical structure, the visual
object would have to be presented against a background which contained
appropriately engineered environmental structures. Only in this way can
the visual object come to be seen as an object belonging to a cultural
World – ours! – which at this historical time is normatively and exclusively Euclidean. In laboratory settings, however, it is usually the case that
the perceptual stimuli are impoverished relative to the settings necessary
for determinate Euclidean vision with the result that the viewer comes
instead under the influence of primitive hyperbolic visual structures
taken by the psychologist to be illusionary. In practice, then, the program
has too little to say about the repertory of perceptual objects as such, and
its pragmatic character. It mistakenly takes parallax, the depth cue for
hyperbolic vision, to be the depth cue for Euclidean vision, and mistakenly regards hyperbolic vision as merely a partial intermediate stage
on two-dimensional patterns of stimulation of the retina, using an array
of two-dimensional patterns with well-known – or at least, allegedly well-
known – phenomenological effects, realistic and illusionary. It has been
shown, however, that the phenomenological effects of such retinal stim-
ulations are dependent on the background perceptual hermeneutical circle – the background anticipations – brought into play.[18] Among these
indeterminacies is the kind of visual space which a fixed retinal input
brings about. Unless then the perceiver (S1) and the researcher (S3)
share the same *perceptual* hermeneutical circle – and such perceptual
hermeneutical circles are rarely unique – uncontrollable errors can infect
the descriptions of what is thought to be the perceptual objects for which
a neurophysiological coding is sought. Such consequences drawn from
the existence of a hermeneutical or pragmatic dimension to visual
perception are usually ignored or overlooked in psychobiological
theories about perception.

Phase 2

The scientific program[19] is a scientific study by S3 of the processing of
retinal inputs by the brain and retino-cortical system of S1.
    The computational program from the start focusses on the retina, i.e.,
the physical interface between the perceiver and the world surrounding

the perceiver. All visual information falling on the retina is encoded on a mosaic of receptors each of which responds in a more or less linear way to the intensity of optical energy falling on it within a certain frequency range. The result is a two-dimensional image on the surface of the retina called a grey-level image, this is composed of a mosaic of small areas or blocks of different levels of excitation. The grey-level image can be represented on paper by a two-dimensional diagram composed of pixels of definite size to each of which a definite measurable level of excitation is assigned. A variety of such grey-level images is in fact constructed by the retina, with different pixel-sizes, and differing in the ranges of electromagnetic frequencies to which the receptors respond; there are grey-level images in the blue, the green, and the red. As the multiplicity of grey-level images does not concern us, I shall speak for short of the grey-level image, as if there were only one.

Possessing a grey-level image cannot be sufficient of itself for perceiving, because it is merely a physical state of the neurological system. Mental states are introduced into the computational program *via* the identity principle: to each neurophysiological state relevant to perception – these are spoken of as "words", "symbolic descriptions" or "representations," – there corresponds a mental act, in this case, a perceptual act, usually taken to be an act of perceiving, which is completely determined relative to content by the representation, and has no independence of the physical brain state. In this view, the content is always a feature of the physical world, that is, of the world as described by the physical and biological sciences – unless, of course, the processes of perception should break down. The representations are neurophysiological states, let me call them "neurophysiological signifiers," or "signifiers" for short. The task is then seen to be that of determining these signifiers, and the repertory of signifiers is the neurophysiological "vocabulary." The domain of the signified is presumed to be that of science. Note that among the systematic ambiguities of the identity principle is one which fails to distinguish between a perceived object and a merely perceivable object, between what is given through the context and what – if anything – is given independently of context. Note also that the identity principle is incompatible with the analysis given in this paper, because it formally excludes the hermeneutical dimension of perception. This is the case because the hermeneutical process is never final or fully specific, and because the identification of a signifier (e.g., a candidate for letter A) is intimately and essentially dependent on whether it can perform the appropriate function within the signifier-signified system (e.g., as the alphabetic A).

The repertory of signifiers is determined in three ways: (1) by neurophysiological methods, (2) by psychological methods, and (3) by computational methods; these last propose to achieve understanding by making a robot computer capable of receiving grey-level images and responding behaviorally to them as humans do; the robot would analyse the contents of its grey-level images with the aid of a computer program (like that believed to operate on the neurophysiological level), and would correctly identify (for the most part) the objects presented in and through it, and would provide evidence of this "understanding" by appropriate behavior.[20]

1. Neurophysiological Methods

Neurophysiological methods study the kinds of optical structures which when presented to the retina produce activity in the cells of the striate cortex and adjacent areas associated with retinal stimulation. Hubel and Wiesel[21] (and others) using microelectrical probes capable of recording the activity of individual cells in the striate cortex of cats and macaque monkeys were able to show that this is composed of columns of cells perpendicular to the surface of the cortex to which they gave the name "hypercolumns." The individual cells of any hypercolumn respond to different structures in the optical stimulation falling on a corresponding area of the retina, its hyperfield. For example, features such as bars of light of different lengths, widths, intensities, and orientations, moving with different speeds, tend to activate single cells of a hypercolumn whenever the hyperfield is exposed to the correspnding feature. The responses of the hypercolumn cells then serve to pick out structures present in the grey-level image. These structures are the candidates for the *embodiment (implementation* in Marr's terms) of the role of feature signifier.[22]

A feature signifier is not just a physical thing or event, it is an element of a symbolic system or algorithm, like *langue* (in the structuralist sense); for a thing or event to be a signifier then it must, says Marr, be the embodiment (or implementation) of a term in a language-like algorithm.[23] Let me make the well-known distinction between a *type* and a *token*: a token is any individual which fulfills a type description. Type descriptions can be of two kinds, they could refer to physical characteristics alone or also to functional characteristics. Alphabetic letters, words, signs, algorithmic terms all belong to the latter kind. A letter, for instance A, has a structural shape, but not everything which happens to have this shape is an alphabetic A, but only such as play an alphabetic role in the spelling of words. Moreover, although it has a physical shape which is normative for the letter A, it is not the case that every A has physically this shape. Think of the squiggle some people make when they sign their names! In such cases, one's knowledge of the functional context in which the token-candidate for the letter A occurs tells one that the squiggle is truly a token of the letter A, despite the fact that its actual shape could be almost anything you please. From all of this one concludes, that not everything that has the physical shape of an A is a token of the alphabetic term A, and not every token instance of the alphabetic term A has the physical shape of an A. This paradoxical situation makes it difficult to design a machine that is as good as a human in identifying under arbitrary circumstances the letters of the alphabet or in general the elements of any algorithm.

Marr who, unlike other psychobiologists, anticipated such consequences,[24] nevertheless attributed the difficulty of identification merely to the complexity of the task.[25] But if what is desired is a capacity which matches the human capacity, the problem is probably insoluable in principle for reasons which will be given below.

The perceiver S1 recognizes the letter A, sometimes from shape and context, and sometimes from the context alone. To identify an A from its shape, S1 must embody in his/her brain states – an unconscious embodiment, like a machine's, is sufficient – the alphabetic sign system (of which the letters are the alphabetic terms). But if S1 identifies an A from the context alone, S1 must in addition be embodied in yet a higher system of signs (or a higher algorithm), that to which the former belongs as a term. To the extent that S1 exhibits powers of discrimination regarding what is and what is not alphabetic, this further embodiment must already be there. Just as the first embodiment of S1 and S3 in the alphabet is the *Vorhabe* which makes it possible for S3 – and then for S1 – to articulate the terms and structure of the alphabet, so the second embodiment is the *Vorhabe* which makes it possible for S3 – and then for S1 – to articulate an understanding of alphabetic systems such as is necessary to discriminate between true and spurious instances of letters. A scientific theory of alphabetic systems is then possible because some S3 can use the alphabetic performance of S1 as data falling under a hermeneutical circle about alphabetic systems pre-experienced in the ability S3 shares with S1 to discriminate between alphabetic instances and non-alphabetic instances of letter-shapes. Such a scientific research program into alphabets will have its parallels in the scientific study of perception.

In order to orient the discussion that is to follow, let me make at this time a summary of the points I want to make and defend: (1) a machine (such as, say, the machine used by banks to machine-read checks) embodies the alphabetic system of signs and can correctly for the most part identify individual letters by certain chosen physical features (say, of machine-typed letters); but it has not (and could not) use this embodiment to generate the system of letter-types which comprise the alphabet. (2) S1 embodies the alphabet and uses it hermeneutically (having learned with the help of some S3 to construct an alphabet, that is, a readable technology of letter signs) to articulate a system of the letter-types of the alphabet. (3) S1, as has been shown, also embodies a higher algorithm of which the alphabet is a term, this embodiment implies the possibility of a higher science, a science of symbolic systems like the alphabet; this permits S1 (having learned with the help of some S3 to construct an appropriate readable technology of signs) to use in addition the higher embodiment hermeneutically to develop an articulated understanding of the higher system. (4) The process of historical differentation of knowledge goes on in this fashion and can go on indefinitely to higher and higher orders of sign systems and algorithms. Every move to a higher order, clarifies and perfects what is already being done in an imperfect way at the lower level. (5) What counts for knowledge then is historically determined by the specific purposes achievable by the systems already articulated, and challenges to the limits of what counts for knowledge will push the inquiry to the next higher level of inquiry; such would arise, e.g., from the need to identify letters (and other linguistic signs) in less and less familiar contexts. (6) The capacity to know (say, letters), and so on indefinitely – which we surely have – does not imply that we can exercise this capacity other than by the progressive, step-wise, and historical acquisition of knowledge as described above. This process is never complete, and consequently at no historical moment will we ever have perfect clarity about what we know about the world. (7) The process just described is existentially hermeneutical; that is, the embodiments referred to above belong in the case of humans to what has already been called *Vorhabe*, this is the pre-conceptual union of subject and object from which the hermeneutical process starts and which ends in articulated knowledge of an objective domain. (8) In this perspective, what differentiates machines from humans is not the fact that each embodies sign systems and algorithms, but that humans with their embodiment possess a general ability not possessed by machines, and which makes human knowing – and the machine-building which depends on this knowing – a historical process.

Returning to the account of the experimental program, the scientific part of the program studies of the retino-cortical system for its capacity to generate identifications of signifiers by modelling with a computer the network of excitatory and inhibitory connections of the receptor cells of the retina with neighbouring horizontal, amacrene, bipolar, and ganglion cells; such processes correspond, for example, with convolution, thresholding, and deconvolution performed on the excitation levels of the grey-level pixels in a computer analogue of the grey-level image. In this way, elementary features of the grey-level image, such as short or long lines, moving lines, corners, edges between light and dark patches, etc. are studied for their potential role as feature-signifiers.

2. Psychological Methods

The program also uses psychological tests to aid or confirm its identifications of potential candidates for the role of signifier structures in the grey-level image.[26] The tests use well-studied psychological after-effects of prolonged intense stimulation of the retina by means of grills and other patterned shapes. Such grills, for example, of oriented stripes of light and dark bands, are used to study neurological connections in the retina. The assumption is that if there is a feature signifier for a given pattern of stimulation, then the associated cells of the neurological network will suffer fatigue when subjected to prolonged intense exposure to the pattern. By testing the subject psychologically immediately after such exposure, the effects of fatigue can help specify the range of feature signifiers brought into play. Useful neurological information can be inferred from lightness and brightness studies, and from the study of illusions such as the Craik-Cornsweet-O'Brien Illusion.

It is crucial to ask, however, how is the gap to be bridged between potential candidate-tokens and signifier-types, that is between the physical feature as a potential feature-signifier, and its identification as a feature-signifier and token of a symbolic or algorithmic type? Think of a mark which might or might not be the letter A! In the program, the gap is bridged by a computational matching process based on purely physical features. Such a procedure, as I pointed out above, will make mistakes; some true signifiers will be missed, and some events will be identified as signifiers which are not in fact so. Practically, such failures would account for some "noise" or unintelligible material in perception. However, human perceivers such as S1 have the special capacity already described, that they can at times (using their embodiment in a higher algorithm or symbolic system) validly overrule the decision of a physical matching process. Decisions of this kind are said by Marr and Frisby to be "conceptually driven".[27] They imply a level of discrimination and decision in S1 higher than that of any deliberate or indeliberate matching process based on physical properties alone. The discovery of this higher level brings no surprise. It witnesses to the possibility of an articulated understanding of perceptual systems as such, and to the possibility of the kind of scientific understanding that this program, despite its philosophical weaknesses, hopes to achieve.

In its search for symbolic systems (taken to be exclusively algorithmic) of psycho-neurophysiological networks, the computational program is, I believe, very much on the mark. If perception involves essentially systematic activities of neurophysiological networks, then these systematic activities must be understood at least as symbolic systems (perhaps even, as algorithms), and have a language-like structure. Such an analogy with language is in the first instance with *langue* in the structuralist sense of the term, *langue* being a diacritical system, that is, a system of signs which define one another by mutual opposition within a system.[28] An analogy with language is made by Frisby and Marr. Frisby, for example, writes, "the word 'vocabulary' is appropriate because the feature symbols [present in the grey-level image] are rather like 'visual words' which stand for features in the scene."[29] Marr also makes the connection with semantics.[30] But neither Frisby nor Marr provide an appropriate philosophical rationale for their use of these terms, and the analogy with language remains in the order of metaphor. It is my view that the computational program lacks a sufficiently general context, and that the missing parts of the context are those needed to articulate the hermeneutical structure of perceptual understanding in a way that accounts for the presumably essential role played by symbolic and algorithmic language-like systems in perception.

3. Computational Methods

The most sophisticated part of the program is the computational part.[31] This attempts to design and construct a computer model of the human visual system by building up algorithms of feature and object signifiers that refer to patterns discoverable in grey-level images. A key concept in the analysis and synthesis of object signifiers is the *primal sketch*:[32] this is a bare-bones structural sketch which simplifies clusters of elementary feature signifiers into single elements, and groups these into structural wholes capable of becoming object signifiers. The analogy is with those properties of line drawings and cartoons which make them instantly recognizable to a viewer. This process is called *segmentation*. By using rules to group and simplify features of the grey-level image into Gestalts, and by using other principles related to brightness cues, texture, movement, stereopsis, color, etc., Marr developed a computer program which to a great extent accomplished this reduction.

According to Marr, the primal sketch leads to a 2 1/2-D representation – the 1/2 dimension is a viewer-centered, non-objective, i.e., non-Euclidean/non-Cartesian depth; this is a rough initial model of a visual object, intermediate between the flatness of the grey-level image and the objective 3-D physical world presumed to be the goal of vision.[33] The 2 1/2-D sketch is not Euclidean; in fact, it turns out to look something like the way that a physical object would appear in hyperbolic vision according to my account of hyperbolic vision.[34] Marr interprets it, however, as an intermediate stage of vision, non-objective (in the scientific sense), and deficient because it does not mirror the physical object. He seems to assume that the passage from the 2 1/2-D sketch to an objective representation is a "natural" one, that is, not dependent as a pre-condition on the cultural transformation of the environment. However, it is likely that Marr simply had not given much thought to this question.

For reasons to be given below, I argue that the 2 1/2-D sketch of Marr is not a deficient, intermediate stage in vision, rather it is itself a primitive pragmatic form of a human vision which has its own "text" and rationale, independently of the objective Euclidean way of vision. According to the account of visual spaces expounded in my book, the latter became entrenched in modern Western culture surely not more than six hundred years ago.

According to this account, the primitive perceptual World of all peoples, including Western peoples, was composed probably of hyperbolic, not Euclidean, perceptual objects; these were the real, natural, but primitive horizons of the Life-World. I believe that toward the end of the 14th century in Northern Italy, a certain kind of cultural transformation took place particularly of the urban environment which changed the public manner of perceiving its World, the Western community went Euclidean. The ancient and medieval World probably regarded the spatial forms given by hyperbolic vision as by and large normative for reality, Aristotelian and Platonic cosmologies suggest this, but the modern World was different, it took reality criteria instead from Euclidean vision and its objective horizons. Such a visual transformation could only have taken place where the carpentered environment contained the public clues necessary to sustain that 'reading,' such as buildings and furniture of regular, geometrical, and modular design.

In the modern World, the horizons of the earlier World have gone into hiding, they are still there vestigially, of course, since they are the product of primitive and natural vision, and they can still be experienced. Representations of these spaces are also found in the pictorial art of the earlier period. But the modern World stripped these perceptions of any authority and relegated them to the class of visual illusions. I take these hyperbolic perceptual horizons to be as justifiable as the Euclidean, as manifestly given, and to possess equal right to be called reality. They were real in the past, we have merely forgotten our past, and in our forgetfulness we blind ourselves to manifest aspects of daily experience.

In a hermeneutic phenomenology which critically appropriates the horizons of our contemporary World and recovers what has been forgotten and suppressed, such hyperbolic horizons would be re-instated, they would stand side by side and complementary with Euclidean horizons of the real. Each would have its own "text," and each would fulfil a different (but complementary) set of human visual interests. Consequently no contradiction would be involved. Scientific realism as an exclusive position is then false; what science does is only to add new realistic horizons to others which precede scientific inquiry.[35]

Phase 3

The act of vision is the process of S1's seeing or coming to see the perceptual objects from which the retinal (and other) inputs came; this process can be studied in three ways, (1) from the first-person point of view of S1 – this would be a phenomenological study, or (2) from the third-person point of view of S3 – this would be a scientific study, or finally (3) from the point of view of philosophy, here of hermeneutical phenomenology – this would be a philosophical account that looks beyond the phenomenology to a covert dependence on text-like structures in the World, brain, and the retino-cortical system, and beyond the scientific account toward its interpretation in the phenomenon of perception.

The final stage of the program attempts of account for vision in the sense of object recognition: according to the computational program there is a repertory of stored object signifiers, a "vocabulary" of stored "structural descriptions" in the brain. What kind of thing is a stored object signifier? Is it a cell which when activated signifies the presence of the signified object (writers speak of an – apocryphal – grandmother cell[36] which is activated by indications of the grandmother's presence)? Or are these object signifiers rythms of cell activity, like musical themes, or spatial patterns of network activity, like words in a written text? We do not know much about this in the present state of knowledge. Some individual cells, however, seem to respond preferentially to quite complex shapes, such as that of a hand in left or right profile.[37] But are they responding to the particular shape as a symbolic or algorithmic term for hand or for hand-shape, or merely to the particular shape as a figure of a certain kind? The answer is not known. It may well be nevertheless that there is a language-like algorithm of perception which is a set of dedicated individual cells within the brain. After all, if the perceiver does possess the algorithmic types, then there must be an appropriate set of type signifiers. It should be noted that the type signifiers cannot coincide with a set of its own tokens, and so would not usually *look like* any of its tokens, a signifier for a hand in the grey-level image or in the brain would not look like a hand (it will, of course, be "read" or "interpreted" as a hand, but it will not be anything like a scale-model of a hand).

How is object recognition achieved? Recognition, says Frisby, is "probably achieved by finding a match between a structural feature of the grey-level image and a stored structural description of the object".[38] Setting aside the problem of how a repertory of such structural descriptions comes to exist in the perceiver, let me focus on the question, how are we to understand its use? The question has two levels: (1) How does the signifier come to be identified as a '*parole*' 'spoken' by the World? and (2) How does the '*parole*' acquire its meaning?

1. The first suggestion – and one already rejected in the preliminary discussion – is that the signifier comes to be identified as a '*parole*' 'spoken' by the World by reason of some physical feature, such as the spatial structure, it possesses. Frisby, for instance, suggests that a structure from the grey-level image is matched with one from a stored "vocabulary" of signifiers. Returning to the discussion of the letter A, normatively the letter A has a definite shape, but physically its shape can be almost anything, because, in addition to shape, the symbolic context can be used to identify it; consequently, a signifier of the letter type A need not have the physical shape of any particular letter A. What is true for an elementary feature signifier is also true of an object signifier. No matching of shapes or other physical properties is sufficient to justify the acceptation of a particular input as a token of a certain type, nor indeed is it necessary. Higher controls must enter into S1's determination, as Frisby admits.[39] A '*parole*' then cannot be identified as such except through

the pre-recognition that a system of signifiers is being used; only under such conditions can a particular candidate-signifier come to be identified – whether correctly or incorrectly – as indeed a signifier. Such conceptually driven processing reveals the necessity for a hermeneutical power in the perceiver and one which is required for the performance of an act of perception as such. This hermeneutical dimension is clearly not a physical or biological property.

2. The identity theory would have us believe that from the scientific point of view – that of S3 – the identification of a signifier is denotationally (or materially) equivalent to the identification of a meaning.[40] What the preceding argument has shown is that signifiers cannot be correctly identified without appeal to the context within which these play the role of terms of a symbolic system, such as an algorithm. Now the terms of a symbolic system do not exist physically, only token-candidates exist physically. But the ability of a token-candidate to play the role, say, of an algorithmic term is more than the possession of any physical property that it can have. Consequently, S3 has to accept that when S1 uses a particular candidate-signifier as a signifier, i.e., to stand for an algorithmic term, it is not just because the candidate-signifier is (according to some threshold measure) like what it should be physically – for it can be otherwise and yet become truly a signifier – but that S1 decides to complement any physical deficiences it may have, and use it as a signifier. *Once S3 admits that doing this is in keeping with good standards of perceiving for all, then it is not possible for a scientific account to contradict the following claims*: (1) symbolic types exist, such as the terms of an algorithm; (2) although symbolic tokens are physical entities, symbolic types are not; (3) S1 (and of course S3) possesses a hermeneutical power which is involved in the identification of particular tokens of each symbolic type; and (4) this hermeneutical power likewise is not physical.

It follows then that the identity theory must be false.

The terms of an algorithm or symbolic system stand for types, these are concepts or meanings, which, though themselves non-physical, permit some individual physical entities to be classified as tokens of such types. To the extent that algorithms are discovered among the activities of (S1's) neurophysiological networks, meanings are involved; S3 knows these meanings, because they enable him/her to read 'texts' or '*paroles*' expressed in the terms of the algorithm. For S3, however, the 'reading' is already laden with the perceptual meanings the algorithm has for S1, for the study is about the scientific conditions for S1's having such meanings. In other words, the postulated neurophysiological algorithms which S3 hopes to discover by the scientific part of the inquiry are precisely those which have the perceptual meanings S1 attributes to them. The algorithms as they are discovered come laden then with perceptual meanings. For S1 the algorithmic '*parole*' to which he/she is responding is hidden in the first-person experience of seeing a perceptual object, while for S3 the algorithmic '*parole*' is known as the term of an algorithm whose meaning is 'read' in the third person; these meanings should be identical in principle, for otherwise either the research problem is not solved, or a new problem is made. Throughout the inquiry then there must be a common set of background anticipations for perceptual meaning which S1 *and* S3 share, such constitutes the presence of a perceptual hermeneutical circle shared by both S1 and S3.

In the practice of experimental research, this condition is often violated. The penalty for a violation is the possible introduction of an uncontrollable kind of error into the results.

VI. MACHINE VISION

What finally is the assessment of the computational program? And what of its attempt – or promise – to achieve machine vision, that is, to make a computer controlled machine capable of extracting from grey-level images of its environment information about the objects in its environment?

Firstly, the program subscribes ostensibly to the three principles listed at the beginning of this paper, these are *scientific realism, the identity principle*, and *the machine principle of vision*. What can be said of each?

To summarize, I hold the principle of scientific realism to be false. In its place I put the principle of the primacy of perception. This claims that perception is existentially hermeneutical in character, that is, in perception there is always a perceptual pragmatic hermeneutical circle mediated by text-like structures in the environment and in the brain, and *parole*-like structures in the sensory-cortical system. The principle of the primacy of perception also applies to scientific entities, these make their being perceptually manifest through readable technologies which mediate the presence of these entities to hermeneutical perception, provided of course that for each real scientific entity such a technology can be constructed. The arguments for these positions are not given here, but in my recent book.

I have tried to show above that an identity theory of the sort first proposed by H. Feigl and as customarily assumed by neurophysiologists is

false. The argument against the identity principle is based on the nature
of algorithms: if the brain and retinal coding for acts of perception constitutes an algorithm – as it is correctly I believe assumed – then perception
can never be explained simply by a processing system of physical inputs
from the grey-level image; algorithmic meanings must be present a priori
to and operative within such processing in the way the hermeneutical circle is operative in the interpretation of a text, or better, in the understanding of a *parole* or spoken word. Human knowers are interpretative
knowers, and what is interpreted is a *Vorhabe*, a structured state of
physical union between (what will emerge as a duality of) subject and
object; such an account escapes the monism of the identity principle and
the dualism of Mind and Body.

What finally of the machine principle? Although the goal of "a seeing
machine built to match human performance"[41] in all its generality is, I
believe, impossible in principle, this should not rule out the possibility in
specific ranges of cases of building a machine that extracts from grey-
level input images the information necessary to execute correctly identifying behaviors like those human viewers use. Both the possibility and
its limitation follow from the leading methodological principle of the
computational program, namely, that acts of visual perception are connected necessarily with unconscious neurophysiological algorithms in
the brain and retina. Such perceptual algorithms constitute something
like *langue* in the structuralist sense, within which are expressed "stored
descriptions" which are like texts, and "feature" and "object descriptions" derived from the grey-level image which are like *paroles*.

One wonders of course why one would want to construct a machine
limited by the input of a grey-level image like the grey-level image of the
human retina. It would be easier to design machines which use inputs
from a wider variety of sensors, responsive to the full range of energy
fields and frequencies, and not only to those to which the human retina is
sensitive. The motivation is of course to test the hypothesis that humans
are just computational machines. Deploring the popular belief that man
is more than a machine, Frisby says that only when a "computer is given
its own eyes, and its own capacity for explicit symbolic scene description," will it reveal its true potential to act in a human way.[42] The goals of
the program then are unfortunately entangled with a certain philosophy
– or better a missionary gospel – which pervades the language and style
even of its scientific expression. The character of that philosophy is clearly set forth by Uttal in his *Psychobiology of Mind*; as he says, "modern
psychobiology . . . is mechanistic, realistic, monistic, reductionistic,

empiricistic, and methodologically behavioristic . . . and in profound logical and conceptual conflict with contemporary religious doctrine concerning human mortality".[43]

That philosophy I hold to be by and large false, and methodologically unsound for the conduct even of the scientific part of the program. One might be tempted to excuse the false philosophy because of certain hard-headed methodological virtues it is thought to exhibit, for example, of refusing to give non-physical entities a place in the physical world. However, the way it does this is by introducing confusion in thought and language; the identity principle proceeds by stipulating changes in the weights of venerable philosophical terms, robbing them of their subtility, and depriving thinkers of the ability to discourse about knowledge in other than crude ways. At any rate, one does not need the identity theory to fight dualisms of matter and spirit. A hermeneutic phenomenology, for example, opposes dualisms – it also opposes all identity principles, materialistic and idealistic – and it does so by introducing the elements of a sophisticated analysis of the embodied human knower. Such an analysis one finds, for example, in Heidegger's *Being and Time* or in Merleau-Ponty's later work *The Visible and the Invisible*.

Nevertheless, once one has taken the computational theory out of its reductionist frame, and shaken it loose from its particular ideological constraints, one finds that it affirms in practice and is guided by a most non-reductionist principle, that of the existentially hermeneutical character of perception, namely, that necessary to perception are language-like structures in the brain, in the environing world, and in the grey-level image, and connected with these, higher-order conceptual schemes which direct and interpret the processing of the retinal input.

A language-like structure of the kind referred to does not need to be known by the perceiver before being used. In fact, to require such would lead to an infinite regress, for such a language-like structure is an empirical object and to know it one would (on this hypothesis) need to use another language-like structure, which on the same hypothesis could not be known without first knowing yet a third language-like structure, and so on to absurdity. Consequently, the language-like structure used by a perceiver does not itself need to be known by the perceiver in order to be used in acts of perceiving. However, sometimes it is known in advance. For example, in the important case where the *parole*-like structure is artefactual – the input, for instance, of a readable technology – then it could be known in advance of perception. But such knowledge, as I have said, is not necessary. For this reason, the hermeneutical aspect of perception is existential – that is, it belongs to the being of the act – rather than methodological, for a methodological hermeneutic generally presupposes that the system of signifiers is known.

The language-like algorithms in the brain and retina are, of course, paralleled by similar structures in the information carrying energy fields that traverse the environment; among these are some, as I have said, which have their origins in cultural artifacts such as readable technologies; the latter are cultural artifacts which are reflected in new language-like structures of the brain and in the environment connected with hyperbolic vision, there are others produced by cultural artifacts, and among these artifacts is importantly the carpentered environment essential I believe for Euclidian vision. The computational program has little to say about the effects of culture on perception, except to be wary of them, since the combination of scientific realism and the identity theory entail that perception is mistaken or subject to illusion if it fails to represent an "objective" world – that is, the physical world as science is thought to picture it without reference to or distortion by cultural influences. Since this is neither a correct account of what perception does, nor a correct account of what is in the World, some confusion necessarily results in the execution of the scientific program.

In a hermeneutical phenomenology, the language-like algorithms in the brain respond to *parole*-like inputs from the retina which have their origin in parallel language-like structures in the ambient information carrying energy fields. The *parole*-like structures effect a physical union between subject and object even before the distinction between the two is articulated perceptually. The physical union is a kind of "fore-having" (*Vorhabe*), prior to, and necessary for the articulation of distinction. The articulation of distinction is done through the application of a conceptual framework (*Vorsicht* in Heidegger's terminology), *via* some clue to the applicability of the framework to this situation (*Vorgriff* in Heidegger's terminology). The triple of *Vorhabe*, *Vorsicht*, and *Vorgriff* constitute what is called the hermeneutical circle. All of these elements occur in the articulation of the computational program, but of course under different names and uneasily introduced into a philosophy of research which really has no way of understanding their relationship. *Vorhabe* occurs in the union of subject and object in the retinal image; *Vorsicht* as the inescapability of higher-order conceptual aspects to the process of object recognition; and *Vorgriff* occurs in the processing of the neurophysiological input as for example in the testing of an image for edges, boundaries, and surfaces with the consequent movement towards building a spatial sketch of the object.

What are the possibilities and limitations of machine vision? These follow from the account given above. Humans, like machines, process the physical embodiments (or implementations) of algorithmic terms, but humans, unlike machines, can come to articulate in a natural language the symbolic or algorithmic systems they embody. All algorithms for humans have their origin and source in the human Life-World. Unlike machines, human knowers come historically to embody ever higher and higher symbolic systems, each level making more precise the indeterminacies of the lower levels. Although no set of rules could ever be drawn up for humans or for machines to bridge ideally the gap between input signal and symbolic system, humans have the capacity, which machines do not have, of perfecting themselves progressively in powers of recognition according to any given set of interests by ascending to ever higher levels of analysis.

Humans use the processes of the hermeneutical circle to articulate conceptually and linguistically the terms of the symbolic systems they embody (or implement) in their pre-conceptual structures; then once a symbolic system is articulated, humans can design a machine to embody the symbolic system. Although machines can in principle be programmed to embody symbolic systems and algorithms at any level, unlike humans they cannot perform the hermeneutical processes necessary to articulate in some natural language appropriate to themselves the symbolic system as such they embody in their structures. People can form concepts, machines cannot.

The source for symbolic systems and algorithms with which to program machines is in natural language; natural language speaks out the World for people, and this is historical, pragmatic, and value-laden. The algorithms machines are programmed for serve primarily then the human interests and purposes which constitute that World. Although machines can be made to model any of the algorithms which humans possess, understand, and conceptually articulate, they do not serve themselves, they have not a history apart from humans, and serve no values apart from the human World.

Moreover, the gap between input and algorithm can be bridged by machines only where specific rules, such as matching by physical features, provide an appropriate means for the identification of terms. But appropriate for what or whom? Here the pragmatic essence of the human World enters. The appropriateness in question is sufficiency for some human purposes. These purposes are both *human* and *specific*, specific, that is, to a historical place, time, and human situation. The machine then will always be no more than a mediation of specific human purposes; it will not have a self-determining "life" of its own in any human sense.

State University of New York,
Stony Brook

AUTHOR: Patrick Heelan was born in Ireland in 1926. He has degrees in mathematics, physics, philosophy, and theology from universities in Europe and the United States. He has had faculty appointments at Fordham University, Boston University, and the State University of New York at Stony Brook, where he is currently Professor of Philosophy. His publications have focused on topics related to phenomenology and science, as reflected in his recent book on *Space-Perception and the Philosophy of Science* (Berkeley: University of California Press, 1983).

NOTES

Research for this paper was supported by NSF Grant No. SES–8204465.

1 See David Marr, *Vision: A Computational Investigation into the Human Representation and Processing of Visual Information* (San Francisco: W.H. Freeman, 1982); John Frisby, *Seeing: Illusion, Brain, and Mind* (New York: Oxford University Press 1980); Z. W. Pylyshyn, "Computation and Cognition: Issues in the Foundation of Cognitive Science," *Behavioral and Brain Sciences* 3 (1980), 111–169. Marr's program is presently being carried forward at the Artificial Intelligence Laboratory of MIT by T. Poggio, S. Ullman, K. A. Stevens, and others.

2 For a full, methodologically and philosophically explicit, account of contemporary research in psychoneurology, see William R. Uttal, *The Psychobiology of Mind* (Hillsdale, NJ: Erlbaum, 1978), particularly, pp. 681–695; for example, "modern psychobiology, most of its students agree, is mechanistic, realistic, monistic, reductionistic, empiricistic, and methodologically behavioristic" (p. 687). Cf. Marr (1982), p. 336, and Frisby (1980), pp. 157–158.

3 See Marr (1982), pp. 3, 99, 105, 354–355; other passages, e.g., p. 31, say that visual representations must be "useful" and not "full of irrelevant information," which suggests a more pragmatic purpose for vision and one which could well collide with the principle of scientific realism.

4 The identity principle referred to is that enunciated by H. Feigl in "The Mental and the Physical" (H. Feigl *et al.* (eds.), *Minnesota Studies in the Philosophy of Science*, Vol. II. [Minneapolis: University of Minnesota Press, 1958], pp. 370–497), and re-affirmed by recent psychobiologists, such as, e.g., Uttal (1978), pp. 60–62. The identity asserted is one of "denotational equivalence." Its essence is to deny the existence of anything, such as mind or mental processes, which cannot be completely described by the physical or biological sciences. Within this perspective, mental representations and processes are nothing but neurophysiological states, e.g., brain states, taken with their correspondences with real or possible worldly states. Marr seems to adhere to this position: for him, a mental representation is a purely neurophysiological state, only representations of this kind are accessible to the scientist, such states are states of a

computer-like (i.e., information-theoretic) representation, seeing is a computational process, etc.; for him, in short, when the scientist (S3) has come to understand someone else's (S1's) acts of perceiving, it is sufficient that he understand the computational processes involved in the information processing of the retinal inputs, and the neurophysiological hardware by means of which these processes are accomplished, and *nothing else is relevant*. I shall argue in this paper that Marr should have recognized that the computational processes as they are performed by S1 differ in some important respects from computational processes of a machine. The reductionist illusion follows from two false assumptions: that only S3 has realistic knowledge (scientific realism) and that S3 can have knowledge without assuming the relation that S1 has.

5   See Marr (1982), p. 331, and Frisby (1980), pp. 156–158.

6   Among the important philosophical works analysing these questions are Hubert Dreyfus' *What Computers Can't Do* (New York: Harper & Row 1982, 2nd. ed. 1979) which is critical of the current orthodoxy, and Fred. I. Dretske's *Knowledge and the Flow of Information* (Cambridge, MA: MIT Press, 1981) which defends it. This author finds Dreyfus' approach compelling, and this paper owes much to what he has written. See also the references given in these two books.

7   Maurice Merleau-Ponty, *The Visible and the Invisible* (Evanston, IL: Northwestern University Press, 1968), Chapter 4; and Martin Heidegger, *Being and Time* (London: SCM Press, 1962).

8   Berkeley and Los Angeles: University of California Press 1983. Hereafter referred to as SPPS. See also my "Natural Science as a Hermeneutic of Instrumentation," *Philosophy of Science* **50**, no. 2 (June 1983), 181–204, and "Perception as a Hermeneutical Act," *Review of Metaphysics* **37**, no. 1, whole no. 145 (September 1983), 61–75.

9   The distinction is made by the de Saussure school of linguistics and used by structuralists. See for example, John Sturrock, *Structuralism and Since: From Levi-Strauss to Derrida* (New York: Oxford University Press, 1979), p. 8, and Paul Ricoeur, *Hermeneutics and the Human Sciences* (Cambridge; Cambridge University Press, 1981), "What Is a Text: Explanation and Understanding," pp. 145–164. For *Vorhabe*, see Heidegger, *Being and Time*, p. 191.

10 See SPPS, Part I. Marr's "viewer-centered 2 1/2-D sketch" shares many of the characteristics of vision in hyperbolic visual spaces. He is concerned, however, with the dynamic of vision to achieve objectivity in Euclidean vision.

11 See SPPS, Chapters 10, 11, and 14.

12 See Rudolf Arnheim, *Art and Visual Perception* (Berkeley and Los Angeles: University of California Press, 2nd and rev. ed., 1947), pp. 289–290; also SPPS, p. 58 where this point is called the "true point."

13 See SPPS, Part I, also Chapters 14 and 15.

14 See SPPS, Chapters 8 and 12; also see Heelan, P., "Perception as a Hermeneutical Act." The distinction between text-like and *parole*-like structures in perception is not made in the cited references but is made here for the first time.

15 For the concept of *readable technology*, see SPPS, Chapter 11; also P. Heelan, "Natural Science as a Hermeneutic of Instrumentation."

16 See Marr (1982), p. 112.

17 See the discussion of Marr's 2 1/2-D sketch below.

18 See Don Ihde, *Technics and Praxis* (Dordrecht and Boston: Reidel, 1979), and SPPS.

19 The material in the text refers to the accounts given by Marr, Frisby, and Uttal in the works cited above.

20 See Frisby (1980), pp. 156–158, and Marr (1982), pp. 329–332.

21   See Uttal (1978), pp. 466–474, and 480–482 for a general account of this field; the basic references are also given there. See also Frisby (1980), p. 159.
22   See Marr (1982), pp. 22–29, 340–343.
23   Marr (1982), p. 345.
24   Marr (1982), p. 341.
25   See Marr (1982), pp. 19–26 for his account of *algorithm*, and of the distinction between *representation* and *implementation*.
26   See Frisby (1980), pp. 106–122, 156–158, also pp. 141–155, and Marr (1982), pp. 331.
27   Frisby (1980), p. 117.
28   See de Saussure, *Course in General Linguistics*.
29   Frisby (1980), p. 26, see also pp. 8–25, and Marr (1982), pp. 343, 351, 357–358.
30   Marr (1982), pp. 355–358.
31   See Frisby (1980), pp. 89–105 and 123–140, and Marr (1982), pp. 329–332 and *passim*.
32   See Marr (1982), pp. 37, 42, 52, and 330, and Frisby (1980), pp. 112–113.
33   Marr (1982), pp. 277–289.
34   SPPS, Part I.
35   SPPS, Chapters 14 and 15.
36   See Marr (1982), p. 15, and Frisby (1980), p. 121.
37   See Uttal (1978), pp. 459–461 where the references are given and analysed; also see Marr (1982), pp. 13–14.
38   See Frisby (1980), p. 157.
39   See Frisby (1980), p. 117.
40   See Uttal (1978), p. 687.
41   Frisby (1980), p. 157.
42   Frisby (1980), p. 158; see also Marr (1982), pp. 360–361.
43   Uttal (1978), p. 387.

EARL R. MACCORMAC

MEN AND MACHINES: THE COMPUTATIONAL
METAPHOR

ABSTRACT. In the 20th century the interpretation of the human mind and brain as a computer has replaced the 18th century metaphor of "man as a machine." This paper traces the development of the computational metaphor with some attention to its 18th-century roots, and then argues that its employment does not necessarily lead to mechanization of thinking and the autonomy of technique. An awareness of the metaphor and, therefore, hypothetical status of the computational metaphor helps prevent technique from escaping intentional human control.

The interaction between men and machines not only changes the organization of society, but also alters the concepts which men use to think, especially those concepts employed to reflect upon human nature itself. Jacques Ellul describes the impact of machines as extending far beyond the adaptation of man to the machine.

When I state that technique leads to mechanization, I am not referring to the simple fact of human adaptation to the machine. Of course, such a process of adaptation exists, but it is caused by the action of the machine. What we are concerned with here, however, is a kind of mechanization itself. If we may ascribe to the machine a superior form of "know-how," the mechanization which results from technique is the application of this higher form to *all* domains hitherto foreign to the machine; we can even say that technique is characteristic of precisely that realm in which the machine itself can play no role.[1]

The employment of technological metaphors to explain the nature of man illustrates one way in which technique extends the process of mechanization beyond the machine. Soon after the advent of machines, the metaphor, "man is a machine," resulted from the discovery of the similarity of the operation of the two entities. This mechanical metaphor, popular in the 18th century, led men to think of themselves as machines. As long as men realized that they were not literally machines, but only like machines in some aspects and different from them in others, man controlled the technique that brought about the process of mechanization of thinking. When the metaphor came to be treated literally, however, and the identification of machines and men became unquestioned – collapsing them into an analogy and dissolving the dissimilarities between the two entities – then the process of metaphoric mechanization

157

Carl Mitcham and Alois Huning (eds.), Philosophy and Technology II, 157–170.
© 1986 *by D. Reidel Publishing Company.*

becomes autonomous. As Ellul suggests, technique reigns, cutting man off from nature and his cultural tradition, thereby robbing man of his own humanity.

Technique has become autonomous; it has fashioned an omnivorous world which obeys its own laws and which has renounced all tradition. Technique no longer rests on tradition, but rather on previous technical procedures; and its evolution is too rapid, too upsetting to integrate the older traditions.[2]

In the 20th century the computational metaphor which interprets the human mind and brain as a computer has replaced the 18th century metaphor of "man as machine." This discussion will trace the development of the computational metaphor, with some attention to its 18th century roots, and then argue that the employment of the computational metaphor through a process of mechanization of thinking does not necessarily lead to the autonomy of technique. An awareness of the *metaphoric* and, therefore, hypothetical status of the computational metaphor will prevent technique from escaping intentional human control and by controlling technique man can retain his relationships with nature and tradition.

Under the computational metaphor, the brain can be viewed as a computational device similar to a computer, and the mind emerges as a series of programs by means of which the brain functions. Human thinking does not necessarily reduce to brain functions; rather, human thinking and brain functions combine to produce a computational process. The "hardware" of the brain operates under the control of the "software" of the mind to produce a computation which has traditionally been called cognition.

Zenon Pylyshyn describes the core of the computational metaphor in the following terms:

The view that cognition can be understood as computation is ubiquitous in modern cognitive theorizing, even among those who do not use computer programs to express models of cognitive processes. One of the basic assumptions behind this approach sometimes referred to as "information processing," is that cognitive processes can be understood in terms of formal operations carried out on symbol structures. It thus represents a formalist approach to theoretical explanation. In practice, tokens of symbol structures may be depicted as expressions written in some lexicographic notation (as is usual in linguistics or mathematics), or they may be physically instantiated in a computer as a data structure or an executable program.[3]

The computational metaphor for cognition represents tangible evidence of a success of an interactionist view of metaphor. The advent of the modern computer brought with it the metaphorical suggestion that these machines think; the discipline of artificial intelligence was developed by those computer scientists, philosophers and psychologists who accepted the metaphorical suggestion that computers engage in mental activities similar to those of humans. In an interaction metaphor, both parts of the metaphor are altered. When it is claimed metaphorically that "computers think," not only do machines take on the attributes of human beings who think – we ask whether computers have intentions and feelings as well as the ability to make rational deductions – but "thinkers" (human beings) take on the attributes of computers. And that is exactly what has happened in the case of the computational metaphor: The mind of a human being is now described in terms of the attributes of a computer, the neuronal states of the brain as if they were like the internal states of a computer, and the mental processes of thinking as if they were algorithmic. Computers are like minds in many respects, they can store data, recall it, manipulate it, learn to recognize new patterns, and even create new cognitive patterns. Human cognition is like machine computation, Humans can manipulate symbol strings according to rules in language and mathematics. Although computers are faster and more efficient than humans in many of their computations, most of the differences between the two remain on the side of humans who have emotions, possess more creativity, and are intentional in many of their actions. Those who deny intelligence to computers emphasize the uniqueness of these human functions, while those who affirm artificial intelligence in computers downplay the differences by dismissing the significance of human emotions for computers and by claiming that computers have intentionality.

What makes the identification of human beings with computers a metaphor and not just an analogy? The computational metaphor could not be understood if it were not possible to recognize similarities between humans and computers. And the differences do not proclude the formation of an analogy, since most fruitful analogies are not isomorphic with a one-to-one correspondence among *all* of their parts. But metaphors differ from analogies in that, although they presume analogies of features and parts of their referents, they also possess a strangeness arising from their juxtaposition of referents. This makes it necessary to explore more carefully the basis of this "strangeness." The strangeness in this case arises from the features of dissimilarity between men and machines that create a semantic anomaly for metaphoric expression.

A theory of semantic conceptual anomaly asserts that the difference between metaphor and non-metaphor, especially analogy, rests upon the conceptual recognition of the semantic anomaly of metaphor and its interpretation as meaningful. Emotional tension exists as a symptom of this recognition rather than as the origin of it. Not all semantically anomalous constructions are metaphors; only those semantic anomalies which we can interpret as suggesting new insights and new possible meanings are metaphors. Strange juxtapositions of referents that produce semantic anomaly are not necessarily ungrammatical, for if they bring about metaphoric conceptual understanding, they are quite normal; they are the initiation of a process of semantic change that may terminate when semantic markers change and new lexical entries enter dictionaries.

The illumination of human nature through the formation of mechanical metaphors did not arise as a new phenomenon in the 20th century, for they abounded in the 18th century. La Mettrie, for example, published his famous *Man a Machine* in 1748. Many who have not read this work imagine that La Mettrie compared man to a mechanical device like a watch, which indeed he did. But he also recognized that "man is so complicated a machine that it is impossible to get a clear idea of the machine beforehand, and hence impossible to define it," and so he resorted to a variety of other metaphors as well. [4]

Consider first La Mettrie's strictly mechanical metaphor.

The human body is a watch, a large watch constructed with such skill and ingenuity, that if the wheel which marks the seconds happens to stop, the minute wheel turns and keeps on going its round, and in the same way the quarter-hour wheel, and all the others go on running when the first wheels have stopped because rusty or, for any reason, out of order. It is not for a similar reason that the stoppage of a few blood vessels is not enough to destroy or suspend the strength of movement which is in the heart as in the mainspring of the machine; since, on the contrary, the fluids whose volume is diminished, having a shorter road to travel, cover the ground more quickly, borne on as by a fresh current which the energy of the heart increases in proportion to the resistance it encounters at the ends of the blood vessels? [5]

But La Mettrie also likens the recall of ideas to a gardener who in knowing plants "recalls all stages of growth at the sight of them." [6] The images produced in the brain, he compares to a "magic lantern." [7] Even the "soul" is described as an "enlightened machine." [8] And in comparing the human body to machine, La Mettrie becomes fascinated with the biological part of the metaphor, so that he begins to speak of the brain as having muscles for thinking and declares that to know man better one must not only look to machines but to animals as well.

Thus, the diverse states of the soul are always correlative with those of the body. But the better to show this dependence, in its completeness and its causes, let us here make use of comparative anatomy; let us lay bare the organs of man and of animals. How can human nature be known, if we may not derive any light from an exact comparison of the structure of man and of animals?[9]

This is the message hidden to those who know of La Mettrie's metaphor of man as a machine only by reputation. La Mettrie's "machine" is a blood and guts machine that can be illuminated not only by the mechanical parts of artifices, but also by comparison with animals. Perhaps the most fascinating proposal La Mettrie made was his suggestion that it should be possible to teach apes to learn human language and thereby further knowledge of human nature. Drawing an analogy with successful teaching of language to deaf-mutes, La Mettrie concludes:

But, because of the great analogy between ape and man and because there is no known animal whose external and internal organs so strikingly resemble man's, it would surprise me if speech were absolutely impossible to the ape.[10]

La Mettrie's suggestions presaged Washoe and Lana by almost 250 years for, along with the computational metaphor, comes the twin metaphor of "man as an animal."

In contemporary version of *Man a Machine,* Michael Arbib's *The Metaphorical Brain* explicitly invokes these twin metaphors as the basis for understanding man.

We want to understand how people think and behave, and in particular we wish to understand the role of the brain in thought and behavior. In some ways the brain of a man is like the computer of a robot, in others it is more akin to the brain of a frog. Our aim here is to convey an understanding of the brain in terms of two main metaphors: the cybernetic metaphor, "Humans are machines," and the evolutionary metaphor, "Humans are animals." We shall not downgrade the differences, but we hope to learn much from the similarities.

Thus, when we call this book *The Metaphorical Brain* we do not imply that the understanding of the brain that it affords will be any less "real" than that afforded by other books – rather we are simply making explicit the aid that metaphor provides us, as well as lessening the risk of misunderstanding that results when an implicit metaphor is mistaken for reality.[11]

As a product of evolution, the biological aspects of man must be described by any metaphor or series of metaphors that attempts to explain the nature of man. Where Arbib uses the twin metaphors of "humans are machines" and "humans are animals" to account for the biological nature of man, Pylyshyn employs only the computational metaphor of "cognition is computational," and assigned the animal nature of man to an instantiation of what he called the "functional architecture" of the mind. Pylyshyn speaks of computation and mind on two levels: (1) the theoretical requirements for computation and mind; and (2) the biological structures and processes of the brain that carry out these computations. These correspond to the software and hardware of a computer. But even with this distinction, Pylyshyn has difficulty accounting for intentionality and consciousness that humans exhibit in forming self-conscious goals that alter their mentality.

More startling, however, than Pylyshyn's attempt to include the biological component in the functional architecture of a computational device, is his insistence that the computational metaphor be taken *literally*.

Given that computation and cognition can be viewed in these common abstract terms, there is no reason why computation ought to be treated as merely a metaphor for cognition, as opposed to a hypothesis about the literal nature of cognition. In spite of the widespread use of computational terminology (e.g., terms like "storage," "process," "operation"), much of this usage has had at least some metaphorical content. There has been a reluctance to take computation as a *literal* description of mental activity, as opposed to being a mere heuristic metaphor. In my view this failure to take computation literally has licensed a wide range of activity under the rubric of "information processing theory," some of it representing a significant departure from what I see as the core ideas of a computational theory of mind.[12]

Pylyshyn wants us to conceive of mental activity not just *as if* it were computational like the algorithms of a computer, but to identify cognition *as* computation. This means not that human beings are identical with computers, for Pylyshyn has carefully formed a hierarchy of levels of theory with computation as theoretical level that can be instantiated in either machines or humans, but that the computational algorithms for both computers and minds are the same. Pylyshyn's claim of a literal status for the computational metaphor means that humans must change the way in which they theorize about themselves and the world. He invokes the analogy of the acceptance in the 17th century of Euclidean geometry as the nature of space. Only in the time of Newton were the axioms of Euclid taken as a literal description of the physical world. This acceptance "profoundly affected the course of science," and acceptance of the computational metaphor will similarly affect theories about cognition – positively, of course.

Accepting a system as a literal account of reality enables scientists to see that certain further observations are possible and others are not. It goes beyond merely asserting that certain things (sic) happen "as if" some unseen events were taking place. In addition, however, it imposes severe restrictions on a theory-builder, because he is no longer free to appeal to the existence of unspecified similarities between his theoretical account and the phenomena he is addressing – as he is when speaking metaphorically. It is this latter degree of freedom that weakens the explanatory power of computation when it is used metaphorically to describe certain mental functions. If we view computation more abstractly as a symbolic process that transforms formal expressions that are in turn interpreted in terms of some domain of representation (such as the numbers), we see that the view that mental processes are computational can be just as literal as the view that what IBM computers do is properly viewed as computation.[13]

Pylyshyn ignores, however, the consequences of taking Euclidean geometry literally that result in beliefs about the absoluteness of length that were shattered by the advent of relativity theory in the 20th century. What guarantee does Pylyshyn offer that converting the computational metaphor into a literal description will not impede cognitive science by constraining how people think about the problem into too narrow channels of thought? Metaphors can be most dangerous when one forgets that they are metaphors; one can become beguiled by familiarity rather than by corroborating evidence into accepting a metaphor as literal. Pylyshyn asks theorists to accept the computational metaphor as literal, not because of corroborating evidence, but because he believes acceptance will produce better theories that are more constrained, and therefore, more focused.

Whether or not to take the computational metaphor literally illustrates one of the major problems facing any theory of metaphor – that of drawing the line between the "literal" and the "metaphorical." Until recently, many linguists, philosophers, and scientists viewed metaphor with disdain as an ungrammatical device characteristic of sloppy thinking rather than a legitimate theoretical tool. In the eyes of these critics, mystics seeking to express the rapture of the moment of union, or poets yearning to express their love or anguish, could resort to metaphor since intuitions and feelings are not subject to precise presentation. But when a scientist resorts to metaphor he is invoking a mushy, imprecise figurative use of language; he should improve his theory to the point that he can present it in more precise terms. Yet the growing recognition that theories require metaphors to be both hypothetical and intelligible brings with it the need to differentiate between the metaphorical and the non-metaphorical or literal.

Some theorists of metaphor claim that all language is metaphorical, and that no such thing as literal language exists. They do admit that many metaphors lose their strangeness and become "dead metaphors," but even these, they claim, retain their metaphorical status by combining two referents that are not identical. In this sense, the sense of representation, all symbols are metaphors because they present the meaning of an object, event, or idea not necessarily present at the moment of utterance. If one admits that all language is metaphorical, then in any particular case like that of the nature of cognition, one is faced with a choice among metaphors. Why choose the computational over other possible metaphors? And, contrary to Pylyshyn, there can be no chance of converting the computational metaphor to a literal statement, since there exists no difference between the two. Pylyshyn might choose a metaphor that was less metaphorical than others, but he could never seek the literal. Such a view entails some harsh consequences, not only for making sense of a metaphor – since one has no standard of literal language by which to make the distinction – but also for a theory of truth. Since corroboration of metaphor can never convert it into a literal one, individuals are faced with a relativistic linguistic realm with no anchor points for that which could be experienced as literal.

However, if a theorist claims a distinction between the literal and metaphorical, then he must make good on this assertion by producing a plausible criterion for the delineation. Such a criterion must be not only linguistic but cognitive, because it will have to show how the literal can be experienced or perceived as literal, and the metaphorical as metaphorical. The cognitive notions of resemblance, similarity, and difference will all be involved as parts of a knowledge process that makes distinctions possible. And to present such a knowledge process will inevitably entail the invocation of metaphors, so that the development of the criterion will be rooted in circularity. The only partial escape from this possible paradox is to differentiate levels of discourse; when speaking about distinctions between the literal and metaphor in the context of cognitive processes, it will be necessary to use a meta-level of language. Because the theory of metaphor on the meta-linguistic level is itself metaphorical, this does not necessarily entail that distinctions between the literal and the metaphorical on the object-language level do not exist.

Any theory of metaphor that maintains a distinction between the literal and the metaphorical will also have to explain how metaphors differ from everyday language, and how metaphors die and become part of ordinary discourse. Metaphors serve as catalysts for linguistic change; the metaphors of one generation become the banal expressions of another. Metaphor exists as a quite normal creative human cognitive process that combines concepts not normally associated to produce new insights.

For instance, John McCarthy, the reputed originator of the appellation "artificial intelligence," argues that the ascription of mental qualities to machines is perfectly legitimate and should not be prohibited.[14]

To ascribe certain "beliefs," "knowledge," "freewill," "intentions," "consciousness," "abilities" or "wants" to a machine or computer program is legitimate when such an ascription expresses the same knowledge about the machine that it expresses about a person. It is useful when the ascription helps us understand the structure of the machine, its past or future behavior, or how to repair or improve it. It is perhaps never logically required even for humans, but expressing reasonably briefly what is actually known about the state of a machine in a particular situation may require mental qualities or qualities isomorphic to them. Theories of belief, knowledge and wanting can be constructed for machines in a simpler setting than for humans and later applied to humans. Ascription of mental qualities is most straightforward for machines of known structure such as thermostats and computer operating systems, but is most useful when applied to entities whose structure is very incompletely known.[15]

McCarthy's argument hinges on the word "same." But when does an ascription express the *same* information about a person as machine? To a thermostat, McCarthy ascribes the simple belief statements of: "The room is too cold," "The room is too hot," and "The room is OK." Yet this does not entail that the thermostat understands the concept of "too cold," which humans certainly do understand. If "belief" means only specific actions or dispositions to act, then the thermostat certainly does possess the three beliefs ascribed to it by McCarthy. If "belief" includes understanding and assent to a proposition, then it remains doubtful that the thermostat possesses "beliefs" in the *same* way that humans do. The metaphorical ascription of human traits to computers, or computer attributes to humans, raises the question of just what parts of the metaphor are the same for both.

To take the computational metaphor literally means that only certain features of human and computer computation will be the same. With metaphorical ascriptions one must decide just how far the similarities between the two referents of the metaphor can be extended. When McCarthy affirms that machines possess beliefs, we can legitimately wonder whether "belief" includes among its proper attributes "understanding" and "self awareness." On the one hand, the notion of a "belief" must be limited in its attributes if it is to apply to a thermostat. On the other, if a machine possesses a "belief" in the sense of a disposition to act, then it need not necessarily possess other attributes that human beings possess who also have dispositions to act (beliefs).

Metaphors allow human beings to extend their knowledge by juxtaposing referents not normally associated, and by suggesting relations between some attributes of each. But metaphors can also be dangerous, by enchanting us into thinking that what they suggest really does exist, or by leading us to assume that the attributes normally possessed by either referent are possessed in the same way by the other. If humans and computers both possess "beliefs," then a person may be lead by the metaphorical usage to assume that the properties of human "belief" should be limited to dispositions to act, since they are so limited in computers.

That is precisely what happens, for instance, in A. M. Turing's famous thought experiment called the "imitation game." An interrogator faces the challenge of deciding which two people separated from him is a woman and which one a man. He questions them and receives answers via a telecommunicator. If, when a computer is substituted for one of the participants in the other room, the interrogator cannot tell the difference, then the machine is said to have passed the "Turing Test." In any question involving the qualities of a computer, if one cannot tell the difference between the output of the machine and that of a human, then one is justified in ascribing the human attribute to the computer.

Consider also the issue of whether computers can think. In the computational metaphor, many of the similarities between humans and computers are obvious. Both entities can add, subtract and multiply; both can make certain kinds of decisions; both can store information and retrieve it; both can learn to recognize new patterns; both can process language. But can both "think"? If "thinking" is defined in terms of the preceding incomplete list of functions performed by both, then certainly computers "think." Indeed, one can take the point of view of the computer and claim that thinking only takes place when an entity follows formal rules. Since much of what passes for thought among human beings rests on haphazard association rather than the instantiation of rules, it might well be argued that computers "think" more often than humans, and that only on those occasions when a man or woman emulates a computer by strictly following formal rules, can he or she be said to be thinking rationally.

Some critics of artifical intelligence have responded that, since human beings who manifest intelligence can feel pain and computers cannot, therefore, computers cannot be intelligent beings. But this argument, just like its counterpart, assumes that if something is to be possessed of intelligence, then it must necessarily possess all of the other attributes that humans possess. Daniel Dennett, by contrast, has argued that to ask if a computer feels pain when it simulates human intelligence is like asking whether a computer experiences wind storms when it also simulates hurricanes.[17]

On the other side, just as humans can be misunderstood through the computational metaphor, so can the machine. Metaphorical personification, which has probably existed since the advent of human speech, has become extensive among computer scientists and everyday users. The naming of computers began in laboratories, and later appeared in science fiction novels and movies. A recent introductory work on computers entitled a section of Winograd's computer program (SHRDLU) "WHAT SHRDLU KNOWS."[18] Primitive man often personified natural objects by giving them a divine status; perhaps we have shifted the deification from nature to technology.

The computational metaphor can be interpreted not only as a creative linkage between normally disassociated referents, but also as a basic presuppositional insight or intuition that undergirds an entire theory. Under such circumstances it can be called a "basic metaphor."[19] Metaphors can at the same time be employed to express a particular feeling or to suggest an individual possibility. In this second situation it can be termed a "conveyance metaphor." Consider now some of the problems posed by the use of the computational metaphor as a basic metaphor for the investigation of mind and brain.

Struck by the algorithmic functions which computers can so efficiently execute and which, before the advent of the computer, only humans were known to perform, the first form of the computational metaphor was packed into the term "artifical intelligence" and the expression that "computers think." This basic metaphor, resting upon a startling suggestion when it was first proposed in the 1950s, combined "intelligence," which before had been confined to humans and animals, with "artifact." From the earliest occurrences mentioned in the *Oxford English Dictionary* to the present, "intelligence" has been associated with mental faculties and understanding, so that one of the usual attributes of "intelligence" has been "biological," as pertaining to humans and animals. Machines have been normally interpreted as having non-biological attributes. Positing the computational metaphor suggests both the examination of computing machines to see just how much in fact they are like humans, and the examination of human beings to see just how much in fact they correspond to computing machines. As a basic metaphor, the computational metaphor proposes to treat human cognition *as if* it were computation, and computation *as if* it were human cognition. When the metaphor was first proposed, it was widely recognized as a metaphor, and not a literal assertion, because people were aware of too many differences between human thinking and the functioning of computers. Gradually, more and more similarities were found, and the strangeness of the metaphor, its tension of suggestiveness, diminished.

Often the process of investigation, which begins with the recognition of a basic metaphor as the presupposition of the entire enterprise, follows the line of positing conveyance metaphors that grow out of the basic metaphor. McCarthy moves in this direction by attributing numerous human traits to machines. He extends the attributes beyond those of cognition to attitudes, feelings, and consciousness. He also extends the notion of computer to devices like the thermostat. The extension of the basic metaphor through other metaphors to a wider range of possible experiences tests the comprehension of the identification of the referents.

This procedure of extending our knowledge through the postulation of basic metaphors arising from intuitions about man and the world has been taking place for centuries. From Plato's presumption of an unchanging realm of Forms to Einstein's presuppositions that the world is orderly and mathematical, philosophers and scientists have recurrently, implicitly or explicitly, invoked basic metaphors as foundations upon which to construct their explanations. The invocation of the computational metaphor as a basic metaphor was done explicitly and with awareness that this presuppositional act brought with it tentativeness and speculation – characteristics Pylyshyn thinks lead to a theory construction that is too unfocused.

By recognizing the computational metaphor as a basic metaphor rather than as the literal expression desired by Pylyshyn, one can avoid transforming a hypothesis into an unreal concrete entity. The improper conversion of technological metaphors into literal statements contributes to the creation of an autonomous technique. When a person thinks that metaphors like the computational metaphor describe what actually is, he engages in the projection of objectivity and reality onto an abstract hypothesis. Forgetting the differences between men and machines dehumanizes men and personifies machines, thereby distorting both. Men and machines do have similarities, but they also possess dissimilarities. Metaphors that identify the two do so successfully when they produce not only analogy but also dis-analogy in the production of semantic anomaly. But so successful and widespread has the computational metaphor become that, through familiarity, one is seduced into forgetting its anomaly. When a person begins to think of the mind as literally a computational device, referring to it as "a computer made out of meat" (Minsky), then he fulfills the prediction of Ellul. The mechanization process creates a fictitious autonomous technique that dehumanizes man by cutting him off from integral relationships to nature, and from an evolutionary relationship to the cultural tradition.

Through an examination of technological metaphors such as the computational metaphor, humans can become aware of the limits of their conceptualization of what is real. A careful examination of the similarities and dissimilarities of minds and computers prevents the all too easy identification of the two in the creation of a false claim about the literal character of the computational metaphor. Such reflective self-consciousness can help undermine Ellul's pessimistic conclusion that technique will become not only autonomous but also inevitable.

We have completed our examination of the monolithic technical world that is coming to be. It is vanity to pretend it can be checked or guided. Indeed, the human race is beginning confusedly to understand at last that it is living in a new and unfamiliar universe. The new order was meant to be a buffer between man and nature. Unfortunately, it has evolved autonomously in such a way that man has lost all contact with his natural framework and has to do only with the organized technical intermediary which sustains relations both with the world of life and with the world of brute matter. Enclosed within his artificial creation, man finds that there is "no exit;" that he cannot pierce the shell of technology to find again the ancient milieu to which he was adapted for hundreds of thousands of years.[20]

Scrutiny of technological metaphors may indeed by only a small defense against the development of a technical autonomy, but it offers at least one exit from the conceptual prison described by Ellul. And an examination of the dominant basic metaphors of other historical epochs can indicate as well the fortuitous shifts in conceptual hypotheses that have undergirded different human views of reality. The supernatural world of the Middle Ages gave way to the Renaissance view of a man-centered universe. While the mechanization of the present age results in the creation of basic metaphors such as the computational metaphor, when this is understood as a hypothetical image it allows the human beings to escape from the almost overwhelming conceptual authority of ideas which Ellul so greatly fears.

Davidson College

AUTHOR: Earl MacCormac was born in 1935 and has received degrees in engineering, divinity, and philosophy. He is Charles A. Dana Professor of Philosophy at Davidson College and Adjunct Professor of Industrial Engineering at North Carolina State University. He has contributed articles to philosophy journals and his *Metaphor and Myth in Science and Religion* was published by Duke University Press in 1976.

NOTES

1   Jacques Ellul, *The Technological Society* (New York: Knopf, 1964), pp. 6–7.
2   *Ibid.*, p. 14.
3   Zenon W. Pylyshyn, "Computation and Cognition; Issues in the Foundations of Cognitive Science," *Behavioral and Brain Sciences* **3,** no. 1 (March 1980), p. 111. Later in this article Pylyshyn argues that the computational metaphor can be treated as a literal expression, as he does in his book *Computation and Cognition: Toward A Foundation for Cognitive Science* (Cambridge, MA: MIT Press, 1984).
4   Julien Offray de La Mettrie, *Man a Machine* (LaSalle, IL: Open Court, 1912), p. 89.
5   *Ibid.*, p. 141.
6   *Ibid.*, p. 107.
7   *Ibid.*
8   *Ibid.*, p. 128.
9   *Ibid.*, pp. 97–98.
10 *Ibid.*, p. 101.
11 Michael A. Arbib, *The Metaphorical Brain* (New York: Wiley-Interscience, 1972), p. vii.
12 Pylyshyn, p. 114.
13 *Ibid.*, p. 115.
14 Pamela McCorduck, *Machines Who Think* (San Francisco: W. H. Freeman, 1979), p. 96.
15 John McCarthy "Ascribing Mental Qualities to Machines," in Martin Ringle (ed.), *Philosophical Perspectives in Artificial Intelligence* (New York: Humanities Press, 1979), p. 61.
16 A. M. Turing, "Computing Machinery and Intelligence," *Mind* **59,** whole no. 236 (1950), reprinted in Alan Ross Anderson (ed.), *Minds and Machines* (Englewood Cliffs, NJ: Prentice-Hall, 1964).
17 Daniel C. Dennett, "Why You Can't Make a Computer that Feels Pain," in *Brainstorms* (Montgomery, VT: Bradford Books, 1978). Dennett also believes that "pain" is a very confused concept.
18 Margaret A. Boden, *Artificial Intelligence and Natural Man* (New York: Basic Books, 1977), pp. 134ff.
19 Stephen Pepper, *World Hypothesis* (Berkeley: University of California Press, 1943). I have extended Pepper's notion of a "root metaphor" to a "basic metaphor" to avoid the metaphysical connotation of the term "root metaphor."
20 *Ellul,* p. 428.

JOSEPH MARGOLIS

INFORMATION, ARTIFICIAL INTELLIGENCE, AND THE PRAXICAL

ABSTRACT. Contrasts bottom-up and top-down strategies for analyzing the concepts of information and artificial intelligence. On the top-down view, information is conceptually dependent on the conditions of human interpretation; whereas on the bottom-up view, information is independent of such conditions. Three kinds of bottom-up strategies are distinguished: physicalistic, Leibnizian (Dretske), and homuncular (Dennett). The top-down approach favors a technological or praxical conception of information and artificial intelligence, and is thus more reasonable and promising.

The developing importance in our time of the concept of artificial intelligence is an implicit tribute to the power and distinction of contemporary technology. But in a larger sense, it merely confirms the perennial complexities of the meaning of human interventions in the world and of the uniquely species-specific range within which consensual and idiosyncratic categories are found or made to fit the world. This indissoluble linkage between the purposive activity and work of man and what might be called the legibility of nature (including, *a fortiori*, human culture and history) is, on the least tendentious reading, what is meant by the *praxical*. Thus construed, the praxical is entirely neutral to the important conceptual quarrels that have been generated by comparing and interpreting Marx, Heidegger, Lukacs, Adorno, Althusser, Dewey, Habermas – not to mention Godelier, Vazquez, Markus, Eagleton, and an army of more recent theorists. But admitting the praxical does entail (even in its most neutral form) rejecting all forms of naive or direct realism – in effect, all forms of the thesis of the cognitive transparency of nature, in particular, of essentialist interpretations of the laws of nature. The reason is simply that, on the admission, our very understanding of the world is seen to be a function of the contingent, varied, and clearly transient forms of human intervention. As we shall see, this conceptual linkage between the paradoxical (or the technological) and the rejection of a cognitively transparent world is critical in weighing the merits of alternative theories of information and artificial intelligence.

The concept of artificial intelligence is, also, rather more inclusive and more abstract than that of computers, Turing machines, and the like – which are simply species of the covering genus (or species of machines of

Carl Mitcham and Alois Huning (eds.), Philosophy and Technology II, 171–186.

some sort) that, by suitable improvement, might come to justify ascriptions of intelligence. The point of this fussy caution about terms concerns liberating our speculation about the import of artificial intelligence from the special constraints of particular kinds of machines, prejudices about the actual functioning processes of the human brain, and privileged convictions about the analysis of human intelligence in terms of machine modelling. In fact, disputes about the nature and analysis of artificial intelligence and about the processing of information (in at least an initially anthropomorphically familiar form) common to human and artifactual systems (even extended to physical systems) essentially rehearse (and must rehearse) the puzzle of the so-called World Knot, that is, the place of mind in nature – now, one believes, with all the advantages of bypassing Cartesian dualism.[1]

It would be a mistake, however, to suppose that to construe the functioning of mind in terms of informational processing appropriate to the most generously abstract conception of artificial intelligence would be to gain a *principled* advantage over purely reflexive studies of human intelligence itself (although there can be no doubt that empirical research gains are bound to be made) – not merely because artificial intelligence is modelled in a decisive way on the human (even if, at the same time, the human is taken to be modelled by a Turing machine, say), but because artificial intelligence (as opposed to extraterrestrial intelligence) *is part and parcel of the actual work and working of human technology and human intelligence*. This is to say that artificial intelligence is modelled in whatever the human can be modelled, because it is the human; not merely that the information processed by artificial intelligence must be modelled propositionally, as must also the human, since we really have no viable alternative models of such processing (*contra*, for example, the failed expectations of B. F. Skinner and D. M. Armstrong).[2] (Perhaps in time, we shall even discover that artificial intelligence can also be extraterrestrial as it is now human.)

Surprisingly, this simple conceptual maneuver resolves in a powerful way several strategic controversies regarding the theoretical significance of the technology of artificial intelligence. For example, it shows at a stroke that Hilary Putnam cannot but have been completely mistaken when, in a early, extremely well-known and much-discussed paper (which offers a view Putnam now frankly pretty well has altered), he announced that

The various issues and puzzles that make up the traditional mind-body problem are wholly linguistic and logical in character . . . [in fact] it is no longer possible to believe that the mind-body problem is a genuine theoretical problem, or that a 'solution' to it would shed the slightest light on the world in which we live . . . [it] is nothing but a different realization of the same set of logical and linguistic issues [as are raised by the "identity" or "nonidentity" of logical and structural states in a machine'].[3]

There are problems raised by Putnam's view that are merely local to the mind-body problem – for instance, that pain is somehow characterized in a purely abstract or functional way, and that mental properties are entirely functional and yet can be assigned a causal role. Viewed solely in terms of our present concern, it is clear that Putnam's maneuver cannot but be vacuous, in the strict sense that the mind/body problem is nothing but the logical/structural problem of information-processing machines since (or, if and only if) that problem is nothing but a manifestation of the generic mind/body problem (no matter how unfortunately skewed in terms of Putnam's peculiar views of psychological states).[4] The implications of this *reductio* are well worth pursuing, because they help to fix the full conceptual import of the notion of the technological.

Putnam holds, further, that "everything is a Probabilistic Automaton under *some* Description,"[5] which is to say (only) that any finite, informationally qualified *segment* of a system can have its informational properties generated by a Probabilistic Automaton – *not* that informational properties or processes can be reduced to the non-informational properties or processes of *any* physical system in which information is thought to be incarnate, embedded, or realized in some particular way, *or* that the ecologically rich and openended *capacities* of humans can as such be modelled by machines. Here, we come directly to the problem of the choice between top-down and botton-up strategies so dear to the speculations of the theorists of artificial intelligence.

There are only two ways of construing the choice between top-down and bottom-up strategies of analysis: one is as a matter of sheer convenience, on the thesis that the two are ultimately equivalent in every important respect that we might wish to preserve – hence, that all complex phenomena can in principle be *composed of*, or *generated from*, the foundational elements of a given system; the other is as an expression, respectively and disjunctively, of nonreductive and reductive accounts of the mental or cognitive or intentional or informational or cultural or historical or praxical or technological – in terms of the physical of purely extensional order or of some suitably analogous order. To construe technology in terms of the praxical, in our deliberately neutral fashion, is just to pose the confrontation of these two strategies: because the very notion of human intervention, of work and activity, both first identifies what we wish to understand (which provisionally favors the top-down strategy) and what conceivably would give us the most unified, simple, and extensive account of the whole of the natural world (what ultimately is promised by the most extreme bottom-up strategy).[6]

Although it is hardly logically necessary, top-down strategies tend to oppose all forms of direct realism (the transparency of nature: essentialist natural laws, the propriety of the correspondence theory of truth, foundationalism, logocentrism, the philosophy of presence, and the like). This is part of the point of emphasizing the praxical sources of human understanding and, in particular, the obvious failure of any known form of analysis effecting a reduction of human language and real linguistic abilities to some set of sublinguistic structures and processes. Correspondingly, again without logical necessity, bottom-up strategies tend to favor some version of the transparency of nature, hence either to neutralize the apparent obstruction of the *sui generis* phenomenon of language (along Wilfrid Seller's lines, for instance) or in various ways to embed the peculiarly propositional structure of language of information in natural processes themselves (for instance, in the manner of Noam Chomsky or Fred Dretske).[7]

In fact, the bottom-up strategy takes three quite distinct forms: if it is physicalistic, it tends to treat the top-down as fully equivalent to the bottom-up and no more than a choice of convenience; if it is Leibnizian, in ascribing at a suitably elemental level attributes of an informational or cognitive sort, it may, yet need not, be reductive, but even if it is, it need not attempt specifically to reduce the linguistic to the sublinguistic (or, for that matter, the biological to the physical); finally, it may treat pertinent bottom-up strategies as no more than *relationally* contrived forms of *factoring* out (not composing or generating) phenomena properly specifiable only at a certain top-down level of discrimination (in particular, at the level of human *praxis* itself). The third of these strategies may fairly be labeled *homuncular*, although its best-known advocate, Daniel Dennett, firmly believes that it can and must give way to the first.[8] Jerry Fodor's version of cognitivism is strongly tempted by the first strategy but remains, at least for heuristic purposes, content with the second[9]; and Chomsky's and Dretske's accounts are flatly versions of the second strategy considered in a fully realist manner. Roughly, the competing intuitions are these: the most complex cultural phenomena are nothing but congeries of atoms or monads (the bottom-up strategy); or, alternatively, such phenomena are emergent and *sui generis* and can only be factored in their own terms (the top-down strategy).

Now, the point of these distinctions is to provide a background against which we may most effectively assess the import of the conception of information and the technology of artificial intelligence. *If*, for instance, as has already been suggested, artificial intelligence is part and parcel of human technology, then a bottom-up strategy of the first sort cannot be effective in principle without an *independent* reduction of the praxical – in particular, linguistic – features of human interventions in nature. Assuming the Leibnizian model to be basically unsatisfactory and no more than metaphorical (since it treats propositional structures as natural or real sometimes even in the inanimate world – and as not even needing to be accounted for in terms analogous to those that support ascriptions of psychological states[10]), there may well be no other plausible bottom-up strategy (at least at the present time) than the homuncular. A brief scan of Dennett's (homuncular) and Dretske's (Leibnizian) strategies, however, should convince us that there are no serious prospects at the moment for pursuing any bottom-up strategy with respect to artificial intelligence stronger than what we have here termed the homuncular.

A word of explanation is in order, however, before we turn to the evidence. Dennett of course introduces the homuncular thesis (quite correctly) as a version of a *top-down* analysis – hence, as *not* intended, at the level at which it is first introduced, to replace or eliminate or reduce without remainder the phenomena thus factored. But somehow, in the middle of the argument, he *does* construe the homuncular analysis as suitably replacing, without remainder, *all* phenomena identified at a given molar level (that is, intentional or linguistic of psychological phenomena). At this point, then, Dennett reinterprets his top-down strategy as, in effect, a fully bottom-up strategy of the second sort (of a sort that can only be introduced independently and non-relationally); and, as is well known, he further maintains that the homuncular-level account (which still exhibits at the sub-molar level the informational or intentional complexity of the molar) can itself be eliminated in favor of a deeper bottom-up account that eschews all intentional or informational attributes (a strategy of the first sort).

The key to the failure of Dennett's argument is simplicity itself. For, as a top-down strategy, homuncular terms are entirely *relational*, that is, introduced only as factoring sub-functions of some functioning, already conceded molar system – hence, as logically incapable of replacing the molar. Thus Dennett says:

The information or content an event within [a given] system has [it has] *for the system as a (biological) whole.* . . . The *content* (in this sense) of a particular vehicle of information, a particular information-bearing event or state, is and must be a function of its function in the system . . . of which it is a part.[11]

But he also says, inconsistently with the above (or, at least on independent grounds, which he fails to supply):

Any psychology with undischarged homunculi [for whom 'internal representations' function as such, that is, informationally, and who are not theoretically replaced (discharged) by 'agents' described in purely physical terms, without reference to representations] is doomed to circularity or infinite regress, hence psychology is impossible.[12]

If these difficulties are genuine, then we have to that extent confirmed the sense in which a study of artificial intelligence cannot yet be shown to provide a principled or independent advantage for the analysis of human technology and human praxis; for it must itself be informed – by way of relational or homuncular distinctions (that is, by introducing sub-molar functions as sub-functions of the molar) – by whatever categories are deemed suitable at the level of molar (or human) functioning. This, in effect, returns us to the vacuity charged to Putnam's early thesis.

Turn now to Dretske. Dretske's theory is surely one of the first fully developed philosophical analyses of cognitive states – neutral to the distinction between the human, the animal, and the artificial – informed by contemporary communication theory (in particular, Claude Shannon's[13]). That Dretske's theory is essentially what is here being called Leibnizian is clear at once from the very opening remarks of his account. In the Preface to his *Knowledge and the Flow of Information*, Dretske explicitly says:

In the beginning there was information. The word came later. The transition was achieved by the development of organisms with the capacity for selectively exploiting this information in order to survive and perpetuate their kind. . . . [I]nformation (though not meaning) [is therefore] an objective commodity, something whose generation, transmission, and reception do not require or in any way presuppose interpretive processes. . . . Meaning, and the constellation of mental attitudes that exhibit it, are manufactured products. The raw material is information.[14]

On Dretske's view, there is no isomorphism, extensional equivalence, or even mutual dependence linking information and meaning.[15] Dretske does recognize, following Warren Weaver's reading of Shannon, that mathematical communication theory ignores the question of the informational *content* of signals in order to study the quantitative effectiveness and capacity of informational channels as such.[16] Nevertheless, in recovering a conception of informational content congruent with that theory, Dretske is entirely prepared to treat what he calls the "objective commodity" of information as fully propositional – that is, he holds that what a system anywhere in nature, actually functioning as or actually housing a signal, "contains" as information it "contains" independently of any "receiver's [say, any human's] actually learning something from that signal."[17] Information is, therefore, both "objective," *there* in the real world, independent of interpretation, *and* propositional or intentional, that is, expressible in the form "*s* if *F*' carrying "*de re* informational content . . . *of* or *about s* that it is *F*'; or, alternatively, determined by "a relation between what is expressed by an open sentence ('. . . is *F*') and some individual *s*."[18] What is Leibnizian, here, in a strong but tendentious sense is marked by the following considerations (granting of course that Dretske takes relevantly favorable and unfavorable stands on each): first, that propositional content is normally and rightly construed both as parasitically modelled on and hardly more than a special abstraction from the functions of natural-language sentences; secondly, that the "objective-commodity" characterization unconditionally favors a thoroughly extensionalist account of information – hence, favors an extensionalist account of the causal processes it is said to depend on as well as of the logical behavior of the set of sentences by which the content of any informational signal can be fixed; and thirdly, that the characterization is expected to be adequate for formulating a competent and plausible (bottom-up) theory of human cognitive states (knowledge and belief), that are otherwise known to pose extremely deep perplexities of an intensional (that is, non-extensional)sort.

Dretske's theory is committed to a bottom-up strategy, is foundationalist in a strong sense, and, therefore, seeks to escape as far as possible the implications of a praxical and technological account of information. Hence, its failure would confirm the peculiar importance for a ramified theory of technology of an analysis of artificial intelligence. Broadly speaking, the counterstrategy to Dretske's approach would emphasize: (a) that the informational import of any system of artificial intelligence is a dependent function of human informational processing simply because it *is* a manifestation of such processing – hence, conceptually dependent on human interpretation; and (b) that human informational processing, in particular, natural-language communication, is ineliminably complicated in intensional ways – hence, capable only of conceding extensional reduction in a dependent, piecemeal, or (more generally) homuncular form.[19]

There are two principal maneuvers that insure (as far as we now understand the matter) the probably irreducibly intensional complexity of human language and of any informational processing embedded in the entire run of linguistic behavior: the first is to press the failure (and the implicit ignorance entailed in that failure) to provide a completely extensional account of reference, propositional attitudes, the distinction between meanings and beliefs, the functional connection between the meaning or sense of predicates and their extensional scope, the truth-functional connectives of natural languages, concepts, and the relation between the rules of linguistic usage and contextual constraints on such usage; the second is to press the import of the indefensibility of foundationalism and of the inescapability, within the space of that admission, of the historicized and praxicalized nature of all human undertakings to understand, describe, and explain the features of the world we encounter – hence, to defeat all prospects of totalizing over all possible worlds.[20] The counterstrategy, therefore, is a top-down strategy that seeks to demonstrate that there can be no realist theory of information or of objective information freed from the intensionally complex features of human information itself: this is just what is captured by construing artificial intelligence as a form of human information processing *and* by exposing the merely figurative nature of the Leibnizian conception of information.

Having set these considerations in place, it is a straightforward matter to demonstrate the inherent and peculiar weaknesses of Dretske's important account (without pausing to consider local problems about his theory of cognition). First of all, though he is prepared to characterize information propositionally or intentionally in the manner already sketched, Dretske offers two very telling further characterizations: "if a signal [he says] carries the information that *s* is *F*, then it must be the case that . . . [t]he signal carries as much information about *s* as would be generated by *s*'s being *F*" (the so-called "communication condition")[21]; furthermore, he says, "the informational content of a signal is a function of the *nomic* (or law-governed) relations it bears to other conditions."[22] Now, the first of these characterizations is either a vacuous translation of some putative state of affairs (*s*'s being *F*) *in* informational terms *or* an undefended thesis that genuinely "objective" information transparently, that is, without distortion or interpretation, *represents* some such state of affairs. Information, then is not anthropogenic, though it seems to be anthropognomic – which is to say, Leibnizian. On the praxical view, it is both anthropogenic and anthropomorphic (that is, linguistically modelled). Knowledge, consequently, is defined by Dretske in a singularly extreme way: "*K* knows that *s* is *F* = *K*'s belief that *s* is *F* is caused (or causally sustained) by the information that *s* is *F*."[23] The nomic universals of causality, therefore, insure that, because (by definition) information perfectly mirrors the states of affairs it represents, genuine knowledge *can* capture that information.

Nevertheless, on Dretske's view, although genuine information preserves *all* extensionally derivable propositions linked either to given lawlike relations or linked analytically to information first posited (in virtue of which "a physical structure has no *determinate* or *exclusive* informational content"[24]), knowledge and believe *do* exhibit a "higher order of intentionality" than does objective information itself.[25] Knowledge (*a fortiori*, belief) need not preserve all such derivable information *in* preserving what it does preserve. This correctly concedes part of the familiar paradoxes of intensionality. But it claims further – untenably or at least without justification – that: (i) all genuine factual knowledge is foundational, essentialist, or informed by a natural world entirely and cognitively transparent at least to that extent; (ii) that knowledge is intentional (or intensional) only in the sense of entailing a limited selection from (but otherwise no distortion or essential skewing of) an exclusively extensional order of information; (iii) that the structures of the real world (including those of human language and culture) *can* be perfectly represented informationally in a completely extensionalist manner; and (iv) that all causal relations, notably those involving the linguistically complex interventions of man, can be characterized uniformly, that is, as behaving extensionally and as subsumable under universal covering laws. Certainly at the present time, (i) is rejected as false by nearly all philosophical theorists of nearly every stripe; (ii) is mortally linked to the fate of (i); (iii) is obviously false or undemonstrable short of a physicalist or at least materialist reduction of language (involving for instance a demonstration of the thoroughly extensional structure of natural language); and (iv) depends on the fate of (iii).

These are surely among the most controversial doctrines of the entire philosophical tradition. They are certainly bold and would, if correct, utterly obviate a technological or praxical conception of information (*a fortiori*, of artificial intelligence): because the latter conception *is*, in effect, committed to the thesis that *information is a function of human interpretive schemata* – hence, subject to whatever complications follow from that concession. Without prejudice to the findings of any sustained inquiry, it is reasonably clear that these complications would probably include the denial or defeat of the four claims just posited. Hence, the theory of information is really the battleground for a fundamental contest between opposed orientations respecting technology already drawn up in terms of much earlier, still troublesome, and entirely unresolved quarrels.

Dretske cannot, therefore, have meant to construe his own thesis about information as concerned only with providing a translational idiom for identifying any and all states of affairs in the natural world; he meant rather to introduce a distinct and objective "semantic" category (information) *not* dependent on, or equivalent to, the category of meaning, and not dependent on the concept of knowledge (which, on the contrary, itself depends on the concept of information).[26] Furthermore, on Dretske's view, the probability that s is F, given that a signal r "carries the information that s is F . . . is 1"; and, as a result of depending on information thus construed, "knowledge [Dretske says] is an absolute concept" – though, contrary for instance to Peter Unger's not dissimilar view, this information-theoretic concept is said not to lead or to need to lead to skepticism.[27]

The upshot is that information about given states of affairs perfectly represents those states of affairs; and factual knowledge, caused by such information, is, though the idea of knowledge "*is* an absolute idea," nevertheless a state that is not "unobtainable" – for its absoluteness depends on the information it "inherits," not on the "certainty" it may or may not inspire.[28] In this sense, Dretske's thesis about information *and* knowledge is foundationalist, Leibnizian, and thoroughly extensionalist.

The program, however, cannot support its own weight. It cannot, at least on the current evidence of the status of the doctrines on which it depends; and perhaps it cannot, in principle, because of the compelling force of the praxical conception of understanding – which undermines it. Here, it may be useful to rehearse very rapidly some of the key failures (or at least some of the unresolved difficulties) on which the success of the various forms of the bottom-up strategy (physicalist, Leibnizian, or homuncular) ultimately depend.

For one, W. V. Quine has argued that

When . . . constructions on sentences are limited to quantification and truth functions [that is, logical transformations involving the simplest sentences and complex sentences constructed from them], one law that is easily proved by . . . [mathematical] induction is that of extensionality, "that is," substitutivity of coextensive terms preserving truth.[29]

Quine means to apply this thesis in a fully sanguine way to the work of empirical science, reassuring us that "Surviving idioms of an *extraneous* sort – indicator words [e.g., 'this,' 'this water,' and the demonstrative pronouns], intensional abstracts [as in treating classes, attributes, relations, and the like as objects], *or whatever* – can remain buried in larger wholes which behave *for the nonce* as unanalyzed general terms."[30] Nevertheless, whatever formal demonstration can be given regarding extensionality, it hardly follows that the terms of natural-language discourse developed in a cognitively engaged way (say, within the constraints of an historicized and praxicalized inquiry) *can* be reliably shown to yield to such regimentation: the pretty expression, "for the nonce," betrays an unexamined assumption. In particular, Quine's attempted elimination of the so-called referential opacity of sentences involving "believe" and similar verbs of propositional attitude (that is, verbs that appear to thwart the extensionality thesis) fails simply because the problem requires a *cognitively* motivated solution, *not* a purely formal one that begs the very cognitive question at stake.[31] No one has as yet supplied the answer needed. In the context of Dretske's thesis, it has yet to be shown that human information processing can be accounted for in extensionalist terms. Quine's convictions about this are curious for another reason, namely, that he himself is quite opposed to all forms of foundationalism and essentialism – which, in the context of *Word and Object*, might not unreasonably be construed as intuitively committed to a praxical thesis. Apparently, he is persuaded that our inability to discover "some fundamental set of general terms on the basis of which all traits and states of everything could in principle be formulated" somehow *does not* render in the least doubtful or uncertain the extensionalist project itself.[32] But his confidence rests on an obvious lacuna. Consequently, so does Dretske's.

Again, there is no clear way in which either to eliminate the reality of mental states and linguistic behavior or to reduce them in the physicalist manner. Rudolf Carnap offered the most radical version of the reductive thesis, but then backed away from it without ever giving it up.[33] Wilfrid Sellars offered the most radical version of the thesis that the roles of persons could be assigned, without any realist imputation, to the otherwise satisfactorily closed world of physical science, but he never provided the argument.[34] If, however, human cognitive states are real, causally effective, informationally significant, *and* intensionally irreducible, then a great part of the technological or praxical conception of information (hence, also, of the analysis of artificial intelligence) would be vindicated.

At the very least, we could not then claim – as in somewhat different

ways Dennett and Dretske do, more or less following Quine – that inten-
sional complications really concern only the *sentences* by which the properties of the inten*t*ional are marked. For example, Dennett challenges
the intentional thus: it appears, on at least one criterion, he remarks, that
"to change the description [of a putative object] is to change the object.
What sort of thing [he asks] is a different thing under different descriptions? Not any object. Can we not do without the objects altogether and
talk just of descriptions?"[35] But this presupposes, without argument,
that it cannot be the case that there are real phenomena that cannot be
satisfactorily fixed in an extensionalist way; hence, that our only option is
to eliminate such objects altogether or else to preserve them under (and
only under) extensionalist conditions. But if *human* informational processing is real and its intensional complexity irreducible, then both the
physicalist and Leibnizian strategies are indefensible. Dretske, for his
part, simply distinguishes between inten*t*ional phenomena (informational phenomena, say, phenomena *about* or *of* given states of affairs) and
the so-called inten*s*ional, that is, the formal property of failing Quine's
extensionality condition "if [that is, when and only when] we are speaking of the *sentences* used to describe such [intentional] phenomena."[36]
But the problem, precisely, is that real human mental states and behavior
are linguistically affected in such a way that *they* are intensionally problematic. *This* is essential to the praxical conception of information and
completely overlooked or unresolved by all bottom-up strategies –
whether crypto-top-down or not – *a fortiori*, by all "objective" accounts
of information.

Add to those considerations the holism of the mental, that is, the relational way in which mental ascriptions are made under the constraint of
an intentional model of rationality, the causal efficacy of the mental, the
consequent difficulty of identifying and reidentifying mental phenomena
extensionally, the threatening heteronomic status of the natural world
construed in terms of physical laws, the possibility (therefore) that not all
causal contexts behave extensionally, the further possibility that not all
causal contexts can be shown to fall under covering laws,[37] and it becomes abundantly clear that all bottom-up strategies for analyzing information, artificial intelligence, and human cognitive states can, at the present
time, be hardly more than a profound conceptual prejudice – that the
top-down approach is considerably more reasonable, methodologically
more straightforward, even unavoidable. But to admit that is, precisely,
to admit the provisional superiority of a technological or praxical
approach to these matters.

Of course, the systematic import of these admissions cannot be narrowly confined to the topic of artificial intelligence. On the argument, that question itself proves to be a particularly strategic version (but no more) of the global question of how to understand the way in which the nature of human existence – of culture and technology – affects our conception of the methodology of science, hence also our conception of an adequate ontology and epistemology. Our survey has been deliberately restricted to the principal strong versions of current analytic views of information and artificial intelligence. By exhibiting their characteristic weaknesses – out of their own mouth, so to say – we have shown the reasonableness of shifting the primary focus of philosophical research programs from a bottom-up to a top-down orientation.[38]

We have not shown that bottom-up programs are self-contradictory or conceptually impossible or anything of the like. That is hardly to be expected. But the failure of physicalistic accounts to carry their reductive intent through to something approaching completion (Carnap, for instance), the stubbornly metaphoric and foundationalist presumption of the Leibnizian view under pressure from current reasonable expectations for a deeper analysis and a conceptual defense (Dretske, for instance), and the inability to convert the homuncular into a distinct bottom-up strategy (Dennett, for instance) greatly strengthen – particularly on the nearly irresistible evidence of the *sui generis*, emergent nature of human language – the sense (what, robustly, may be admitted to be rhetorical or socialized concern) in which top-down strategies are overwhelmingly more reasonable that bottom-up ones. Furthermore, wherever the bottom-up approach seems provisionally feasible (a point of considerable dispute, apparently, among the proponents of quantum and relativistic physics[39]), an overriding top-down orientation can provide a suitable accommodation. What it cannot do – and what the bottom-up alternatives have never been able to show they can manage in their own favor – is admit that wherever the linguistic, cultural, historical, praxical, technological work of man directly affects or forms or informs the phenomena that we describe and explain (*even* the phenomena of stellar space, for instance), a bottom-up account could in principle be adequate.[40]

The single most important consequence of this shift in orientation is the perceived need to recast the entire theory of what is to count as *method* in science and rational inquiry. A few heterodox theses have already been suggested (for instance, that causal contexts need not behave extensionally, or that causal processes, even if regular, need not fall under universal covering laws). The point is that the breach in the stonewalling versions of analytic philosophy demonstrates that the rigor of such philosophy need not be committed to essentialist, foundationalist, cognitivist, extensionalist, universalist, reductionist, physicalist, anti-historicist, anti-praxicalist theories; hence, that there is every reason to consider how to reconcile the strongest features of contemporary Anglo-American and Continental philosophical currents in a fresh way. That, at any rate, is what reflecting on the technology of artificial intelligence and information promises.

Temple University

AUTHOR: Joseph Margolis (born 1924) received a Ph.D in philosophy from Columbia University and has taught at Long Island University, University of South Carolina, University of Cincinnati, and University of Western Ontario. Currently he is Professor of Philosophy at Temple University. He has written widely on the philosophy of art, ethics, and the philosophy of psychology.

NOTES

1   The issues of the philosophy of mind and the cognitive sciences are explored in some detail in Joseph Margolis, "Conceptual Links between Cognitive Psychology and Philosophy of Psychology," *Cognition and Brain Theory* 5 (1982); *Philosophy of Psychology* (Englewood Cliffs, NJ: Prentice-Hall, 1984); *Persons and Minds* (Dordrecht: D. Reidel, 1978).

2   See B. F. Skinner, *Verbal Behavior* (New York: Appleton-Century-Crofts, 1957); and D. M. Armstrong, *Belief, Truth and Knowledge* (Cambridge: Cambridge University Press, 1973).

3   Hilary Putnam, "Minds and Machines," *Philosophical Papers*, vol. 2 (Cambridge: Cambridge University Press, 1975,), pp. 362, 384. Cf. also, Hilary Putnam, "Mind and Body," *Reason, Truth and History* (Cambridge: Cambridge University Press, 1981).

4   See further, Ned Block, "Troubles with Functionalism," in C. Wade Savage (ed.), *Minnesota Studies in the Philosophy of Science*, vol. 9 (Minneapolis: University of Minnesota Press, 1978).

5   "The Nature of Mental States," *op. cit.*, p. 435.

6   See, for example, Herbert Feigl, *The "Mental" and the "Physical": The Essay and a Postscript* (Minneapolis: University of Minnesota Press, 1967); Mario Bunge, "Emergence and the Mind," *Neuroscience* 11 (1977); and Mario Bunge, "Levels and Reduction," *American Journal of Physiology* 103 (1977).

7   See Wilfrid Sellars, "Philosophy and the Scientific Image of Man," *Science, Perception, and Reality* (London: Routledge and Kegan Paul, 1963); Noam Chomsky, *Rules and Representation* (New York: Columbia University Press, 1980); F. I. Dretske, *Knowledge and the Flow of Information* (Cambridge MA: MIT Press, 1981).

8   See Daniel C. Dennett, *Content and Consciousness* (London: Routledge and Kegan Paul, 1969); and *Brainstorms* (Montgomery, VT: Bradford Books, 1978). For a detailed account of the problem, see Joseph Margolis, "The Trouble with Homunculus Theories," *Philosophy of Science* **47** (1980).

9   See Jerry A. Fodor, *The Language of Thought* (New York: Thomas Y. Crowell, 1975); *The Modularity of Mind* (Cambridge MA: MIT Press, 1983).

10   See Jerry A. Fodor, *Psychological Explanation* (New York: Random, 1968).

11   Daniel C. Dennett, "Toward a Cognitive Theory of Consciousness," in Savage, *op. cit.* On the distinction of the top-down and bottom-up, see Daniel C. Dennett, "Artifical Intelligence as Philosophy and as Psychology," in M. Ringle (ed.), *Philosophical Perspectives in Artificial Intelligence* (Atlantic Highlands, NJ: Humanities Press, 1978).

12   Dennett, "A Cure for the Common Cold," *Brainstorms*.

13   See C. Shannon and W. Weaver, *The Mathematical Theory of Communication* (Urbana: University of Illinois Press, 1949).

14   Fred I. Dretske, *op. cit.*, p. vii. There is a convenient summary of the book in Fred I. Dretske, "Précis of *Knowledge and the Flow of Information*," *Behavioral and Brain Sciences* **6** (1983).

15   Cf. *Knowledge and the Flow of Information*, Chapter 2.

16   *Ibid.*, p. 41.

17   *Ibid.*, p. 57.

18   *Ibid.*, p. 66.

19   (b) may be fairly construed as the import of Alfred Tarski's program of extensional semantics but *not* of Donald Davidson's, which professes to follow Tarski's. See Alfred Tarski, "The Semantic Conception of Truth," *Philosophy and Phenomenological Research* **4** (1944); and Donald Davidson, "In Defense of Convention T," in Hugues Leblanc (ed.), *Truth, Syntax and Modality* (Amsterdam: North Holland Publishing, 1973). Cf. also, Ian Hacking, *Why does Language Matter to Philosophy?* (Cambridge: Cambridge University Press, 1975), Chapter 12; and Gareth Evans and John McDowell (eds.), *Truth and Meaning; Essays in Semantics* (Oxford: Clarendon Press, 1976).

20   For a brief introduction to these problems, see Joseph Margolis, "Pragmatism without Foundations," *American Philosophical Quarterly* **21** (1984); "Relativism, History, and Objectivity in the Human Studies," *Journal for the Theory of Social Behavior* **14** (1984); and " 'The Savage Mind Totalizes'," *Man and World* **7** (1984).

21   *Op. cit.*, p. 63.

22   "Précis of *Knowledge and the Flow of Information*," p. 58.

23   *Knowledge and the Flow of Information*, p. 86.

24   *Ibid.*, p. 174 and remainder of Chapter 7.

25   *Ibid.*, particularly pp. 172–174.

26   *Ibid.*, Chapter 2.

27   *Ibid.*, pp. 65, 109–110. Cf. Peter Unger, *Ignorance* (Oxford: Clarendon Press, 1975).

28   *Ibid.*, p. 110 and 252n4.

29   W. V. Quine, *Word and Object* (Cambridge MA: MIT Press, 1960), p. 231.

30   *Loc. cit.* Italics added.

31   See Joseph Margolis, "The Stubborn Opacity of Belief Contexts," *Theoria* **43** (1977); Quine, *op. cit.*, pp. 141–156.

32   Quine, *op. cit.*, p. 231.

33   See Rudolf Carnap, "Psychology in Physical Language," trans. G. Schick, in A. J.

Ayer (ed.), *Logical Positivism* (Glencoe: Free Press, 1959); and Joseph Margolis, "Schlick and Carnap on the Problem of Psychology," in Eugene T. Gadol (ed.), *Rationality and Science* (Vienna: Springer-Verlag, 1982).

34 See Wilfrid Sellars, "Philosophy and the Scientific Image of Man," *Science Perception, and Reality* (London: Routledge and Kegan Paul, 1963); and Margolis, *Persons and Minds*, Chapter 1.

35 *Content and Consciousness*, p. 29.

36 *Knowledge and the Flow of Information*, p. 75.

37 See Donald Davidson, "Mental Events," in L. Foster and J. W. Swanson (eds.), *Experience and Theory* (Amherst: University of Massachusetts Press, 1970); and Joseph Margolis, "Prospects for an Extentionalist Psychology of Action," *Journal for the Theory of Social Behavior* **2** (1981).

38 This is to trade, of course, on Imre Lakatos's notion of a scientific research program; but it is not to accept his somewhat conservatively adjusted Popperian reading of that notion. There *is* no strongly verificationist *or* falsificationist procedure in place in philosophy in general and there *is* no generally accepted sense in which, within whatever generous limits, the boundaries of philosophical progress can be even roughly formalized. But to say that is not to deny that the professional work of philosophy proceeds in accord with recognizable practices of clear rigor; or to deny that, in doing so, it remains hospitable to both moderate and radical methodological innovation. See Imre Lakatos, *Philosophical Papers*, vol. 1, John Worral and Gregory Currie (eds.), (Cambridge: Cambridge University Press, 1978).

39 I owe a glimpse of the importance of this quarrel to discussion with Patrick Heelan and Donald Hockney, who may however not favor my use of the reference.

40 See further, Joseph Margolis, "Relativism, History, and Objectivity in the Human Studies."

PART III

ETHICAL AND POLITICAL ISSUES
ASSOCIATED WITH
INFORMATION TECHNOLOGY
AND COMPUTERS

ALBERT BORGMANN

PHILOSOPHICAL REFLECTIONS
ON THE MICROELECTRONIC REVOLUTION

ABSTRACT. Microelectronic technology maintains a modern split between revolutionary machinery and conventional commodity. This is mirrored in the division between labor devoted to the construction and maintenance of technological artifacts and leisure devoted to the consumption of commodities. On the labor side, microelectronics by way of automation will eliminate much degrading work and increase affluence. On the leisure side, the distraction and passivity of typical technological pastimes will be aggravated. In the longer historical perspective, the microelectronic revolution is not revolutionary at all, but only intensifies tendencies which have been at work for two centuries.

The microelectronic revolution is of epochal significance – so at least we are told in cover stories of *Newsweek* and *Time* and by representatives of science and industry.[1] According to the National Academy of Sciences, "the modern era of electronics has ushered in a second industrial revolution. . . . Its impact on society could be even greater than that of the original industrial revolution."[2] And *Newsweek* tells us that as "the industry likes to picture the future, the new technology offers potential solutions to humanity's most intracable problems – the allocation of energy resources, food enough for all and the worldwide improvement of health care, to name just a few."[3] There are of course many more indications that the developments and impacts of computers are stirring up enormous curiosity, excitement, and apprehension.[4]

What does philosophy have to say about this most advanced and perhaps most consequential issue of modern technology? It may come as a surprise or challenge when I say that in the encounter of philosophy and technology philosophy is more of a defendent than a judge. By this I mean that mainstream philosophy in this country has had very little directly to say about modern technology generally or about computers in particular.[5] One indication of this neglect is the absence of technology as a topic in the standard introductions and anthologies through which the profession initiates novices into the major concerns of philosophy. A defender of orthodox philosophy might reply that philosophers cannot talk about everything but must attend to the fundamental problems of reasoning and knowledge, of freedom and determinism, mind and body, of God and ethical standards.[6] All other problems, one might say, will

189

Carl Mitcham and Alois Huning (eds.), Philosophy and Technology II, 189–203.
© *1986 by D. Reidel Publishing Company.*

fall into place once the fundamental ones are solved. Philosophy of technology would then be a kind of applied philosophy.

But this defense of philosophy must leave us dissatisfied. For one thing, the fundamental principles that we are invited to apply to technology and microelectronics have not been worked out cogently, and likely they never will be. For another, it seems unlikely that a phenomenon, which is as deeply fascinating and disquieting as microelectronics, and especially one application of it, namely computer technology, can be illuminated merely by applying some fundamental philosophical principles.

One should not think, however, that the philosophical neglect of microelectronic technology is just an oversight or a matter of professional arrogance. It is an indication, I believe, of the peculiar way in which modern technology is both obvious and novel. Modern technology is obvious because it is so pervasive in its extent and so articulate and purposeful in its structure. It seems too evident to require philosophical investigation. And yet modern technology constitutes the sharpest break in the hundreds of thousands of years of human development. Within two centuries it has totally transformed the face of the earth. When seen in this broad perspective, modern technology appears to be so sudden and radical a transformation of the world that philosophical curiosity is daunted and deflected.

If such considerations puzzle and concern us, we may be on the way to an answer, for wonder is the beginning of philosophy. To begin, then, we may wonder how one might bring the character of microelectronic technology into proper relief. Let us take the contemporary computer as the first instance of microelectronics. We can then try to grasp its present significance in two ways. The first is to see the computer in various well-defined contexts. The second is to relate the computer to the modern center of gravity. Let us call these the partial and the central reflections on the microelectronic computer. Let us try to find our way to the central concern by way of the partial ones.

Partial reflection might first consider the immediate historical contexts of the computer and then the illuminating force that computer technology has in regard to various philosophical problems. Historically speaking, one might call the computer the extreme issue of a number of traditional human pursuits. The first of these is the concern with formal systems, with rigorous logic and mathematics, a concern that goes back at least to Aristotle and Euclid. I am not competent to judge whether the formal systems that are embodied in the machine and programming languages of computers rival or exceed those of traditional logic and mathematics in ingenuity, depth, or elegance. They certainly surpass them in size and in the degree to which we can control and manipulate them. That is, we can carry out formal operations by means of computers which formerly were beyond the attention and life span of humans. Computers allow us to do this because they are powerful calculators. Thus they are the culmination of a second traditional concern, namely the endeavor to construct machines that disburden us of the tasks of remembering and reckoning, and perhaps more than that, to construct machines that show in a rigorous and inspectable construct what it is to compute.

But the workings of a computer are too rich to be captured entirely under the titles of storage and calculation. Third, then, computers process information and manipulate symbols with an intricacy and autonomy that are reminiscent of the ability to solve problems and to make decisions.[7] Here too, however, there are ancient and simple precursors – namely, regulators, governors, and automata that have existed at least since Hellenistic times. Fourth and finally, considered as human artifacts, computers are the most intricate that have been produced. The endeavor to shape material reality to human purposes has achieved its finest and densest point in the integrated circuits of silicon chips. To use a suggestive if ambiguous metaphor, the human spirit has infused material reality most intimately in certain microchips. Or put more soberly, the number of functional features per unit of matter is greatest, among artifacts, in integrated circuits.

These four historical perspectives are misleading insofar as they may reduce to a matter of degrees what is a difference in kind. Computer technology represents such an extreme in progress and perfection that one may well consider it to have broken through traditional confines of understanding and achievement and to have dissolved ancient puzzles and beliefs. The computer, one might say, is not just a powerful tool for the solution of problems; it constitutes in itself, as an object, the clarification of such old mysteries as intelligence, freedom, and creativity. Speculations of this kind constitute the other half of the partial reflections on the character of computer technology. I am unable to do these problems justice; but I must say, to keep our reflections on the appropriate path, that the hopes just expressed about the significance of the computer seem unrealistic. It is true that computer technology has stimulated, enriched, and to some extent clarified our understanding of the mind-body relation, of intelligence, of the modes and levels which fix our comprehension of reality. But computer programs have not illuminated the nature of, say, intelligence, the way models of physiology and epidemiology have explained late medieval pestilence. In the latter case there is something like final and universal clarity. It is a matter of fact that this is not so in the case of intelligence – although to render this fact theoretically perspicuous is a deep and challenging problem.

It could also be that to expect of the computer that it provide or constitute a solution to traditional and difficult problems is to press it into an inappropriate service. Let the tradition ponder its own problems. I may not know what freedom is, but we do know what a computer is; and perhaps when seen in its own right, it does center and illuminate our time. Daniel Bell has made a suggestion of this sort. "Technological revolutions," he says, "even if intellectual in their foundations, become symbolized if not embodied in some tangible 'thing', and in the postindustrial society that 'thing' is the computer."[8] Bell does not pursue this idea very far or very deeply, but it is a suggestive remark. It reminds us of the cathedrals as the tangible things in which the high Middle Ages had come to be focused. Such a cathedral was for the Medievals the embodiment of the real world, of Solomon's temple and of the heavenly Jerusalem; and so it constituted a center and ordering force for the visible world. The cathedral was the tangible presentation of the world's hierarchy which assigned everyone and everything a rank and place. It presented the beginning, the middle, and the end of all things. And finally it invited everyone to share in this order through the power of its artistic presence and through the services and rituals in which it came fully to life.[9]

Does the computer have an analogously focal significance in our world? It does have a crucial organizing function according to Bell. "[T]he computer," he says, "has been the 'analytical engine' that has transformed the second half of the twentieth century."And, more pointedly, "[T]he computer is a tool for managing the mass society, since it is the mechanism that orders and processes the transactions whose huge number has been mounting almost exponentially because of the increase in social interactions."[10] but even under the most favorable interpretation, these remarks make the computer not the embodiment or symbol of order but merely its implement. Concentrating on the miracles of implementation one overlooks the question of the character and the validity of the order itself. Whatever that order, it is difficult to see how the computer could be its ruling and orienting presence as the cathedral was of the medieval order. As physical objects, the characteristic components of contemporary computers, namely integrated circuits, are paradigmatically inaccessible and insensible. They do not disclose their workings to our eyes, nor would they disclose their functions to the probing and intrusions of a layperson. Their functions, when considered in the abstract and down to the details of machine language and design, are as a rule forbiddingly complex to all but a handful of people in this country. And in their overt functions, where the integrated circuits connect or interface, as they say, with ordinary folk, they are so variable that they seem to represent everything and nothing. What do PacMan, a microwave oven, and a chess computer have in common as objects in an everyday setting? On the other hand, if we disengage the computer from its uses in the ordinary world and consider it as a computer properly so-called, it is merely the embodiment of a formal system or at most a neutral, ambiguous tool. The computer, then, is too inaccessible, forbidding, variable, or abstract an object to constitute the center of gravity of our world.

But it is evident from the stable and pleasant character of the advanced technological societies that our world in its paradigmatic and privileged regions does exhibit a definite order. And from the excitement microelectronics has generated in these very regions, we can gather that there is a crucial tie between the paradigmatic technological order and the developments in microelectronics. These developments in fact are commonly considered revolutionary.[11] Here we have one clue about a proper issue for philosophical reflection on microelectronics to address, the claim that microelectronics has ushered in a revolution. Another clue follows from what we have found so far. The central significance of microelectronics for our time cannot be captured when we attend to this or that aspect, objectified and exemplified in a computer; rather we must trace the ways in which microelectronics has entered the inconspicuous fabric of everyday life and how it affects that fabric.

Let us ask then how revolutionary the developments in microelectronics truly are, and let us approach the question by looking at a microelectronic object that is now widely and unobtrusively used, namely a digital watch. We can bring its peculiarity into relief by comparing it with a spring-driven one. The latter kind of clock was first built in the early 16th century. Thus it was thoroughly familiar a century later to men such as Bacon and Descartes. In the late 17th century, Newton provided the scientific insight which provided a precise and general explanation of its workings. There have been many refinements in the construction of mechanical watches since, yet they would all be readily intelligible to a Bacon, Descartes, or Newton. But what would they say when shown a digital watch? Even if we gave them hundreds of watches to dissect and examine, their inner workings would remain impenetrable. To understand a digital watch at its functional level the way they comprehended mechanical ones in their functions they would have to recapitulate 300 years of revolutionary science or do graduate studies in modern logic, mathematics, physics, chemistry, and engineering. When it comes to the structure and working, what I want to call the machinery, of a digital watch, it is surely no exaggeration to say that it is separated by a revolutionary gap from the machinery of the spring-driven watches that were current only ten years ago.

But in another sense the digital watch is not revolutionary at all. This becomes apparent when we ask how difficult it would be to teach someone like Newton not how to comprehend but how to use a digital watch.[12] This would take only a few minutes. In fact, it is easier than teaching someone how to read the dial and hands of a traditional watch. Thus what a digital watch procures, namely time indication, is familiar and accessible. Of course it procures it much more commodiously, i.e., in digits, with more precision, more variety, greater completeness, and with less bulk, without the need to wind it up, to turn it past the 31st of November, or to take account of leap years. It appears then that in a technological device such as a digital watch we must sharply distinguish between machinery and commodity. While the machinery typically undergoes revolutionary changes which remove it ever more from the comprehension of the common person, the commodity generally develops continuously and makes ever slighter demands on the user's competence or care. This tendency is quite general and can be seen to cover all of the technological items that surround us.

It requires a little practice to recognize how pervasive and progressive the pattern of the technological device has become. The food, the news, the music that we consume are all pleasantly and effortlessly at our disposal, and so are less tangible commodities such as health, safety, sex, and excitements of all sorts. And precisely when we hesitate at such claims and think of lonely, frustrating, or painful moments in our lives, we ought to recognize also that technological machineries are being designed and constructed that mean to give us perfect health, unfailing safety, and ever-pleasant sex. These endeavors have the support and unchallenged, if uneasy, support of society. But all these commodiously available goods are procured by more and more complex and discrete machineries which, just because of their complexity and discreteness, are ever more removed from our competence. As we are moving more deeply into a Cockaigne of consumption, we allow ourselves to be more and more disfranchised from competent and insightful citizenship in the technological society.

We do, of course, pay our dues as citizens of technology through our labor. And in our labor we are to some extent in touch with the machinery of technology. In fact, the division in the technological device between machinery and commodity is impressed on our lives as the division between labor and leisure. Normally we attend to the production and maintenance of the technological machinery in labor and devote our leisure to the consumption of commodities. Given the depth of this split, the developments in microelectronics may have very unequal consequences in these two realms. From our reflections so far it seems that the effects on our leisure or consumption will be slight. People's lives did not change significantly when they abandoned their spring-driven watches and bought digital ones or when they bought programmable microwave ovens.

But could it not be that these locally negligible changes will have a profound global effect, and that it is the latter which the prophets of revolution have in mind? I have already quoted a pronouncement to this effect, one that promises an easing of energy constraints, and the liberation from global famine and disease. On the labor side, as *Business Week* tells us, liberation is promised from "work that is hazardous, dirty, or monotonous."[13] Regarding leisure, the promise is one not only of liberation but of enrichment as well. According to James Albus of the National Bureau of Standards: "The robot revolution will free human beings from the pressures of urbanization and allow them to choose their own lifestyles from a much wider variety of possibilities." Similarly British Agriculture Minister Peter Walker: "Uniquely in history we have the circumstances in which we can create Athens without the slaves." And finally Isaac Asimov: "Robots will leave to human beings the tasks that are intrinsically human, such as sports, entertainment, scientific research."[14]

How are we to judge such promises? First we must realize that they are not revolutionary at all but have an ancestry of 350 years. They were formulated as the practical version of the Enlightenment by Bacon and Descartes, holding out a future of freedom and wealth on the basis of the new natural sciences.[15] They constitute the promise of technology which began to be enacted in the Industrial Revolution and has since accompanied the progress of technology as the official rhetoric and as a justifying and animating force.[16]

Seeing that there is now a record of two centuries on how we have fared

with the realization of technological liberty and affluence, one naturally
wonders what verdict the evidence supports. The answers differ depending on whether one considers the centers or fringes of technology, and
whether one looks at labor or leisure. In the technologically advanced
societies, the conquest of famine, disease, and illiteracy has been undeni-
ably successful, and just as obviously there has been no such conquest in
the developing countries. Regarding more specifically the quality of
labor and leisure in technological societies, the promise surely has not
been fulfilled. Labor, to be sure, has been rendered relatively safe, more
pleasant in its surroundings, and much more lucrative. But typically it has
been degraded all the same, stripped of initiative, responsibility, and
skill.[17] The typical quality of leisure activities appears to be low as well, if
one is prepared to measure it by any standard of excellence at all. Our
intuitions or apprehensions that most discretionary time is spent watching television and very little on activities such as participatory sports, the
theatre, museums, making music, correspondence, or reading books is
borne out by the evidence of social research. Accoding to one study, the
total time devoted to these latter activities is on the average only a fifth of
the time spent on watching television.[18] But as we have heard, the microelectronic revolution is to gain us final admission to a world where
food is universally adequate, disease conquered, literacy accomplished,
where people spend their time in fulfilling work and ennobling leisure.
How likely is this to happen?

The belief that rising affluence in the industrial countries will bring relief to the Third World rests on the assumption that our failure to help the
starving peoples overcome famine is due presently to insufficient wealth
on our part. But this is doubtful at best. In 1950 the standard of living in
this country was incomparably higher than that in the developing countries. By 1972 average real family income had more than doubled.[19] During the same time, foreign aid as a share of the federal budget has declined by a factor greater than five and now hovers between one and two
percent.[20] Effective foreign aid is difficult to achieve. But clearly our determination to achieve it has not grown as a function of rising affluence.
There is, of course, a good possibility that high technology will by its own
dynamics come to envelop the entire globe and so extinguish famine and
disease. But it would do so slowly and over the graves of millions who
have died of hunger and illness.

I have already indicated that the affluence we have retained for
ourselves has done little for the normal quality of labor and leisure. Will
the microelectronic revolution within the technological society admit us

to the promised land? To answer the question on the side of labor we must first ask why typical labor has been degraded until now. The main reason lies in the endeavor of technological societies to construct a powerful and reliable productive machinery. Machines and engines, of course, are eminently tireless and highly productive. Their construction and improvement has been the primary vehicle of productivity gains since the Industrial Revolution. But in general it was impossible until now to build machines that ran largely or entirely by themselves. Humans were needed for selecting, feeding, fastening, and steering. As Marx saw, the logic of machines has been primary, and human work was divided and paced according to the demands of the machines.[21] The human component of the productive machinery is inevitably imperfect, however; it cannot compete with a robot. "Not only," so we read in *Time*, "can the robot work three shifts a day, but it takes no coffee breaks, does not call in sick on Mondays, does not become bored, does not take vacations or qualify for pensions – and does not leave Coca-Cola cans rattling inside the products it has helped assemble."[22] Thus, through the microelectronic revolution the degradation of work comes to its conclusion in the elimination of work. The elimination of "hazardous, dirty, or monotonous" work is merely an aspect of the larger phenomenon which will lead to the loss of more and more skilled work.[23] How much work will be lost? Some estimates say that up to 75 percent of the factory work force will be displaced.[24] In short, the most likely and significant consequence of the microelectronic revolution for the world of labor is, in light of the history and logic of technology, not the improvement of work, but unemployment.

Still, since automation is undertaken in the name of productivity gains, it may well raise the standard of living; and though there may be much less work, there could be much more affluence as well with which to fill the growing leisure time. The realm of leisure and consumption turns out to be the court where the microelectronic revolution must finally rest its case, and here too the revolution is emphatically traditional because leisure and consumption have always been technology's highest court of appeal. The microelectronic case may be unprecedented, however, since its products are said to be radically novel. This clearly is the implication in the lead paragraph of the *Newsweek* article on microelectronic. It says:

A revolution is under way. Most Americans are already well aware of the gee-whiz gadgetry that is emerging, in rapidly accelerating bursts, from the world's high-technology laboratories. But most of us perceive only dimly how pervasive and profound the changes of the next twenty years will be. We are at the dawn of the era of the smart machine – an "information age" that will change forever the way an entire nation works, plays, travels and even thinks. Just as the industrial revolution dramatically expanded the strength of man's muscles and the reach of his hand, so the smart-machine revolution will magnify the power of his brain.[25]

Such pronouncements, however, are simply promises that yield little insight into the flavor and texture of the new microelectronic world. But here too *Newsweek* has intrepidly pressed ahead and given us a glimpse of the microelectronic everyday. It is the preamble to the article where a microelectronic citizen speaks to us as follows:

Welcome! Always glad to show someone from the early '80s around the place. The biggest change, of course, is the smart machines – they're all around us. No need to be alarmed, they're very friendly. Can't imagine how you lived without them. The telephone, dear old thing, is giving a steady busy signal to a bill collector I'm avoiding. Unless he starts calling from a new number my phone doesn't know, he'll never get through. TURN OFF! Excuse me for shouting – almost forgot the bedroom television was on. Let's see, anything else before we go? The oven already knows the menu for tonight and the kitchen robot will mix us a mean Martini. Guess we're ready. Oh no, you won't need a key. We'll just program the lock to recognize your voice and let you in whenever you want.[26]

The sketch is short of course and may seem shallow and glib. But in its essentials it is like the more studied scenarios in *The New York Times* or like the sweeping and breathless account one finds in Toffler's *The Third Wave*.[27] What does the picture tell us? Let us look at the individual features. (1) The smart machines will be "friendly," i.e., easy to use. (2) We will consider them indispensable. (3) They will allow us to do the following: (a) We will be able to be evasive or rude on the telephone by way of electronics, rather than through our children or personally. (b) We will be able to turn off appliances at a distance so saving ourselves the trouble of having to traverse entire rooms. (c) We will be disburdened from having to plan our menus and from having to mix drinks for guests with our own hands. (d) We will be spared the possibility of losing the house key or having it stolen.

It is clear that technological liberation from the duress of daily life is leading more and more to a disengagement from skilled and bodily commerce with reality. Our leisurely contact with the world is being narrowed to pure consumption, the unencumbered taking in of commodities which requires no preparation, provides no orientation, and leaves no significant trace. Perhaps the account above fails to do justice to the riches of information, entertainment, and games that the new electronics will present us with. But these too will be consumed, i.e., they will not make the demands of commitment, discipline, or skill.[28] They will be more diverting due to a greater variety and closer fit with our individual tastes. But since they will fail to center and illuminate our lives, their diversion will more and more lead to distraction, the scattering of our attention and the atrophy of our capacities. It is already apparent that the new video technology is not used by people as the crucial aid which finally allows them to develop into the historians, critics, musicians, sculptors, or athletes that they have always wanted to be.[29] Rather the main consequence of this technological development appears to be the spread of pornography.[30]

Let me now summarize and clarify what we have found about the relation of microelectronics and computers to the modern center of gravity. First, the computer as an instance of microelectronics is not itself that center of gravity or orientation. And unlike pretechnological cultures, the technological society is not structured or focused by any one central and eminent thing. Rather it owes its character to a pervasive and incisive pattern which becomes paradigmatically evident in a technological device. How then is microelectronics related to the ruling paradigm of the technological device? Devices that incorporate microelectronics, programmable or not, constitute the most advanced forms of such devices, both on the commodity and the machinery side. Such devices procure hitherto unavailable commodities, and they provide traditional ones in more refined, effortless, secure, and ubiquitous ways. These commodities rest on machineries which are more discrete and intricate and so less accessible and intelligible than preceding ones. But strictly within the device pattern, microelectronics is not revolutionary at all. It is merely the most advanced stage of a generally familiar and well-established development.

Does this latest move in the history of technology constitute a qualitatively new phase when considered in a broader cultural and social perspective? It is my hope that it will, but this new phase will only come into its own if we are able to pass through the common hopes that have been tied to the promise of technology. It is the promise of liberty and prosperity on the basis of scientific knowledge. We should not reject this program simply but recognize and restrain the pattern according to which the innocent ambiguity of its youth was confined and resolved. We must learn to see that genuine freedom and wealth cannot be achieved in a life which is shaped by the division into mindless labor and distracting consumption. There is an uneasily dawning recognition in this country, that this kind of life occludes the eloquence and depth of the world and atrophies our profounder faculties. The microelectronic revolution is in some ways a last and feverish attempt to deny this. Perhaps it is the crisis we must pass through to recover our health.

The reform of technology and the recovery of its promise are large and demanding tasks. But a critique of technology must appear hopeless and aimless without an indication of the positive steps that can be taken. These I can only sketch. Yet the brevity of my remarks is balanced by the fact that the reform of technology is not a utopian proposal that must be painstakingly unfolded to an uninitiated audience. Rather the reform of technology is already underway, inconspicuously and often uncertainly, but clearly just the same once the pattern of technology which needs reform is recognized.

Reform must begin on the leisure side of technology, and it can begin there more easily as well. In leisure everyone has a large measure of discretion. From the preceding critical remarks on consumption in the mature phase of modern technology, it is clear how the discretion is to be used. We should make a clearing in the clutter of devices and commodities to make room for things and practices that engage us as fully human and bodily beings and engage us in their own right. I mean focal concerns such as the preparation and celebration of meals, the exercise of sports, the experience of nature, or the making of music. It is obvious that microelectronic devices can be helpful to these concerns, but they are not crucial. Those things and practices that, as far as I can see, have orienting, engaging, and sustaining force are all of essentially pretechnological origin though they assume a new splendor and radiance when exercised in a technological context. In this way, one might say, they become metatechnological.

If our leisure practices recover a measure of soundness and orientation, we can hope to reshape the world of technological labor and machinery. This will become possible because sounder leisure will naturally curb our luxuriating consumption and stretch our goods, resources and capital. Thus we obtain the leeway for economic reform. The latter will be necessary since the soundness of leisure in the end requires the restoration and security of work. Here the microelectronic revolution contains a genuine promise. Through automation it can indeed rid us of "hazardous, dirty, or monotonous work," and we can render the indispensable technological background of our lives more efficient and reliable. But the automated economy must be constrained in two ways. One is the exclusion or decrease of frivolous commodities which, as argued, would follow naturally from the reform of leisure. The second is the establishment of a local and labor-intensive economy whose welfare would have to be an explicit and agreed-upon goal of public policy. The primary problem would of course be that of guarding or strengthening the second economy against the encroachments of the automated economy. In return the local and labor-intensive economy would not only provide rewarding work for all, at least in the long run, but also help in establishing again local cultures which reflect their natural setting and the peculiar heritage of their citizens.

Such reforms would also lead to the recovery and rehabilitation of technology. Microelectronic technology, no matter how sophisticated and awesome in itself, is demeaned when put in the service of distraction or obscenity. When, on the other hand, our lives are centered in practices of engagement, the dignity of technology is restored in three ways. First the liberating and enriching force of technology attains an unambiguously beneficial place. Second, the intensive, limited, and often strenuous nature of focal practices sustains our sensibility and admiration for the easy and wide-ranging power of technological devices. Third, the discipline and calm that are nurtured by focal practices may give us the strength and time to study and understand the scientific and engineering principles on which technological devices rest. Being scientifically and technologically literate, we would at last attain full citizenship in the technological society.

University of Montana

AUTHOR: Albert Borgmann was born in Germany in 1937. He studied in Germany and the United States and has taught at the University of Illinois, DePaul University, University of Hawaii, and the University of Montana, where he is currently Professor of Philosophy. He has written extensively on the problem of technology, and his book *Technology and the Character of Contemporary Life* has just been published by the University of Chicago Press.

NOTES

1 *Newsweek* (June 30, 1980) and *Time* (December 8, 1980).
2 Quoted in Colin Norman, *Microelectronics at Work: Productivity and Jobs in the World Economy* (Washington, DC: Worldwatch Institute, 1980), p. 5.
3 See *Newsweek* (June 30, 1980), p. 51.
4 A landmark among these indications was the decision by *Time* magazine to celebrate the computer in place of a person as "Machine of the Year." See the January 3, 1983, issue.

5   Notable exceptions are Hubert L. Dreyfus, *What Computers Can't Do*, 2nd edition (New York: Harper & Row, 1979); and Daniel C. Dennett, *Brainstorms* (Montgomery, VT: Bradford Books, 1978). An indication of the common philosophical neglect of technology is the absence of any reference to technology as an object of philosophical concern in Richard T. DeGeorge, *The Philosopher's Guide to Courses, Research Tools, Professional Life, and Related Fields* (Lawrence: Regents Press of Kansas, 1980).

6   These headings are taken from James W. Cornman and Keith Lehrer, *Philosophical Problems and Arguments: An Introduction*, 2nd edition (New York: Macmillan, 1974).

7   See Herbert A. Simon, "What Computers Mean for Man and Society," in Tom Forester (ed.), *The Microelectronics Revolution* (Cambridge, MA: MIT Press, 1981), p. 424.

8   See Daniel Bell, "The Social Framework of the Information Society," in Forester, p. 509.

9   Cf. Otto von Simpson, *The Gothic Cathedral* (Princeton: Princeton University Press, 1974).

10 See Bell, "The Social Framework," p. 509.

11 Cf. the title of the Forester anthology, cited in note 7 above, and the remarks quoted at the beginning of this essay.

12 Cf. Alvin Toffler's account of how quickly and easily he came to master a simple computer used as a word processor in *The Third Wave* (New York: William Morrow, 1980), p. 205.

13 See *Business Week* (June 9, 1980), p. 63; also *Time* (December 8, 1980), p. 78, and Simon, p. 428.

14 *Time* (December 8, 1980), p. 83; cf. *Newsweek* (June 30, 1980), pp. 51 and 56, and Toffler, pp. 155–67, 380–91, and passim.

15 See Francis Bacon, *The New Organon and Related Writings*, Fulton H. Anderson (ed.) (Indianapolis: Bobbs-Merrill, 1960), p. 16; and René Descartes, *Discourse on Method*, tr. Laurence J. Lafleur (Indianapolis: Bobbs-Merrill, 1956), p. 40.

16 For historical accounts see Hugo A. Meier, "Technology and Democracy, 1800–1860," *Mississippi Valley Historical Review* **43**, no. 4 (March 1957), 618–640; and Euguene Ferguson, "The American-ness of American Technology," *Technology and Culture* **20,** no. 1 (January 1979), 3–24. For a classic contemporary statement see Jerome B. Wiesner speaking in a U.S. Steel advertisement in *The Wall Street Journal* (April 13, 1976), p. 11.

17 See Harry Braverman, *Labor and Monopoly Capital: The Degradation of Work in the 20th Century* (New York: Monthly Review Press, 1974), and Louis E. Davis and Albert B. Cherns (eds.), *The Quality of Working Life*, 2 vols. (New York: Free Press, 1975).

18 See John P. Robinson, *How Americans Use Time* (New York: Praeger, 1977), pp. 102 and 107.

19 See Daniel Yankelovich, *New Rules: Searching for Self-fulfillment in a World Turned Upside Down* (New York: Bantam, 1982), p. 18.

20 *Information Please Almanac*, Ann Golenpaul (ed.) (New York: Simon & Schuster, 1975), pp. 80 and 87.

21 See Bernard Gendron and Nancy Holmstrom, "Marx, Machinery, and Alienation," *Research in Philosophy and Technology*, vol. 2 (1979), p. 120.

22 See *Time* (December 8, 1980), p. 73.

23 See *Business Week* (June 9, 1980), p. 62.
24 *Ibid.*, p. 63; see also Norman, pp. 29–40, and Edmund F. Byrne's "Microelectronics and Workers' Rights" in this volume.
25 *Newsweek* (June 30, 1980), p. 50.
26 *Ibid.*
27 See Part 2 of the *New York Times Magazine* (September 27, 1981), which was devoted to the issue of "Living with Electronics," and Toffler's book referred to in note 12.
28 Toffler in *The Third Wave*, pp. 265–288, argues that the consumer will more and more become a producer as well, thus constituting a "prosumer." But it is clear from the great majority of his examples that "prosuming" is the typically unencumbered and unskilled, if newly busy, consuming which is guided by and rests on an impenetrable productive machinery.
29 Cf. Joe Weizenbaum, "Once More, the Computer Revolution," in Forester, pp. 550–570.
30 See Tony Schwartz, "The TV Pornography Boom," *New York Times Magazine* (September 13, 1981), pp. 44, 120–122, 127, 129, 131–132, 136.

EDMUND F. BYRNE

MICROELECTRONICS AND WORKERS' RIGHTS

ABSTRACT. A description of how microelectronics and robotics are tending to increase unemployment, followed by comparisons between the social policies of Western European countries and the United States with regard to this problem. A conclusion points out the need for a social philosophy of technology that acknowledges workers' rights.

The developed world is moving rapidly into what is generally referred to as a microelectronics revolution, one major consequence of which will be the demise of many traditional jobs and job skills. But not in comparable numbers. The transition is being eased, in some countries more than others, by so-called new technology agreements that protect workers presently employed. But little is being done, outside of Scandinavia, to develop jobs for people who will be seeking them in the future. What jobs there will be, especially for the unskilled or inappropriately skilled, may not provide enough income to support those who are thus employed. In a word, neither the proverbial sweat on one's brow nor even the knowledge stored behind will be sufficient conditions for earning one's bread. What, then, is to be the value of work in the age of the microchip; and what value shall we assign to those whose skills are merely quaint in such a high-technology driven economy?

At issue is a fundamental question of human dignity and social responsibility which until recently has been sidestepped as moot by most theoreticians, perhaps including even Marx, because of the extreme unlikelihood of an economy not based on human toil.[1] The twentieth century has seen the development of various systems of unemployment compensation and welfare maintenance. But in spite of this social cushion, the new technology is worrisome nonetheless to the extent that it involves what some have called "the collapse of work."[2] The methodology of futuristics in this regard as in any other is unreliable. But careful prognostication on the basis of recent developments and discernible trends does suggest that we face quite radical changes in our patterns of work.

Microelectronic technology will probably render entire industries, e.g., the postal service, obsolete. It has already transformed various work *processes*, e.g., tool and die manufacturing, printing and publishing, retail sales, banking, insurance, and clerical work, to a point at which in these sectors comparatively few jobs need to be done by hu-

205

Carl Mitcham and Alois Huning (eds.), Philosophy and Technology II, 205–216.
© *1986 by D. Reidel Publishing Company.*

mans. Many traditionally valued skills are being rendered obsolete, and
fewer new skills are likely to be needed in the sectors affected.[3]

Consider just a few examples of what has already happened. Between
1969 and 1978 eight manufacturers of business equipment, as surveyed
by Olivetti, reduced their employment by 20%. In three years, from 1975
to 1978, Ericsson, a Swedish manufacturer of telecommunications equipment, reduced its production workforce from 15000 to 10000. Between
1972 and 1979 in West Germany 35000 employees in the printing industry lost their jobs, usually to a visual display unit (VDU). Similar changes
have been documented in insurance, banking, and now clerical work.[4]
Even employment in the once labor-intensive production of computers
declined by 50% from 1963 to 1965.[5]

In the production of machine tools, once requiring high-grade human
skills, humans now do little more than monitor and feed information.
Skills still needed are based more on analytic and logical ability than on
workplace acquired experience. Clerical skills, which often include a
range of administrative responsibilities, are now being dissipated by the
word processor, which, not coincidentally, impacts inordinately upon
women in the workforce.[6] Electronic components of an earlier generation were produced by a workforce made up mainly (70–80%) of semi-
skilled workers. New electronic components (large-scale integrated circuits) are produced by a workforce almost equally divided into thirds
among trained engineers and technicians, semi-skilled workers, and unskilled workers.[7] So in the area of microelectronics production we are
witnessing an exacerbation of the division of labor first espoused by
Adam Smith, encouraged by Charles Babbage, and implemented in the
mechanical age by Frederick Taylor.[8]

There is no obvious limit to the range and variety of sectors that may be
similarly affected. Yet it is more than mere coincidence that the sector
first affected is decidedly blue collar. Repetitive work that requires
minimal skills, e.g., in assembling, joining and handling, has been ripe
for the arrival of the robot.[9] A robot may be defined as a programmable,
self-correcting manipulator of versatile automation components.[10]
According to one estimate, there are already some 15000 robots instal-
led around the world, about half of them in Japan and a fourth in the
United States.[11] There are some 6–7000 units in the Soviet Union, but
most of these are technically limited on only 3–4 axes of movement.[12] By
1986 the Russians hope to have added 40000 additional units, and during
the five years thereafter they will be installing sensory robots (see
below).[13] One hundred and fifty companies in Japan (five times as many

as in the United States) produced robots at a level of $400 million in 1980, and expect to be producing at a level of $2.2 billion in 1985, $4.5 billion in 1990.[14] In the United States, robot production was at a level of $50 million/year in 1981, but may expand to $250 billion by the turn of the century.[15]

The robots of the future will have "sensory" capability, in varying degrees depending on the task, both in regard to "touch" and in regard to "vision," and both are now becoming technically and economically feasible. A Mitsubishi robot, for example, "knows" when it has reached the correct object on a work-bench. A Hitachi robot is so touch-sensitive that it can insert a piston into a cylinder with a clearance of 20 μ in three seconds.[16] Selective choice and evaluation of parts will be coming soon. Still in the future is a "thinking" robot that when shown what to do will determine the most efficient way to do it.[17]

The impact of robotization on the workforce is only gradually becoming apparent, but it clearly results in unemployment at least indirectly. General Electric, already a user and intending to become a producer of robots, will have a significant effect on both its own and others' payrolls. So far, GE has limited its workforce reduction to attrition.[18] But it plans to robotize as many as half of its 30 000 assembly-line jobs to achieve 6%/year improvement in productivity. GE's U.S. competitor, Westinghouse, has established a Robotics Division and given it a mandate to robotize "any and all manufacturing areas."[19] The PUMA (programmable universal machine for assembly), a $20 000 robot arm developed by General Motors and Unimation, is expected to displace half of GM's assembly line workers by 1990.[20] Robogate, an assembly line robot developed and installed by Fiat, has not yet displaced many workers. But once sensory robots are installed the Italian manufacturer could, it is estimated, cut manpower 90% before 1990.[21]

In Japan, MITI, the quasi-governmental research arm of Japanese industry, is investing $140 million over a seven-year period to achieve completely robotized assembly of a product, such as an automobile, the design of which could be changed simply by changing the system's software. Hitachi hopes to have smart robots doing 60% of its assembly work by 1985.[22] This company has already opened a $60 million prototype of a flexible manufacturing complex (FMC) that involves five fully automatic manufacturing operations, all interconnected and controlled by a hierarchy of computers, with humans on hand only as safety overseers of lasers used for treating and machining. It is expected that 20% of Japan's total factory output will be FMC'd by 1985.[23] Meanwhile, Fujitsu Fanuc Ltd.

has opened a \$38 million plant to produce other robots and computerized tools automatically, using robots, numerically controlled machine tools, and only one shift of 100 human workers to assemble robot-made parts.

If these developments in particular industries are expanded into larger-scale projections, the likely impact on the blue collar workforce takes on alarming proportions. According to one projection, in the last decade of this century robots will be producing half of all manufactured goods; and, as a result, up to one-quarter of the factory workforce may be dislodged.[24] Another estimate has it that increased use of robots and other electronic devices in U.S. industry will lead to a 30% decrease in use of workers, commonly by introducing an unmanned third shift: what the Germans call "*die Geisterschicht.*"[25] According to yet another techno-seer, "smart robots could displace 65% or more of today's factory work force."[26]

In the face of this rather sudden transformation of the means of production, even *sushin koyo*, the vaunted job security system of Japanese factory workers, has become vulnerable. The manufacturing workforce in Japan dropped from 14.4 million in 1973 to 13.7 million in 1980. The conclusion of a government study that the impact of microelectronics on employment would not be serious was heavily criticized, and the heretofore acquiescent unions have begun to worry.[27] One result: the Federation of Japan Automobile Workers' Unions has entered into an agreement with Nissan that protects the jobs of those presently employed against lay-off or downgrading due to the introduction of robots and microelectronics.[28] But the government is doing little to create new jobs.[29]

Nor is the problem of displacement limited to developed countries. Computer-controlled assembly in the United States and Japan is now competitive with labor-intensive and increasingly expensive production elsewhere ("outsourceing"), and thus it is now less advantageous for electronics manufacturers to depend upon developing Asian countries for low-level assembly operations.[30]

In a word, knowledgeable observers of the robot industry debate not the threat of workers displacement but only its magnitude. There is, however, yet another dimension to the displacement which amounts to a reversal of the thesis that the whole is greater than the sum of its parts. At issue here is the diminution if not outright demise of unions organized historically to protect workers *in their jobs*. The United Auto Workers expects to lose 200 000 of its 1 million members between 1978 and 1990. The International Association of Machinists (IAM) and the International Brotherhood of Electrical Workers (IBEW) will also be hard hit. But microelectronics will eliminate white collar jobs as well. Employment in the U.S. postal service has declined from 70 000 in 1970 to 667 000 in 1981.[31] And as the great giant AT&T moved toward its 1983 break-up into a long-distance service and five regional companies, the Communications Workers of America (CWA) focused its concern in bargaining a contract on the issue of job security.

What was of concern to employees of AT&T has been of no less concern to government leaders in Western Europe. The so-called Nora Report, submitted to the President of France Giscard d'Estaing in 1978, is especially pessimistic. According to Nora, the revolutionary telecommunications system about to be introduced by IBM and its American partners (Comsat and Aetna) is likely to undermine any claim to sovereignty on the part of a small nation state such as France.[32] Nora's warning that small nation states must accordingly develop a collective policy in their own defense has in fact been translated into an agreement between IBM and the EEC countries to standardize many of its computer components. It is unlikely, however, that contracts and agreements such as these can prevent the technological unemployment that will impact, proportionally, no less heavily on middle- and even upper-level management rendered superfluous by the departure of those whom they have been managing.

If technological unemployment does occur in the years ahead more or less as here portrayed, it will so happen not because there are no alternatives but because robots and other microelectronic devices are already perceived as cost effective in the long run and hence a necessary condition for staying competitive in the industries affected.[33] At least one writer would add, however, that it is only by eliminating humans that microelectronic automation can be cost effective, because the greatest expense in incurred in trying to accommodate man in the loop. Says Lawrence B. Evans, an MIT chemical engineer:

The cost of complex electronic circuitry continues to decrease exponentially (by a factor of about 1/2 each year) due to large-scale integration (LSI) semiconductor technology. . . . The real cost of a system is in the hardware for communication between man and that system (displays, keys, typewriters) and this cost is a function of the way the system is packaged. Thus, automation functions and data processing become economic if they can be done blindly, without the need for human communication.[34]

Estimates vary as to just how much less expensive it may be to use robots rather than humans; but that there will be significant savings is widely assumed. As one writer puts it, a Japanese robot in automotive production can do at $5.50/hr what a UAW worker does for $18.10/hr (wages and fringes).[35] An estimate of this sort is typically based on a comparison between costs incurred from labor and costs of procuring and maintaining a robot. Robot providers claim that robot costs will be recouped within a three-year payback period from savings in labor alone. Of course, assumptions with regard to the cost of money, the cost of installation, and the cost of power and maintenance of a robot need to be adjusted up rather significantly in the present economy. But the initial cost of producing the robot may well drop from $50 000 in 1980 to just $10'000i in 1990. So recent estimates are probably at least in the correct order of magnitude. However, cost considerations have on occasion given way to a desire for product quality improvement, e.g., in production of Chrysler's K-car at the Newark, Delaware, plant and of GM's Fleetwood in Detroit, where $8.5 million of robots save only $120 000/ yr. In such instances, of course, a more affluent market is targeted, and cost is expected to be recouped through sales.[36]

What is more noteworthy about these cost-savings calculations from a societal point of view is that only internal costs are being taken into account, not the external costs, direct and indirect, that spill over onto society in the wake of technological upheaval.

In some important respects, Western European countries are far ahead of the United States with programs in place to smooth the transitions made necessary by industrial transformation. To be sure, there have been some enlightened steps forward in various places in America. But as a general rule neither the private nor the public sector seems really prepared to deal with the technological unemployment that is already becoming endemic in Western societies.

That workers should lose their jobs is, of course, nothing new. Nor is it unprecedented that the installation of a new technology is the immediate cause of losing one's job. It is not even new that technology should impact so heavily on one sector of the economy, since much the same sort of transformation resulted from the mechanization of agriculture beginning a century ago. What is new is the wide range of specialized skills that the new technology is rendering economically without value. But these skills have served society well for many generations. They have been acquired as a result, directly or indirectly, of socially supported priorities and programs. It is, accordingly, only the unconscionable or perhaps amnesiac society that would simply abandon the victims of its own devices.[37]

European countries, at least in principle, have more lucrative unemployment benefits than is the case in North America. The percentage of the labor force covered by such benefits is significantly higher in the United States than in Europe, where smaller firms and agriculture tend to be exempted from coverage. Nonetheless, there is substantial evidence that European societies assume greater responsibility for workers displaced by technology, especially but not only in Sweden and other Scandinavian countries.[38] Job-securing new technology agreements, now fairly common in Europe, are not unknown to American unions, notably the UAW and now also the CWA; but Americans in general still have much to learn in this regard.

What a society chooses to do collectively, however, depends greatly on the theories it espouses as to what properly ought to be done and in what manner. But theories about welfare given public support in the United States do not extend to individuals the courtesies recently received by a now resurgent Chrysler Corporation. Economists, resigned to the demise of the balanced budget, debate whether there is such a thing as a Phillips curve to account for unemployment and on occasion wonder if perhaps technological revolution is a factor after all, as Kondratiev and Schumpeter after him maintained early in this century.[39] Nor are economists alone in the at least tacit realization that the challenge we now face surpassed the lore of their trade. As we contemplate the microelectronic transformation of patterns of work, we find comparatively little in our traditional values and institutions that will help us deal humanely with the pervasive unemployment that it will engender. And this is especially the case because of a belief attributed to the Judaeo-Christian tradition that there is a conditional relationship between the means of subsistence and work.

As noted in a wry Haitian proverb, "If work were a good thing, the rich would have found a way of keeping it to themselves." To the contrary, of course, the rich are considered exempt, almost by definition, from the requirement that one work in order to live. Others are often excused if they are in some way disabled, by virtue of age (either too young or too old) or physical condition. In the absence, however, of some socially acceptable excuse, one is led to believe that one's value as a person is a function of one's utility as a worker. This utility, as Marx so well perceived, is based primarily upon market conditions, but as mediated by complex intervening structures due to social and political considerations. In a socialist economy employment tends to be a given, even if absenteeism reduces the amount of work actually performed to a minimum. In a capitalist system, the basic determinant of employment, even in an organized plant, is the needs of the employer, as defined, for the most part, by the employer.

In Anglo-American law, the employer's prerogative in this regard is expressed in the time-honored doctrine of "employment at will," that is to say, that an employee has rights *qua* employee only so long as the employer sees fit to continue that relationship. The harsh reality of dismissal may on occasion be obviated in a unionized work environment through a grievance procedure the ostensible purpose of which is to assess the propriety of the grounds for dismissal. But the grievance mechanism is ordinarily activated only to determine whether one individual rather than another ought to be employed at a particular job. If the employer determines that there is not enough work to go around or enough money to pay for the work, layoffs are taken to be inevitable, with only the terms and conditions thereof subject to negotiation. And if the employer, say, a multinational corporation, determines that all the work done in a particular plant can be better (read: more cheaply) done elsewhere (*via* "outsourceing") or otherwise (*via* automation), then the entire local workforce will be terminated.

This management right of "employment at will" is subject to modifications in several ways, notably either by agreement between the parties or by governmental intervention. In some countries, such as Italy and Sweden, mass terminations simply are not tolerated. In others, labor-management agreements at various levels assure employees access to information relative to contemplated technology changes ("data agreements") and/or opportunity in one way or another to limit the effects on personnel of such changes ("new technology agreements"). These NTAs may involve any aspect of a new technology, from safety in its operation to planning how it should be designed and/or under what circumstances it should be implemented.

Unions in many European countries have negotiated agreements with employers at plant, industry or national level to limit the impact of new technology on the present workforce, albeit not on any successors thereof. A common concern commonly the subject of agreement is the health risk associated with use of visual display units (VDUs). In Norway and Sweden in particular, both statutes and negotiated agreements assure workers a significant voice in determining how computer-based technologies are to be introduced. Especially important to these programs are (1) the establishment of workers representation in the decision-making process and (2) a program to develop computer literacy among the rank and file with the cooperation of academics. Similar but less progressive developments have taken place in West Germany and in the United Kingdom.

Building a checklist of issues published by the Trades Union Congress (TCU) in 1979, British unions have negotiated over one hundred NTAs, mainly at company level or below and mainly for the benefit of white collar (clerical and managerial) workers. About half of the West German workforce is now covered by collective agreements that give special protection in the event of rationalization, e.g., in chemicals, leather and footwear, paper, textiles, metalworking, and especially (a recent focus of controversy) printing. As a matter of fact, it was primarily due to the initiative of the metalworkers' union in West Germany (IG Metall) that that nation's watch-making industry was finally prodded out of its complacency to switch from mechanical to quartz technology.[40]

By contrast, only a comparatively small segment of the workforce in North America has been able to protect itself against obsolescence in the face of microelectronics. Two exceptions, the blue-collar UAW and the more white-collar CWA, have been noted above. And Canadian workers, it should be mentioned, were among the first to recognize and attempt to deal with microelectronically created unemployment. That U.S. labor unions are comparatively lethargic in the face of microelectronics is ironic, since it is they who were most upset about the (premature) threat of automation in the 1960's. What unions are seeking to achieve collectively in Europe, individuals have been achieving in the United States through litigation. In courts in various jurisdictions managerial personnel who have lost their jobs are convincing (similarly situated?) juries that they have endured "unfair dismissal" or "abusive discharge." These verdicts have resulted in damage awards typically in the $200–300 000 range; but they have gone as high as $4 million. The complaints assert breach of express or implied contract (in the employee handbook), breach of implied covenant of good faith and fair dealing, and/or public policy (e.g., in the case of a "whistle blower").[41]

If such recognition of employee rights in the United States should ever be extended beyond the management level to the workforce as a whole, then public policy with regard to unemployment will have achieved something like equilibrium across both oceans. And, as a matter of fact, the agreement entered into between AT&T and the Communications Workers of America contains most, if not all, of the kinds of rights with regard to new technology which Europeans have been demanding for over a decade.

As noted above, however, these NTAs protect at best only the present

workforce of a company or industry. They do nothing directly for job-
seekers in the future who, in many instances, will find employment only,
if at all, in kinds of jobs that have not yet even been created. Whether
there will ever be enough new jobs to go around or whether working will
remain a condition for subsistence are questions still impossible to
answer. But in spite of the social inertia attributable to and instilled by
the traditional work ethic, the time has come to develop ways to distribute wealth more equitably, e.g., by means of a negative income tax.
And this in turn suggests a need for the concerns of social philosophy to
be brought to bear upon our philosophy of technology. For, what the
Germans call "humanization of the workplace" cannot advance without
some consensus about ethical priorities and how these ought to apply to a
society being transformed by microelectronics. These priorities, in turn,
might well be generated from the old but never more appropriate maxim:
TO EACH ACCORDING TO NEED; FROM EACH ACCORDING
TO ABILITY.

AUTHOR: Edmund F. Byrne (born 1933) has advanced degrees in philosophy and in law.
He is Professor of Philosophy at Indiana University – Purdue University in Indianapolis,
and a member of the Indiana Bar. He has published a number of articles on the relation
between law and technology, the humanization of technology, and is finishing a book on
technology and work.

NOTES

1   See Clive Jenkins and Barrie Sherman, *The Leisure Shock* (London: Eyre Methuen,
    1981). See also Josef Wolkowski, "The Philosophy of Work as an Area of Christian-
    Marxist Dialogue," *Dialectics and Humanism*, vol. 5 (Winter 1978), 113–122.
2   Clive Jenkins and Barrie Sherman, *The Collapse of Work* (London:Eyre Methuen,
    1979). See, however, J. L. Missika *et al.*, *Informatisation et Emploi: Menacee ou Muta-
    tion?* (Paris: La Documentation Française, 1981), pp. 73–74.
3   Guenter Friedrichs and Adam Schaff (eds.), *Microelectronics and Society: A Report to
    the Club of Rome* (New York: NAL Mentor, 1983), pp. 115–202; Christopher Evans,
    The Micro Millenium (New York: Washington Square, 1979), pp. 121–145; Tom Sto-
    nier, *The Wealth of Information* (London: Thames Methuen, 1983), pp. 99–122.
4   European Trade Union Institute (ETUI), *Negotiating Technological Change* (Brus-
    sels: 1982), pp. 8–11, 16. See also Ian Benson and John Lloyd, *New Technology and
    Industrial Change* (London: Kegan Paul, 1983), pp. 39–43.
5   Paul Stoneman, *Technological Diffusion and the Computer Revolution* (London:
    Cambridge University Press, 1976), p. 177.
6   ETUI, *op. cit.*, pp. 12–18. Regarding impact on women, see Ursula Huws, *Your Job in
    the Eighties: A Woman's Guide to New Technology* (London: Pluto Press, 1982).

7 ETUI, *op. cit.*, p. 20. The terms 'skilled' and 'non-skilled' translate into American usage the author's terms, 'qualified' and 'non-qualified.'

8 *Ibid.*, p. 33. A comprehensive analysis of this process, from a Marxist perspective, is that by Harry Braverman, *Labor and Monopoly Capital: The Degradation of Work in the Twentieth Century* (New York and London: Monthly Review Press, 1974). See also Benson and Lloyd, *op. cit.*, pp. 31–47.

9 ETUI, *op. cit.*, pp. 8–10.

10 Jasia Reichardt, *Robots: Fact, Fiction and Prediction* (New York: Penguin, 1978), p. 141; Wayne Chen, *The Year of the Robot* (Beaverton, Or: Dilithium Press, 1981), pp. 9–24.

11 D. Smith "The Robots (Beep, Click) Are Coming," *Pan Am Clipper* (April 1981), p. 33; E. Janicki, "Is There a Robot in Your Future?", *The Indianapolis Star Magazine* (Nov. 22, 1981), p. 55.

12 "Russian Robots Run to Catch Up," *Business Week* (Aug. 17, 1981), p. 120.

13 *Ibid.*

14 "The Push for Dominance in Robotics Gains Momentum," *Business Week* (Dec. 14, 1981), pp. 14, 108. See also *Business Week* (April 5, 1982), p. 40; (June 27, 1983), p. 40.

15 Smith, *op. cit.*

16 Reichardt, *op. cit.*, p. 140.

17 "Racing to Breed the Next Generation," *Business Week* (June 9, 1980), pp. 73 and 76.

18 "How Robots are Cutting Costs for GE," *Business Week* (June 9, 1980), p. 68. See also "General Electric: The Financial Wizards Switch Back to Technology," *Business Week* (March 16, 1981), pp. 112–3.

19 "Robots Join the Labor Force," *Business Week* (June 9, 1980), pp. 62 and 64; "GE is About to Take a Big Step in Robotics," *Business Week* (March 8, 1982), pp. 31–32.

20 "GM's Ambitious Plans to Employ Robots," *Business Week* (March 16, 1981), p. 31.

21 "Racing to Breed the Next Generation," *op. cit.*, p. 76.

22 "The Push for Dominance in Robotics Gains Momentum," *Business Week* (Dec. 14, 1981), p. 108.

23 "The Speedup in Automation," *Business Week* (August 3, 1981), p. 61. See also Paul Kinnucan, "Flexible Systems Invade the Factory," *High Technology* (July, 1983), pp. 31–37, 40–42.

24 "High Technology: Wave of the Future or a Market Flash in the Pan?," *Business Week* (Nov. 10, 1980), p. 96: chart, "The Coming Impact of Microelectronics."

25 "The Speedup in Automation," *op. cit.*, pp. 58–59. See, however, "Robots Bump into a Glutted Market," *Business Week* (April 4, 1983), p. 45.

26 "Robots Join the Labor Force," *op. cit.*, pp. 62–63, 65.

27 See Kuni Sadamoto (ed.), *Robots in the Japanese Economy* (Tokyo: Survey Japan, 1981); "Japan: The Robot Invasion Begins to Worry Labor," *Business Week* (March 29, 1982), p. 46.

28 *International Metalworkers Federation News* (May, 1983).

29 "A Changing Work Force Poses Challenges," *Business Week*, Special Issue: Japan's Strategy for the 80's (Dec. 14, 1981), pp. 116–118.

30 "Automation is Hitting a Low-Wage Bastion," *Business Week* (March 15, 1982), pp. 38–39. See also *ibid.* (Dec. 14, 1981), p. 40; Juan F. Rada, "A Third World Perspective," in *Microelectronics and Society, op. cit.*, pp. 203–231; Ira C. Magaziner and Robert B. Reich, *Minding America's Business* (New York: Vintage, 1983), pp. 99–101.

31 "The Speedup in Automation," *Business Week* (Aug. 3, 1981), pp. 62–67; "Technology Challenges Postal Union," *In These Times* (Aug. 17–23, 1977), p. 8.

32 Simon Nora and Alain Minc, *L'Informatisation de la société*. Rapport a M. le President de la Republique (Paris: La Documentation Française, 1978). English edition, *The Computerization of Society* (Cambridge, MA: MIT Press, 1980).

33 See Tom Forester (ed.), *The Microelectronics Revolution* (Oxford: Basil Blackwell, 1980), pp. 159–160, 192–195, 383.

34 "Industrial Uses of the Microprocessor," in Forester, *op. cit.*, p. 144. Originally published in *Science* (18 March 1977).

35 Janicki, *op. cit.*, pp. 54–55.

36 Information based on an unpublished study by graduate students at Carnegie-Mellon University entitled "The Impacts of Robotics on the Workforce and Workplace," Pittsburg, PA., June 14, 1981.

37 A particularly pointed example involves the manufacture of machine tools. See ETUI, *op. cit.*, p. 13.

38 See Roger Kaufman, "Why the U.S. Unemployment Rate is So High," in M. J. Piore (ed.), *Unemployment and Inflation* (White Plains, NY: M. E. Sharpe, 1979), p. 160; Magaziner and Reich, *op. cit.*, pp. 12–18, 143–154; 211–214, 271–276, 333–334.

39 See Benson and Lloyd, *op. cit.*, pp. 49–55; David Wheeler, "Is There a Phillips Curve?" in *Unemployment and Inflation*, *op. cit.*, pp. 46–57.

40 ETUI, *op. cit.*, pp. 50–51. For historical background to this development, see Benson and Lloyd, *op. cit.*

41 "It's Getting Harder to Make a Firing Stick," *Business Week* (June 27, 1983), pp. 104–105; "Fire Me? I'll Sue!" *American Bar Assoc. J.* **69** (June 1983), 719. Compare Paul O'Higgins, *Worker's Rights* (London: Arrow Books, 1976), pp. 62–72; Jeremy McMullen, *Rights at Work: A Worker's Guide to Employment Law*, 2nd edition, with supplement (London: Pluto Press, 1979), pp. 144–184; *Cessation de la Relation de Travail a l'Initiative de l'Employeur* (Geneva: Bureau International du Travail, 1980).

DANIEL CÉRÉZUELLE

INFORMATION TECHNOLOGY
AND THE TECHNOLOGICAL SYSTEM

ABSTRACT. Information technology came into being at a specific stage of technological development and made possible the creation of the contemporary technological system. But the goal of complete knowledge and control, at which information technology aims, is in fact illusory. We can thus anticipate in the future a corresponding increase in crises in the relations between technological systems, the technological system and nature, and the technological system and society.

The aim of this short paper is to provide some insight into the significance of information processing technology in relation to the ensemble of already existing technologies. It is possible and necessary to study information processing, cybernetics, and computers from various points of view – metaphysical, epistemological, political, and so forth. Such studies are quite useful, but they risk being beside the point if, from the beginning, they are not careful to locate information technology in its proper context, that is to say, within the dynamic reality of the modern technological society. One should not limit oneself to examining an isolated technique, because it is the technological context within which it exists which determines its application and significance. One must therefore delineate this relationship, and not until then, not until this preliminary work has been accomplished, can the particular details of information processing or the computer adequately be investigated.

This paper does not exhaust the question of the place of information technology in the contemporary technological system. Authors such as Jacques Ellul and Ingmar Granstedt have done this in greater detail and depth, but consistent with their analysis I will develop two main ideas. The first is that information technologies came into being at a clearly defined stage of technological development and gave impetus to the dynamic of the contemporary technological system by assisting, at least in theory, the bringing about of an extraordinary increase in knowledge, integration, and control. Such developments favored the growth of technology in power and complexity. The second idea is that since, practically speaking, absolute knowledge and control remain an illusory possibility, we can anticipate a corresponding increase in crises and technological contradictions.

Carl Mitcham and Alois Huning (eds.), Philosophy and Technology II, 217–225.
© 1986 *by D. Reidel Publishing Company.*

I. TOWARD A TECHNOLOGICAL SYSTEM

According to Ellul, the growth of information processing technology has brought technology to a new level of perfection. On the whole, since the 1950s the technological society has tended more and more to become a technological system. To appreciate this point, it is necessary to review briefly the character of modern technical change.

As Ellul first argued in *The Technological Society*,[1] the latter 19th century and first half of the 20th witnessed the development of a technical rationality which gradually invaded the areas of economic, cultural, and social life. The customs and symbols of humanity became to an ever greater extent organized in a technological form which was increasingly constraining and which everywhere imposed its requirements. Time and work schedules were measured by clocks instead of the length of the day; vacations were determined by industrial shut-downs instead of the seasons; travel came to be regulated by traffic lights instead of the conventions of a social hierarchy. Technology thus became the principal characteristic, the common denominator of "modern" societies. Modern society tends to be nothing in the final analysis but the foundation for an independent technical growth, and to be judged primarily in terms of its ability to support or contribute to such growth. This is why from the end of World War II one could speak with accuracy of a "technological society."

But as Ellul has also pointed out in *The Technological System*,[2] which constitutes a second-look a quarter century later at the main themes of the first two chapters of *The Technological Society*,

Twenty-five years ago, there was no way to speak of the technological system, because all that could be ascertained was a growth of technology in all areas of human activity. It was an anarchic growth, however; these areas were still kept specific by the traditional human divisions of labor, and there was no relationship between them (p. 101).

At first, as technization proliferated, different technical specializations tended to become more independent, autonomous, and unintelligible to outsiders, so that it became progressively difficult for each to pursue its own objectives without running up against the requirements or objectives of others. Urban water and sewage treatment operations, electric power and telephone networks, various means of transportation (car, truck, bus, train, airplane, boat), all tended to become more and more the fields of specialists and to engage in conflict-prone encounters. Technical operations expanded in quantity, complexity, and speed, and exceeded the possibility of general human oversight, so that increasingly bottlenecks and dislocations occurred. One simple illustration is the conflicts which ensued in many urban settings between the partisans of different modes of transportation as they competed for access to public facilities and support.

The only possible way out of such problems was to come up with institutional arrangements which would set up procedures and relations to mediate the diversity of technical specializations and the different areas in which technicians worked. To this end technological growth brought about an enormous increase in bureaucracy. Yet because of inherent weaknesses and imperfections in the human members of a bureaucracy, the functioning of these procedures and relations were no less imperfect. The result was that more and more it became necessary for each technological process to be planned not only with an eye on the specific task at hand, but also from the point of view of disseminating information about it, and with an understanding of the condition of technology in the larger environmental context. It is from this point that modern technological organization began to grow into a system. As Ellul says, "The various technologies have unified into a system by dint of the information transmitted from one to another and utilized technologically in each sector" (p. 91).

Interestingly enough, John Diebold, an influential management and technology consultant, recently made a similar point. In the new introduction to a reprint of his 1952 book on *Automation*, Diebold argues against using computers merely to increase the mechanization or efficiency of certain specialized functions. Early mass production focused

on discrete actions and . . . ignored the complex implications of moving from one specialized activity to the next. We refined the actions, but really did not address their connections into "transactions."

With computers, however, it becomes possible to "take account of a transaction when it [takes] place, and of its impact throughout the entire enterprise, whether that enterprise [be] government, an educational institution, or business."[3] It is then just a simple extension to move from this creation of what Ellul calls technological "ensembles" to the technological system.

Thus, what characterizes the recent worldwide development of technique has been its integration into a system which can function only if all the parts and subsystems which compose it are related and maintained within the interconnected network necessary for its internal coherence. Such a unity can evolve over a span of time, but at a given moment it is limiting and constraining with regard to the elements which compose it. Building codes, zoning plans, traffic laws, and various government regulations all determine how surface, air, and sea transportation systems interact in any urban setting. This is true for each technical ensemble, but one observes the same thing at a much higher level with the consolidation of diverse industrial ensembles or institutions into a system. The functional coherence of this system demands a relatively stable degree of coordination which reaches beyond personal interests and responds to the diversity of functions undertaken by the subsystems personnel. The coherence of the system, the maintenance of the relations which define it, takes precedent over the development of any particular technical ensemble.

It is at this point that information processing and computers arose as the sole means of meeting the demands for information which have grown at such a dizzying speed as a result of the development of the technological society. Computers emerged in the fifties to guarantee the coordination of technical activities, as more and more ensembles and subsystems were consolidated by the necessity of bringing about savings in the industrial economy. As Ellul has stressed,

Thanks to the computer, there emerged a sort of internal systematics of the technological ensemble, expressing itself by, and operating on, the level of information. It is through reciprocal total and integrated information that the subsystems are coordinated (p. 102).

Only the computers could do this, and thus the basic function of the information processing ensemble

is to allow a flexible, informal, purely technological, immediate and universal intersection of the technological subsystems. Hence, we have a new ensemble of new functions, from which man is excluded – not by competition, but because no one has so far performed those functions (pp. 102–103).

From this point on, computers make *possible* the integration of various technical functions according to an internal logic, which in turn leads to the creation of a new technical universe, a "technical system" operating according to its own rules and regulations and introducing a new dynamism to technical progress. Such a technical system appears to create itself, now that information processing and computers allow a generalized form of interaction that draws on diverse technical ensembles and subsystems.

Yet one can no longer speak of a complete and totally harmonious system. Indeed, we shall have to inquire as to whether the computer is truly capable of integrating the various technical subsystems, because the computer also encourages such an expansion in power, complexity, and speed of innovation that the hope of being able to bring about a global coherence and resolution of problems posed by technological society seems to evaporate.

II. THE VICIOUS CIRCLE

To the extent that technologies are perfected and spread on the basis of, among other things, progress in information, processing and computers, technical and organizational interdependence becomes all the more common and penetrating. Slowly but surely this dependency becomes more widespread, complex, and inelastic. Since modern technologies have reached out into all areas, to all types of activities on a world-wide basis, the need for coherence causes a relentless drive toward a global economic, organizational, and trans-technical integration. If, on the one hand, computers make this integration possible, on the other, the more complex and powerful its own technological components become, the more difficult are harmonious inter-connections between systems. This becomes evident as soon as one considers connecting systems among themselves, or in relation to natural or social realities. Ingmar Granstedt shows this clearly in *L'Impasse industrielle*, which examines technico-organizational problems brought about by modern industrial techniques and the economies which depend on them.[4]

In point of fact, if one examines the area of industrial economy, the development of the technological system can be illustrated by the way the means of production are tied together

in a lengthy, substantive series which extends and complicates itself in order to integrate technologically masses of workers in various regions, countries, and continents, without interruption. This involves vast resources which are quite resistant to modification, since the necessary interconnections cannot be recast from one day to the next. One must transform these complexes, negotiate other arrangements, undo established groups and interests, create new structures, reorganize transportation capacities.[5]

There are thus vast networks of resources on a world-wide scale the interrelations of which are deeply entrenched and unadaptable, so that troubles affecting one sector will have increasing repercussions on the whole ensemble of production. Moreover, to keep functioning, production ensembles must be resupplied at regular intervals and have their products distributed at a steady pace. Industries must be assured that the entire range of their productions are coordinated and that the necessary component parts are of a consistent quality. In addition, to assure that various enterprises work with a minimum of stability over a sustained period, these activities must be planned on a long-range basis, and so on. Moreover the technico-organizational integration toward which technical development in industry moves further depends on vast reservoirs of manpower scattered to the ends of the earth, sources that are obliged to maintain working relations with each other, relations that are pre-planned and inflexible. As this evolution progresses modern man, in the same manner as the means of production, finds himself bound by economic interdependencies that are relentless and resistant to change in the demands they make and over which control is dubious – as is revealed periodically by financial, economic, and environmental crises.

In this manner we see how technical progress ceaselessly extends itself and complicates the context in which the technical ensembles function. For example, each undertaking must adjust to an ever-increasing range of factors in the global environment concerning innovations, markets, raw materials, parts, etc. This is why information must be organized and programmed to include different kinds and levels of data. Thus as techniques increase in scope, it is necessary to increase functional interrelationships, which enlarges the degree of interdependence and thus increases the need for integrated information. One must keep in mind an increasing number of variables and the complexity of their interrelations in order to supply a particular enterprise (or administration) with a true picture, one that accurately portrays the facts at hand. In this manner the need for background documentation and information are multiplied and it is necessary to coordinate and unify the information which those sources both transmit and receive.

Unfortunately, sooner or later the technico-organizational integration of management and production are such that to function adequately each component needs complete information about all the others as well as about itself. Technological progress and the development of the technological system are such that the context of each technique is the sum total of all other technologies.

Informational integration has exceeded the limits of the means at hand and has become available to a host of necessary but anonymous agencies. . . . Faced with the necessity of considering an ever increasing list of background factors on a world-wide basis in order to manage the means of production, new services and new bureaucratic structures are created which specialize in the monitoring of a particular factor such as water pollution, trade balance, technical innovation, patents, domestic affairs, tariff changes, security, economic forecasts for foreign countries, energy costs, the recycling of wastes, etc. . . . At the same time that the institutions of communications are developed . . . informational contexts and interrelations are thus extended in parallel manner around each unit without end.[6]

But there will come a time when this synthesis is no longer possible because the components have become unmanageable. While information acquisition, computation, and analysis remains possible, its interpretation into an integrated, coherent, and meaningful whole which will make various options available, and offer a reliable basis for decision making and control, ceases to be practical. Overwhelmed with information, technicians and those in charge become aware that they are always under informed, that certain elements elude them. The programmed review of the sum of technological interventions always shows itself to be insufficient. From the fact that the human ability to act and communicate is limited, information utilization is pushed beyond the threshold of what is possible, information is no longer pertinent, decision making is adversely affected, and the technical components no longer function in a coherent manner; insufficient information becomes the norm and there is no longer any control of the technico-economic ensembles. Beyond a certain degree of technological complexity and integration of skills and decision making, viable choices can no longer be made; one can no longer explain what is happening. At this stage making larger quantities of information available is no longer useful, and an increase of facts does not equal "information." Stockpiling and analysis of data does not produce information in any humanly meaningful sense.

Thus it is that the integration of all the information exceeds the human capacity for thought. No conclusion other than this is possible, it being unrealistic to imagine that such a function could be excercised by interdisciplinary "collectives" or by super-machines whose programs reflect or exaggerate the limitations of the programmer. We are here approaching boundaries that cannot be transcended.

In such a manner technical progress itself, nurtured and made possible by computers, leads to an enlarging spiral of problem handling which gets out of control. Undoubtedly, the use of each new fact creates a potential for new solutions to problems created by the already existing technology, but the excess of both power and effectiveness which results from these new facts creates a new set of problems. One can, therefore, with adequate justification, speak of a reciprocal reinforcement between the system and the chaotic conditions which have been created, not by an insufficiency but by an excess of organization and rational techniques.[7]

It is within this global context that one must consider contemporary progress in information technology. More than with any other technology, the understanding of information technology is tied in the ensemble of other technologies. This, it seems to me, must be the foundation of further philosophical, moral, or political analyses of its fundamental character. The epistemological or ontological status of information, the relation between computers and human nature, the neutrality or autonomy of information technologies, the ethical issues of privacy and responsibility – all these will be affected by the relationships which have been pointed out. Certainly it is only in considering such interrelationships that the options for the future can be significantly evaluated, and a guide can be established for the serious choices which lie ahead.

Aquitaine Regional Institute for Social
Work and Research,
Bordeaux

Translated by J. Kesel
and C. Mitcham

AUTHOR: Daniel Cérézuelle studied under Jacques Ellul at the University of Bordeaux and under Hans Jonas at the New School for Social Research. His publications include an important review article, 'Fear and Insight in French Philosophy of Technology', *Research in Philosophy of Technology*, vol. 2 (1979). He has also been active in the environmental movement and done research in behaviorial technologies at the Hastings Institute of Society, Ethics, and the Life Sciences. Currently he teaches at the Institut Regional de Formation de Travailleurs Sociaux et de Recherche Social d'Aquitaine in Bordeaux, France.

NOTES

1 Jacques Ellul, *The Technological Society*, trans. John Wilkinson (New York: Knopf, 1964). Translated from *La Technique* (Paris: Colin, 1954).
2 Jacques Ellul, *The Technological System*, trans. Joachim Neugroschel (New York: Continuum, 1980). All quotations with page references in the text are to this volume, which was translated from *Le Système technicien* (Paris: Calmann-Lévy, 1977). In some cases the translation has been slightly revised.
3 John Diebold, *Automation* (New York: AMACOM Books, 1983), p. xvii.

4 Of course, this critique is not the only possible one. Once can also show that development of the industrial economy reaches the limits of the economic order which in its turn
   creates serious social and political displacements, displacements beyond the reach of
   technological regulation. This is François Partant's theme in *La Fin de development*
   (Paris: Maspero, 1981).
5 Ingmar Granstedt, *L'Impasse industrielle* (Paris: Seuil, 1980), p. 87.
6 *Ibid.*, p. 126.
7 Cf. Bernard Charbonneau, *Le Système et le chaos* (Paris: Anthropos, 1973).

NATHANIEL LAOR AND JOSEPH AGASSI

THE COMPUTER AS A DIAGNOSTIC
TOOL IN MEDICINE

ABSTRACT. Diagnosis is not a discrete or limited process, but one which is always operative in the physician-patient encounter. It is neither possible nor desirable fully to compute-
rize or automate this process, even in restricted areas. But it is possible and desirable to set
up a computer-assisted diagnostic network with which the physician and patient can both
interact. Such a system would educate patients and allow them to monitor their own care.
This paper includes as an appendix the outline of a major study along these lines which has
not been published because of opposition arising from a narrow comprehension of self-
interest in the medical profession.

A few years ago we completed a study dealing with the role computers
should play in medical diagnosis and how this role could best be instituted. The basic conclusions of this study, which challenged many
accepted views in the field, may be summarized as follows:

- The diagnostic part of medicine is the weakest link in the chain of
  medical services today.
- A fully-automated diagnostic system is not feasible in the near or
  foreseeable future. Nor is it morally acceptable. Yet most current
  attempts at computerization aim toward this immoral, unfeasible,
  and unattainable goal. Research budgets are drained by false
  promises.
- A computer-assisted (as distinct from a fully automated) diagnostic system is a three-party system which replaces the current two
  party system. The clinical, social, ethical, legal, and economic implications of involving computers in the diagnostic process call for
  careful consideration. Until this is done, opposition to the implementation of computer-assisted diagnosis must be viewed as
  rational, at least in part.
- A comprehensive analysis is theoretically as well technologically in
  order for the rational planning and implementation of computer-
  assisted diagnostic services. Implementation can be planned in a
  manner encouraging growth, both of the geographic area where
  the service is to be offered and of the medical specialties to be
  served.
- All clients of the diagnostic system must be assisted and allowed
  access to the computerized elements. Allowing patient access to

227

Carl Mitcham and Alois Huning (eds.), Philosophy and Technology II, 227–238.
© *1986 by D. Reidel Publishing Company.*

these elements will not only be an excellent educational tool, but will increase the contribution of the system to scientific progress and technological success. Opening the computerized elements to all clients may well improve clinical quality control and thus allow for free education in the democratic process by encouraging the sharing of responsibility by all parties.

A complete description of this study, entitled *Diagnostics Computerized*, is provided as a supplement to the present paper.

In submitting this study of publication we discovered an enormous hostility to it – so much so that even though quite a few editors and publishers showed great enthusiasm, the work remains unpublished. It thus seems reasonable, on this occasion, to try to explain the sources of this hostility while reviewing and restating some of our major arguments.

Let us begin by comparing computer diagnostic service with any other diagnostic tool, such as X-ray photography or computer-assisted tomography (CAT scan). All these instruments are in no way different from the magnifying glass and the stethoscope – they enlarge the diagnostician's vision, they add to the stock of available diagnostic information about the patient. But they do not constitute the process of reasoning that is involved, nor do they form the judgment or verdict that the diagnostician may reach. Considering, then, not the input but the process of reasoning, along with the decision prescribing treatment, to be "diagnosis" in the full sense, we should ask: What are the principles of diagnosis? How is the computer able to contribute to it? And when is this contribution desirable?

There is one principle of diagnosis which is prescribed by virtually every medical text and in nearly all medical schools, which is at the same time either utterly impossible or utterly useless. The principle says: Have every diagnostic encounter lead to as thorough and complete a diagnosis as possible. What does the phrase "as complete as possible" mean in this context?

The very fact that some diagnostic procedures require hospitalization proves that the confines of a diagnostic session in a clinic are sometimes too narrow. It is not even very informative to say "as complete as possible within the clinic," when this can include or exclude simple laboratory tests whose results are available at once or within a day or two, when it can include or exclude a trip to a lab for an X-ray or CAT scan, when it can include or exclude sending the patient home with or without some self-administered medication to return the next day if symptoms persist. To specify any such list of conditions is problematic, yet to establish no such conditions is even more problematic. Even the problem of when diagnosis should lead to hospitalization for diagnostic purposes has no general guidelines – although in some cases it is met with artificial guidelines or pre-determined answers. Yet such pre-determined answers may change with the progress of medicine, and even in quite commonsense ways, so that clearly there are some guidelines, although they are seldom articulated.

A few corollaries may be noted. First, when guidelines are articulated, the articulation may be mistaken or insufficient and thus invite criticism or re-wording, perhaps even reform. This is all to the good. The fear of critical attention is an obstacle on the road to articulation, an obstacle which must be removed. Removal is easier when the taste is rewarding – as it is these days, since it is a necessary step in the introduction of computers into the diagnostic process (although even well-formulated general guidelines are usually far from being as explicit as computers need them to be).

Second, we cannot limit the diagnostic process to the standard clinical diagnostic session. As has already been pointed out, no one really limits diagnosis in this way, although its extension is usually done in an *ad hoc* commonsense manner. To improve the extension we may, again, want to articulate the rules for it, thus once more opening ourselves to criticism. Diagnosis will then be defined as a process of deliberation, including deliberations about whether to invite more input from the laboratory or hospital – perhaps even an operation for diagnostic purposes – as long as this deliberation leads to the prescription of a treatment.

It follows that in every stage of treatment, deliberation is diagnostic, since it inevitably opens (as it should) possibilities for changing the treatment. It also follows that check-ups are diagnostic, even when the treatment they lead to is no treatment at all. More specifically, combining the two corollaries, we need the articulation of rules telling us when to discharge a patient, when to hospitalize (and even operate on) a patient for diagnostic purposes, and all the situations which might arise in between these two extremes.

From this perspective almost all medical encounters include a diagnostic component. Since most modern medical complexes employ computers, it also follows that almost no physician-patient relationship in the modern world is free of some computer influence. Almost all modern diagnosis involves computer-assistance of some sort. Yet the question of computerization, to what degree and to what effect, is thus far barely studied. The claim that computer use is normally quite marginal (as in simple record keeping) is true, but the conclusion that it is therefore insignificant is false.

Along with the general case – which almost always concerns computers, however marginally – we should mention those cases where computers are centrally involved: computer diagnosis par excellence. These are, almost exclusively, cases of artificial intelligence – or so-called "smart systems" – in which a computer simulates the best available medical knowledge and practice in certain restricted areas and thus performs the complete job of the diagnostician. This is a current practice that we find quite dangerous and are inclined to view as plainly objectionable. The dangers include both petrification of the practice and its bureaucratization. Furthermore, a physician may blame the computer and thus relegate personal responsibility to an inanimate object.

People who object to computer diagnosis on the ground that artificial intelligence systems are objectionable overlook two facts: that computer-assisted diagnosis is much less problematic than fully computerized diagnosis; and that computer-assisted diagnosis is not as morally objectionable as diagnosis by fully automated systems. But one may object even to computer-assisted diagnosis – on the grounds that the individual patient is unique, that the computer accelerates the bureaucratization of medicine, and that the computer endangers the privacy of the individual patient.

These three objections are powerful and must be met. But they cannot be met by simply opposing the tide of computerization which we are now witnessing. To secure the reference to the uniqueness of the individual patient we must find *democratic safeguards* against the petrification of computer techniques and against the tendency of physicians to relinquish their own responsibility to machines. And the process of bureaucratization will be controlled not by proscribing the computer (which is a useful tool for making bureaucracy more efficient and responsive), but by centralizing the computer services and setting them up so that they can form a national and international network. Such networks are extremely efficient, but again require special protocols to safeguard the individual and his privacy. This holds for government, commerce, industry, and medicine: we must take the bull by its horns.

Computers can be used also as means of protecting individual persons against maltreatment and abuse. Computer-assisted diagnostic services can help here – provided we require informed consent not only for treatment, or for specific diagnostic procedures, but for all diagnosis. The fiction that diagnosis is confined to some initial clinical encounter is the excuse for not regularly eliciting informed consent and for not allowing the patient to be as active a participant in his own diagnosis as he may wish. Once we insist on on-going informed consent in diagnosis we thereby encourage patients themselves to use the computer-assisted diagnostic services on their own, both for general information and for monitoring their own care.

Computer-assisted diagnostic services should include sub-systems that might be linked into a network, both across medical specialities and across towns, regions, and countries. The network should include general information for the lay public, glossaries and patterns of diseases, simple calculations of compound probabilities that could prevent the errors involved in intuitive assessments of probabilities, and some epidemological information. It could also store the cumulative records of individual patients under some kind of limited access protocol.

Our proposals along these lines explain, we think, the hostility provoked by our original study. The narrow view of the short-term self-interest of the medical profession clashes here with the public interest. As in all similar clashes between the democratic application and development of technology and special short-term class interest, the public deserves to be informed.

All this of course leaves open the question with which we began: How thorough should each diagnostic procedure be? The truth is there are no hard and fast guidelines. We can, however, name the general theory for such guidelines. This is the theory of cost effectiveness. This makes use of the whole field of systems analysis, including decision theory, so that its considerations may be integrated into part of the computer-assistance program that should itself be integrated into the network of public medical services. But this also means that a country in which such an elaborate and powerful system is available must establish some kind of monitor to guide and control the use of computers on the national level. But this seems to be called for on other grounds as well, and (to repeat) in all sectors – government, commerce, industry, and, no less, medicine.

DIAGNOSIS COMPUTERIZED: THE PROS AND CONS OF COMPREHENSIVE COMPUTER-ASSISTED DIAGNOSTIC SERVICES

(a descriptive summary)

Preface. What often stands between a patient and the best available medical treatment is a set of difficulties involved with securing a correct diagnosis. In a world of decreasing supply of, and increasing demand for, medical manpower, the automation of the diagnostic process may appear to be the best solution. Indeed, most attempts to utilize the computer in the medical field aim at full automation. This seems either extremely naive or megalomaniacal. Even if possible, it would certainly be both too expensive and too dangerous. This fact, together with the absence of explicit criteria for when it is useful to implement computers in the service of diagnosis, reinforces the attitude of some individuals and institutions concerning the implementation of even existing partial programs – fear and hostile reluctance.

A comprehensive view is in order for any rational planning. This should not be utopian, but serve as a blueprint for the gradual implementation of partial programs which might eventually be integrated into a comprehensive system. Both patients and physicians must have access to, and be in control of, any computer-assisted diagnostic services, for their own education and exercise of autonomy. Hence, democratic discussions and careful analyses are required for the generation of common-sense rules concerning the limitations, testing, and reform of such a system. This study offers the beginning of such a discussion.

CHAPTER 1. THE PROBLEM SITUATION

A computer system should be designed to assist in medical diagnosis and in improving its quality, if and when possible – without damage to other parts of the diagnostician's performance which, at present, may not be computerizable, and without violating the social and moral aspects of the system of medical diagnosis. Its availability to all beneficiaries is clinically necessary and morally desirable. In order to insure that all these points are fully considered, a system-analysis of the medical diagnostic system is required before recommendations for the rational implementation of computer-assisted diagnostic services are put forward.

CHAPTER 2. THE SYSTEM APPROACH TO MEDICAL DIAGNOSIS

2.1. Introducing the System Approach

The system approach is always problem oriented. It subordinates the parts of a problem to the whole and favors a critical attitude toward even well-understood and uncontroverted matters. Hence, the most general task which the analyst has when approaching a system is to look for the problem, and in doing so to locate the decision-maker, the beneficiary, and the aim of the system under conditions whose variability he likewise has to study. The standard of propriety in the operation of the system may depend on either the beneficiary or the decision-maker or both: there is no general rule here.

2.2. Application to Diagnostics

Our aim is dual: on the clinical level it is the ascription of a disease to the patient; on the research level it is the improvement of medical knowledge. The first aim assumes a given theoretical background, the second an independent monitor on it. Hence we have a number of different factors – such as the diagnostic encounter, the conceptual framework employed there, the research unit at large, and the clinic – and we may take any one of these as the conditions or the environment within which action takes place, and consider the others subordinate, depending on the problem at hand.

CHAPTER 3. THE BENEFICIARY OF THE SYSTEM AND HIS PROBLEM

3.1. Individualistic Ethics

The present study is outside the domain of ethics proper, though it attempts to apply ethics to one given end: to increase the possible efficiency of the responsible diagnostic agent. It can do so by assuming that the patient is the ultimate responsible agent and thus the final decision-maker in the diagnostic process; the question whether consulting the computer is advisable may be left to him or to his diagnostician as agreed upon between them. Any other division of functions will contradict the individualistic principle, that the individual is fully and unconditionally responsible for whatever happens in the realm of his activity. The application of this principle, however, is often problematic.

3.2. The Place of Informed Consent in Diagnostics

Providing background diagnostic information for patients alleviates the problem of informed consent in therapy. Computer-assisted medical diagnosis and monitoring may distribute democratically the burden of responsibility for the improvement of medical research, planning, prevention, diagnosis, and treatment among the various beneficiaries of the system.

3.3. The Client of the Scientific Medical System

The individual client of any medical service plays both the role of patient and the role of a guinea pig; when he is a mere patient, society loses a guinea pig; when he is a mere guinea pig, he could forfeit proper treatment; yet if he is made at one and the same time both a patient and a guinea pig, then society benefits without his suffering more for it. The incentive for voluntary participation of patients and diagnosticians in the large-scale human technological experiment must be such that while adding one's own data one benefits from the existing large-scale pool, i.e., a computer-assisted diagnostic service operating in the free market.

CHAPTER 4. SUBSYSTEMS OF MEDICAL DIAGNOSIS

The implementation of computers in the diagnostic system is dangerous. Means of risk prevention must be built in and regularly monitored. Difficulties include over-stabilization of the system, maintaining the level of the system's efficiency at unreasonable cost, and more. Any recommended change must rely on the analysis of the system as a whole.

4.1. Diagnostic Theory

Viewed as pure science, medicine has no patients or performance criteria. It is preferable, then, in the first instance, to approach medicine as a subsystem of medical diagnostics.

4.2. Rational Diagnostic Technology

The diagnostic means available to medical science and technology are constrained by extra-scientific, human, moral, and legal factors. Total risk-avoidance is impossible under such conditions. Comprehensive computer-assisted diagnostic services may aid diagnostic rationality by making it clear which errors are easily avoidable (with or without the aid of computers), so that committing them constitutes a violation of the demands for rational technology. But this will only be so if the computer service is regularly improvable. Otherwise it will be a useful tool for evading responsibility.

4.3. Rational Diagnostic Method

All effective diagnosis is differential and as such statistical. Medical diagnostic intuition should be assisted by readily available computation.

Yet the temptation to indulge in the thoughtless collection or application of statistics to data must be checked. Hypotheses about diagnostic methods as well as about diagnosticians' strategies are needed, and not a machine whose task is the automatic emulation of some experts (i.e., a so-called "expert system"). The diagnostic system may be improved by integrating readily available formalizable computer elements with the human diagnostician in one system, on the condition that responsibility lies always with the human element, never with the machine.

4.4. Clinical Diagnostic Situations

In the clinical situation diagnosis and treatment intertwine. Knowledge of the range of treatments and their costs advise the physician about the desirability of continuing with the diagnostic process. The costs change according to the clinical situation. Defining the situation as well as filtering relevant information are unformalizable and rely on medical competence. Yet the computer may be introduced at any stage for assistance in the process of diagnosis and for assessments of costs. Enlisting the computer must submit to cost-effectiveness analysis for which the comprehensive computer-assisted diagnostic system may be of great help.

4.5. Rational Diagnostic Control and Public Health Approach

It is possible to monitor and control a given individual element of diagnostic service to one client with the aid of a comprehensive computer-assisted diagnostic service system by means of a few rough cost-effectiveness computations, based on simple decision-theory considerations together with some general information and available statistics. However, using the general to check the particular is dangerous; it always allows the general to win, since the general generates its own statistical reinforcement. A computer program can only ring a bell or flash a red light, never decide. Decisions concerning individual cases are left to the human partners of the computer-assisted diagnostic service. Deviations from a rule should be recorded and monitored, and at times lead to an attempt to reform the rule.

4.6. The Rational Control of Single Diagnostic Encounters

The output of the proposed computer-assisted diagnostic system, then, must be fed back as a control, and all control functions must be fed back into the computer and out to diverse institutions of control and decision (customers, public representatives, and programmers). Clinical computer-assisted diagnostic conferences (CCC) and clinical ethical conferences (CEC) should be introduced as a means of control and improvement of the single diagnostic encounters. Preventive and public health services as well as medical education can take advantage of the computerized system and help control it. The individual's confidentiality must be secured, however, by restricted access to personal data – by code-numbered credit cards, for example.

CHAPTER 5. SURVEYING THE COMPUTER-ASSISTED DIAGNOSTIC SERVICE

5.1. Rational Diagnostic Service Itemized

The computer service can offer data concerning standards, costs, benefits, etc., as well as diseases and their treatment. The patterns recognizable by the computer are imperfect, however – like other empirical messages they are contaminated by noise. The minimization of noise increases the efficiency of the system, yet eliminates the odd occurrence and the rare diagnosis which are so important for the advancement of medical diagnosis and treatment. This can be remedied by keeping as stable only the initial two stages of pattern-recognition – first the elements, and second the patterns – while leaving open to more frequent alterations the two latter stages – third, pattern and element fitting, and fourth, decision. Fifth and sixth, monitoring of implementation and improvement, should remain mixed.

5.2. The Present State of the Computer in the Service of Diagnosis

Traditionally, there are three methods for formalizing diagnosis: (1) ignoring all noise; (2) allowing for noise by introducing alternative options and giving each a statistical weight, and (3) computer simulation. All three aim at the impossible, i.e., at the full formalization of medical diagnostic competence; they are, therefore, severely limited. Their combination, however, may serve as a first approximation for a more rational, system analytical approach toward comprehensive computer-assisted diagnostic services.

5.3. The Future of the Computer in the Diagnostic Service

Computer-assisted diagnostic services are suggested as a regulative idea for partial services to be flexibly integrated in a later stage. Such a plan calls for discussions concerning aims and standards of our general services, to which the system analysis of the medical diagnostic system offers modest beginning.

APPENDIX A. THE SYSTEM APPROACH

The system approach is endorsed both in science and in technology as regulative yet criticizable. It seems that (1) cost effectiveness cannot always be rationally computed, and (2) an analyst's designs must be safeguarded against his own intrusive bias. Hence, it is suggested that perspectives on any given system be developed from different viewpoints and that they be democratically discussed. To that effect the implementor and the monitor should be added as separate entities or functions of any given system in addition to those thus far proposed, namely, the decision-maker, client, and analyst. This is how the system approach could be compatible with individualistic ethics and rational technological planning.

APPENDIX B. DIAGNOSIS

There is no pattern recognition without noise. At times noise originates from the clients of the pattern-recognition process. Medical diagnosis recognizes systematically intruding clients as malingerers. Computer diagnosis without eliminating malingerers constitutes defective computer simulation, yet we cannot eliminate malingering. Computer-assisted human diagnosis remedies the situation by broadening the system to include its human environment: by educating the clients of the medical system in autonomy, they will be allowed to share in the responsibility for improvement of individual diagnostic encounters as well as the general diagnostic service.

APPENDIX C. TREATMENT

Treatment is usually based on very poor pattern recognition. The pattern recognition of treatment as medical is also traditionally poorly defined.

This is why it is preferable to describe both diagnosis and treatment in the abstract. Implementation, however, requires context-dependent considerations concerning the therapeutic commitment – not to harm, but to explain, to get the patient's permission and to treat him adequately. Unfortunately, different physicians propose different explanations, use different standards, and vary widely in treatment. Such a situation is unacceptable and calls for discussion in an attempt to generate responsible consensus. Only then may explanations, standards, and treatments be responsibly computerized. Moral standards and decisions which involve treatment should not be computerized and their implementation must always be openly checked.

Yale Medical School

York University and Tel Aviv University

AUTHORS: Nathaniel Laor, M.D., Ph.D., is Assistant Professor of Psychiatry, Yale University School of Medicine, New Haven. Joseph Agassi, who has degrees both in physics and in logic and scientific method, studied under Karl Popper, and is Professor of Philosophy, York University, Toronto, and Tel Aviv University, Israel. Agassi's works include *Towards an Historiography of Science* (special issue of *History and Theory*, 1963; reprint Middletown, CT: Wesleyan University Press, 1967), *Towards a Rational Philosophical Anthropology* (The Hague: Martinus Nijhoff, 1977), and *Science and Society: Selected Essays in the Sociology of Science* (Boston: Kluwer, 1981).

HANS LENK

SOCIO-PHILOSOPHICAL NOTES ON THE
IMPLICATIONS OF COMPUTER REVOLUTION

ABSTRACT. Reviews the background of previous thought about automation (Aristotle, H. Schelsky), and points out that consumption is not a perfection of human nature. Human beings are prefected more by self-achievement. Argues against the necessity of a computer-based technocracy, and suggests how computers might help provide the means to a more humane (if not utopian) society.

"What (if anything) to expect from today's philosophers" – this was the title of an editorial essay in *Time* magazine almost twenty years ago (Jan. 7, 1966, pp. 24–25). The subject can easily be raised again with regard to the role philosophy might play in dealing with the problems of the so-called "computer revolution" in society.

"The new priests come from the lab," Richard McKeon is quoted as saying. Is this also true with regard to the social problems of the computer society, microelectronic phase – that it is scientists and engineers who pose the problems and have all the answers? Notice that such an idea excludes not only philosophers (the professional "lovers of wisdom") but all amateurs (a word derived, interestingly enough, from the Latin *amare*, to love). The men in the lab in our case are the computer scientists and all kinds of computer technicians. Many "computer freaks" are no more than amateurs. So what, if anything, can social philosophy contribute to the debates concerning automation and computers.

* * *

Let us begin by turning to the first philosopher who ever mentioned our problem. Surprisingly enough, this is Aristotle. In the *Politics* he writes:

For if every instrument could accomplish its own work, obeying or anticipating the will of others, like the statues of Daedalus, or the tripods of Hephaestus, which, says the poet, 'of their own accord entered the assembly of the Gods'; if, in like manner, the shuttle would weave and the plectrum touch the lyre without a hand to guide them, chief workmen would not want servants, nor masters slaves (1253b33–1254a1).

Aristotle clearly believed this would be a positive development. He did

239

Carl Mitcham and Alois Huning (eds.), Philosophy and Technology II, 239–245.
© *1986 by D. Reidel Publishing Company.*

not, perhaps could not, predict the negative consequences of unemployment for a workaholic society – of which Greek society was not an example. Yet Aristotle did foresee the purely objective social implications.
When, over two thousand years later, the Polish emigré philosopher,
Adam Schaff, edited a Club of Rome report on microelectronics and
society – under the revealing title *For Better or For Worse* – he really did
not go beyond what Aristotle had predicted already in antiquity. (Indeed, apparently he was not even aware of Aristotle's statement.)

                              *   *   *

The truth is that most of the newly discovered social implications of the
computer and automation were predicted, outlined, and discussed in our
own time as early as 1957 in a complete if preliminary manner by the
well-known German sociologist Helmut Schelsky. Indeed, Schelsky
identifies not only the problem of unemployment, but the possibility of
ideological distortion, which he estimates to be a greater totalitarian danger than the more immediate consequences of technological development. (This point remains a rather tentative and speculative thesis subject to revision; perhaps events have already proved it mistaken.) A
more important point is that Schelsky does not think that the plain and
simple fact of industrial automation can be called "revolutionary." Only
a combination of critical factors can be labelled a social revolution.
Schelsky argued that all talk of second or third industrial revolutions was
wholly unfounded or at least premature. (Perhaps this judgment is likewise in need of revision.) And as for practical measures, Schelsky recommended piece-meal strategies among all parties: entrepreneurs and corporations, the state, the workers and unions.
    According to Schelsky, one requirement for both research and ratio-
nal social action is proper information and knowledge. This means that
corporations, for example, have a social obligation to inform the public
about planned changes in organization, employment, and production.
The systematic management of job changes by means of retraining programs was proposed as well. Most important, he thought, would be a
reform in technological education. Finally, worker demands as expressed by the unions as well as general public concern for an overall increase
in economic productivity were to be taken seriously.
    So with regard to both social impact and appropriate response, there is
little that is new in recent discussions of microelectronic automation –
although in practice the phenomena in question have or are currently

receiving a rather dramatic emphasis. Nevertheless, neither Schelsky's work nor that of other social philosophers seems to have had much real impact on past or present discussions of the social consequences of technology. So what should philosophers struggle to think or to say about such matters today? – since thinking still seems to be the primary task of philosophy.

*   *   *

One suggestion is that social philosophers should try to go beyond merely piecemeal pragmatic advice – recommending gradual adaptation, a pacing of automation, retraining programs, etc. These are common sense ideas that any society will come up with all on its own, without the help of philosophers.

A second suggestion is that philosophers might well avoid unrealistic specific proposals like that articulated by Schaff, who thinks that "permanent education as a form of universal activity" will solve the employment problem and at the same time help realize the ancient ideal of a universal man – i.e., the universally educated, harmoniously developed individual in the Platonic sense. One need not appeal to Auschwitz or depth psychology to point out the utopian character of such an ideal. Human beings are not the good people that Schaff wants to make of them.

The possibility that a combination of *homo studiosus* and *homo ludens* could replace *homo laborans* will obtain for only a few people. Many if not most do not want to go to school their whole lives. And with regard to other leisure time activities, as Albert Borgmann has pointed out in his own reflections on the computer revolution:

> It is clear that technological liberation from the duress of daily life is leading more and more to a disengagement from skilled and bodily commerce with reality. Our leisurely contact with the world is being narrowed to pure consumption, the unencumbered taking in of commodities which requires no preparation, provides no orientation, and leaves no significant trace. . . . [D]iversion will more and more lead to distraction, the scattering of our attention and the atrophy of our capacities. It is already apparent that the new video technology is not used by people as the crucial aid which finally allows them to develop into the historians, critics, musicians, sculptors, or athletes they have always wanted to be.

Interestingly enough, Borgmann stresses that "those things and practices that . . . have orienting, engaging, and sustaining force are all of essentially pre-technological origin though they assume a new splendor and radiance when exercised in a technological context." His conclusion parallels my own argument in *Eigenleistung* (1983, cf. also 1979), that man in the last analysis is an active and acting being capable of achieving greater and greater skill. He is an achieving and performing being. Real life in its deepest sense is personal achievement and self-development, which is what Borgman means by "engaging activities." As Aristotle wrote in the *Politics*, "But life is action, and not [*only* – my emendation] production" (1254a7).

* * *

Consumption and consumerism are indeed not the perfection of humanity. Paradise is not, at least in the Western tradition, a passive state; it is active, even activistic. Here, clearly, social philosophy and philosophical anthropology must come into play to examine and understand this ideal. Thus philosophers of technology should not only describe but try to explain why it is that pre-technological activities are the more truly rewarding, and then identify the more deeply engaging ones. A simple identification of such pre-technological activities still leaves open the question of why we cannot also act *technologically* in a truly human and humane sense. It is the grappling with issues such as this which may be a special contribution of social philosophy to the computer revolution.

Let me just mention, then, a related issue – that is, the problem of mediation in the technological world. It is not necessary to repeat in detail the whole debate about the so-called administrated world, with all its phenomena of red tape, fragmentation, functionalization, manipulation, alienation, etc. The passive way in which people look at pictures and movies, the vicarious experience created through television, the resultant pseudo-excitement without proper personal engagement – all this is well-known. The electronic media exercise a distracting and abstracting, if not displacing, force in our society which should not be facilely underestimated.

But there also looms a new computer version of the mediated world. Computer modelling has its own abstracting and displacing affects. Computer models are not reality. They constitute an artificial world which can give rise to its own pathologies: compulsive computer hackers and adventurous computer criminals seem to replace the former neurotic "book worm" and library vandal. Social psychologists such as Sherry Turkle have begun to do serious studies of various computer-associated neuroses. And the movie *War Games* signals the increasing public awareness of these issues. Computers do have a seductive power; they are subject to passionate pseudo-identification – perhaps in part because, like a beautiful but unknown woman, they are ready carriers for psychological projections. (One might even become sick from computer love.) However productive or active such computer compulsions may appear, they are nevertheless distracting from personal – that is from person-to-person – engagements. Indeed, pseudo-love projections can become so exaggerated as to be positively dangerous.

*   *   *

Let me return to socio-philosophical issues in a stricter sense. Are not the trends toward "computerocracy" but instances of what I referred to ten years ago as "systems technocracy"? Systems-technocratic tendencies seem to be more and more encouraged by computer and information networks. Indeed, if we were to generalize in a rather gross manner we might come up with some ideal-type descriptions of such trends.

But does systems engineering and the computer revolution inevitably lead to a kind of systems technocracy? Discussions of systems technocracy as a special variation of technocracy could well become more prominent as the information and systems-technological era comes more clearly into focus – with such dangers as large-scale computerized information systems that could be put to centralized autocratic uses. (See, in this regard, David Burnham's fearful exposé, *The Rise of the Computer State*.) Yet such possibilities should not be considered in isolation from democratic forces and certain humanizing possibilities. Rising social awareness and participatory democratic engagement have their own momentum and influence on the direction of computer-enhanced social systems. No doubt societies of the future will be increasingly confronted with technocratic challenges, particularly in the form of information-network threats to privacy. Social philosophers, philosophers of science and technology, along with moral and legal philosophers should not abrogate the discussion to engineers, politicians, and sociologists alone. Systems networks call for interdisciplinary research. This is particularly true with regard to the social implications of the computer revolution – and even more so in regard to the problem of unemployment.

*   *   *

Let me briefly (and finally), then, turn to work and labor, and ask whether the auspices are as gloomy as many think. The truth is that the number of jobs – at least manual jobs – are and will continue to decline.

High unemployment rates can be expected to remain a structural feature of advanced automation production. What socio-philosophical implications can be drawn from this observation or anticipation?

It is not enough to recommend permanent education or to advocate the biblical position which, ironically enough, was written in the Soviet constitution of 1937: ". . . that if any would not work neither should he eat" (II Thessalonians 3:10). From a socio-philosophical point of view it can easily be argued that we will have to abolish traditional alternatives between the options of work or starvation. An affluent industrial society with structural unemployment can and should supply a minimum life subsistence to everyone. A guaranteed income – which need not be limited to "satisficing" – appears to be both prudent and practical in highly productive societies.

This does not mean that reward for achievement should be totally eliminated. Beyond the guaranteed income, achievement could still serve a socially differentiating role and as a relatively "just" means for distributing income and other social amenities. Moreover, the uncoupling of the work-survival relationship will offer new possibilities for exercising volunteer work and non-monetary rewards. People need not and should not in the future judge their and others' social value solely in terms of monetary income. This may seem a bit utopian, particularly in the American context, but we will gradually be obliged to move in this direction.

Furthermore, computers and automation may also help alleviate what Marxist and neo-Marxist social critics call alienated labor. By disjoining the rigid work-survival relationship, the realm of free personal and social engagement will be opened up. Professional work in itself and not just for pay confers a sense of life. We can and perhaps should retain the right to be activists and at times even workaholics. But the overall pattern of social gratification and monetary reward will be de-dramatized, so to speak. Voluntary and freely chosen activities done for their own sakes and values, or for socially creative or even recreational purposes, will receive greater social status and value.

Indeed, it may well turn out that with the help of automation and computers, along with other dynamic technological processes, we will eventually come close to creating a personalized and more socially just society which is no longer based primarily on rugged individualism and a compulsory work ethic. Such a society would leave room for individualism and non-alienated work – self-achievement, even at times of workaholism – yet would still involve a kind of basic human solidarity. It would render competition, especially competition for individual survival, less serious, totally eliminating the idea of a rather mitigated competition as a vehicle for progress and development, though only on top of the basic guaranteed income. Competition, instead of being deadly serious, would become more symbolic, a kind of sport play and means of self-development. Computers – not as a sufficient condition, but as one important contributing factor – may take us closer to the realization of such an apparently utopian ideal. This could also serve the venerable though by no means outdated social values of Christianity. The potential social implications of computers create real historical opportunities which call for serious consideration.

University of Karlsruhe

AUTHOR: Hans Lenk (born 1935), who has written widely on the social philosophy of technology and on the philosophy of science, was a 1960 Olympic rowing champion and has served as a coach. He has taught in Germany, the United States, Brazil, Venezuela, Japan, Korea, and Norway. He is currently Professor of Philosophy at Karlsruhe University and a member of the European Land Use Faculty in Strassbourg, France.

REFERENCES

Burnham, David. *The Rise of the Computer State*. New York: Random House, 1983.
Deutsches Institut für Fernstudien an der Universität Tübingen. *Mikroprozessoren. Die Elektronische Revolution*. Textsammlung und Basistexte. Tübingen: Zeitungskolleg, 1980. See especially chapters 7–12.
Lenk, Hans. *Philosophie im technologischen Zeitalter*. Stuttgart: Kohlhammer, 1971.
Lenk, Hans. *Social Philosophy of Athletics*. Champaign, IL: Stipes, 1979.
Lenk, Hans. *Zur Sozialphilosophie der Technik*. Frankfurt: Suhrkamp, 1982.
Lenk, Hans. *Eigenleistung*. Zurich and Osnabrück: Interfrom and A. Fromm, 1983.
Lenk, Hans (ed.), *Technokratie als Ideologie*. Stuttgart: Kohlhammer, 1973.
Lenk, Hans and Moser, Simon (eds.), *Techne – Technik – Technologie*. Pullach: Verlag Dokumentation, 1973.
Lenk, Hans and Ropohl, Günter. *Technische Intelligenz im system-technologischen Zeitalter*. Düsseldorf: VDI Verlag, 1976.
Schaff, Adam and Friedrichs, Günter (eds.), *Microelectronics and Society: For Better or For Worse*. A Report to the Club of Rome. New York: Pergamon, 1982.
Schelsky, Helmut. *Die sozialen Folgen der Automatisierung*. Düsseldorf: Diederich, 1957.
Turkle, Sherry. *The Second Self: Computers and the Human Spirit*. New York: Simon & Schuster, 1984.

CARL MITCHAM

INFORMATION TECHNOLOGY AND THE PROBLEM
OF INCONTINENCE

ABSTRACT. Incontinence, or the inability to act on the basis of one's knowledge, poses a special challenge to any optimism about the ability of information technologies to contribute to the rational control of modern technology. This paper examines both weak (Aristotelian) and strong (Augustinian) versions of the problem of incontinence in order to explore the extent of this challenge.

"Information technology" is an ambiguous term. It could refer to all technology used to transmit information in the human sense – from written alphabets and signal flags to radios, TVs, and data processing computers. This would make it roughly synonymous, in at least the last three instances, with "communications technology," although this term too has its own ambiguities. Alternatively, "information technology" can refer to those technologies which are designed to manipulate digital signals as analyzed by the mathematical theory of information which has grown out of the work of Claude Shannon and others. The justification for restricting "information technology" to this latter reference is that there is a certain kind of electronic engineering which can reasonably be described as applied information theory – i.e., computer science. In the present paper the term is used primarily in the second sense, but also as a convenient way of referring to any kind of electronic technology which functions to transmit meaningful information in the first or human sense. It thus includes electronic media such as the telephone, radio, TV – which are, in fact, rapidly becoming digitalized.

Implicit in much of the excitement surrounding information technology in the technical sense has been a kind of optimism about the possibilities of control which it seems to promise. The ability to analyze signals and signal interaction at discrete levels raises the possibility of eliminating much ambiguity – and thus contingency or lack of control – in communication. The very title of Norbert Wiener's classic book, *Cybernetics: Or Control and Communication in the Animal and the Machine* (1948), points in this direction. Later Soviet interest in the field is explicitly predicated on similar notions. And many contemporary hopes for the benefits to be derived from the information/computer revolution likewise hinge on similar beliefs.

Carl Mitcham and Alois Huning (eds.), Philosophy and Technology II, 247–255.

There are many questions that can be raised with regard to such an optimism. The fundamental problem concerns the relation between signals and their messages – that is, the problematic of hermeneutics or interpretation – and the slippage this re-introduces into communication and control outside any information technology itself. But even if this issue could be bypassed, there would still remain the problem of incontinence.

"Incontinence" is another word which calls for clarification. Although most commonly used to designate the medical pathology of being unable to retain urine or feces, in scholastic moral philosophy it had a more general reference to the absence of *contentia* or self-control – of which the medical pathology is only a specific instance. Here the term is used to indicate a hiatus between knowledge and action, in an effort to avoid terms such as *akrasia* or "weakness of the will" which prejudge the interpretation of the phenomenon.

Following an argument for incontinence as a problem to be addressed by any ethics of technology, two versions of the problem are described, and the ability of information technology to meliorate each is considered. The conclusion is that, although information technology can in some instances meliorate the problem as understood in the first version, it may well exacerbate the problem as understood in the second.

I. RESPONSIBILITY AND INCONTINENCE

It is helpful to begin by situating the problem of incontinence within the broader context of ethical discussion regarding modern technology. If power or the ability to act increases, then so must responsibility – otherwise power will eventually lead to disaster. Technology increases power. Increased technology, therefore, calls for a corresponding increase in responsibility. Everyone probably would admit the soundness of this argument, and agree that such a formulation is general enough to apply to a wide range of technologies – from nuclear weapons and feats of biomedical engineering to automobiles and chainsaws.

But what are the preconditions for the full exercise of such responsibility? The responsible use of technology depends on
   (a) knowing what we should do with technology, the end or goal toward which technological activity ought to be directed;
   (b) knowing the consequences of technological actions, prior to the actual performance of such actions; and
   (c) acting on the basis of or in accord with both types of knowledge – in other words, translating intelligence into active volition.

Most discussions concerning the responsible use of technology focus on (a) and/or (b), often with reference to the abilities of information technologies and computers to deal with problems in these areas. Insofar as (c) is recognized, it is subsumed under questions of societal organization (cultural lag) or observed as a psychological pathology (alienation, etc.).

More specifically, with regard to (a), knowing and getting agreement about the ends or goals of technological actions present special difficulties for the pluralistic societies of the West, embedded as they are in a scientific and technological cultural with marked positivist leanings. The most widely proposed solutions appeal to some form of democratic control, with at least implicit reliance on radio, TV, telephone, and computer-analyzed public opinion polls, properly protected from the dangers of bureaucratization and authoritarian manipulation. As for (b), the equally difficult problem of predicting the consequences of technological actions seems to rely even more strongly on electronic monitoring and computer modelling to deal with the enormous information complexities of environmental impact statements, technology assessments, and risk-cost-benefit analysis.

At the same time, on a daily basis one encounters any numbers of examples of the problems related to (c), the issue of incontinence. There is the nurse or physician who is well aware that smoking causes cancer and any number of other health problems, yet continues to smoke. There is the automotive safety engineer who knows full well the importance of seat belts, but fails to buckle up himself. And genetic counselors tell horror stories of persons who, even when appraised of the near certainly of passing on disabling or fatal genetic defects, nevertheless choose to bear children. In each case conditions (a) and (b) are clearly met. The individuals in question know what they should do and how to do it. They should pursue health; indeed, they even actively do so in many aspects of their lives. Moreover, they know the consequences of particular actions that are diametrically opposed to the good they desire. And yet, they do not perform those actions which are dictated by such knowledge of ends and means. They know the good but do not do it. Should an analysis of the dimensions of such behavior not form a substantial part of the discussion of ethical responsibility in the exercise of science and technology?

II. *AKRASIA* AND THE WEAK VERSION OF INCONTINENCE

An appraisal of the dimensions of the problem at issue must begin by distinguishing two major versions of incontinence. In the weak version, there is what might be described as the resistance of matter to intelligence. Discussions of this go back to the works of Plato and Aristotle. In the strong version, there is an opposition of intelligence to intelligence. This is associated with theological discussions of freedom of the will and the possible ability of a creature to know God but reject him.

The *locus classicus* for a discussion of incontinence in the weak sense is Aristotle, *Nichomachean Ethics*, Book VII. According to a common historical distinction (based on Aristotle's own remarks), Socrates identifies knowledge with virtue and thus rejects the problem of incontinence, whereas Artistotle argues that this is patently contradicted by the facts of experience. The central difficulty in analyzing *akrasia* or weakness of the will is thus to explain in what sense a person could know the good and still not do it.

According to Aristotle there are four senses in which a person can know but not do the good (see *Nichomachean Ethics* VII, 3): (1) A person can know in the sense of being able to remember, but not presently remembering. (In New York City it is not legal to turn right on a red light – although it is legal in the state as a whole – but one can just space out and forget.) (2) A person can have universal knowledge which in reality includes some particular, but not actually be aware that it does so. (One can know that sugar is unhealthy, without realizing that catsup is laced with sugar.) (3) A person can have knowledge which is obscured by sleep, drunkenness, or some other physiological state. Finally (4), a person can have two different kinds of knowledge, and the lower can overcome the higher by virtue of an accidental feature of his individual state (or social condition).

Cases (1) through (3) cannot, however, really account for the experience of incontinence. In none of these instances will a person experience a struggle to resist temptation. After the fact one may well look back and recognize a failure to act rationally or in accord with the good. Yet at the time, the person who forgets, or fails to recognize how a particular falls under a universal, or is drunk, does not experience a struggle for the good which ultimately fails. It is only case (4) which offers some explanation for this experience.

Aristotle's highly condensed presentation of this fourth case may be elaborated with the following example. Suppose that a person knows both that working in an asbestos factory is bad for his health and that working in such a plant will pay him a high wage. He is in need of work and is offered the job. But he hesitates. If he takes it, he will get his first paycheck in two weeks and begin to pay off his debts. The bad effects of asbestos exposure will be much more distant, to say the least. It will be years before asbestos will take its toll on his health. In fact, the first two weeks of work all by themselves may take no toll at all. Only if they become combined with hypothetical other weeks and years of exposure will they be bad for him. Besides, what is bad is not the work; that, in itself, is good. Given the remoteness of the bad and the immediacy of the good, is it any wonder that people in such circumstances will, after some hesitation, take the job – knowing in some sense, that they should not?

Here is the practical analogue of a principle of Aristotelian epistemology. In science, the Stagirite observes, what is more knowable in itself is commonly more remote from us, while what is less knowable in itself is more immediate to our experience. The order of demonstration (beginning with first principles) is not the same as the order of discovery (beginning with experience). In the case of ethics, likewise, the highest good is commonly more remote from human experience, while a lower good is more immediate.

Notice what this implies. Science is the attempt to replace ignorance with knowledge by bringing the more knowable but less known closer to our range of experience. This is why books are better teachers of science than the world; books can explain first principles. Ethics, likewise, must try to make the less influential but higher good more immediately influential in our lives. Thus, following his analysis of the moral experience of incontinence, Aristotle asks, How is the "ignorance" of the incontinent man to be dispelled?

The incontinent man does not act against knowledge in the truest sense, says Aristotle (agreeing now with a view earlier attributed to Socrates). If he had been as vividly aware of the long-range effect of asbestos exposure as he was of the short-range effects of work (and as able to act on such awareness), then he would not have taken the job. The overcoming of incontinence becomes a function of education and moral training (and perhaps the restructing of society). The "artifice" of the *polis* is a better teacher of ethics than nature. Incontinence loses its force as a conundrum and becomes merely an indicator of the need to transcend nature with culture.

This is the understanding which animates much of the practice in our society relative to raising consciousness about the dangers of certain technologies – often with the use of various information technologies. In order to discourage smoking, its long-range effects are made as vivid and as immediate as possible by means of computer-assisted epidemological studies, warning labels, newspaper articles and books, films, public service TV ads, and glass-encased sections of dead smokers' lungs fashioned into ashtrays. Similarly, as war is made more horrible by advances in weapons technology, argue those who are encouraged by this approach, media technologies drive the horrors of war home to everyone's mind. (The depressing thing is that this neither lowers the number of people who smoke below one third of the population, nor does it seem to have much effect on the number of wars in the world.)

Supplemental to this information-oriented approach are two other strategies. One is to structure the environment so that it "artificially" reflects the long-range consequences of smoking. This brings remote knowledge down to the level of everyday experience. Extra tax burdens are placed on smokers. Laws against smoking in public places are passed. Parents withhold allowances and even resort to corporal punishment. Human beings – or at least some and perhaps most human beings – are not guided by information alone.

A second strategy is the so-called "technological fix," which would disjoin "bad" artifacts and actions from their long-range consequences, so that only the short-range consequences really matter. The American Tobacco Institute is determined to invent a cigarette that does not cause cancer. Short of that, we have a national effort to find a medical cure for cancer.

One problem with the information-oriented response to weakness of the will is that an information-rich society can sometimes aggravate the weakness at issue through an inclination to postpone action in favor of the pursuit of more information, or otherwise undermine the ability to perform difficult and heroic actions. Certainly the former tactic has been used most effectively by the tobacco lobby to thwart and delay anti-smoking legislation; a bias in favor of more information can also be used to protect established elites against rapid social change. As for the latter possibility, George Will has pointed out that if there had been TV news cameras at Gettysburg, the Civil War would have had a different ending.

III. FREEDOM OF THE WILL AND THE STRONG VERSION OF INCONTI-
NENCE

The weakness with Aristotle's analysis of the relationship between knowing the good and doing it is that despite his professed intention to preserve the experience of incontinence, this experience is subverted by his defense of the power and primacy of true knowledge. The gap between knowledge and action is closed by distinctions between different types of knowing, and an affirmation of the power of at least some kind of knowing fully to determine human behavior. The gap opens up only in the presence of a weak or inadequate form of knowledge (or in a social situation which deprives that knowledge of its efficacy). What is commonly referred to as Aristotle's analysis of the weakness of the will is really an analysis of the weakness of certain kinds of knowledge.

The *locus classicus* for a discussion of incontinence in a much stronger sense is St. Augustine's *De libero arbitrio voluntatis*, especially Book III. For Augustine, the issue of the power of knowledge arises in relation to the question of the origin of evil. If God created the world out of nothing and did not give human beings truly independent agency, then God would have to be the remote cause of all evil. But if he *did* give human beings truly independent agency, how can this be seated in the intellect and its act of knowing, since knowledge must always bear on what already is? Insofar as the intellect "chooses" evil, the choice (as Aristotle rightly observed) comes about because the intellect somehow fails in its act of knowing. Yet if this limitation is built into the intellect by its creator, once again God must be ultimately responsible. Faced with this difficulty, Augustine turns to the will as the source of evil and the cause of a gap between knowledge and action, which can only be called a stronger form of incontinence.

It is crucial to note that Augustine in no way "solves" the problem of incontinence in this stronger sense. He cannot explain how it is that the will could freely choose to do what is known to be a lesser good. He simply tries to acknowledge what he sees as a fact of experience. To explain how it happens would in fact once again subordinate the will to the intellect. If it is a truly independent element, the will must be unintelligible to the intellect. All Augustine does is recognize how the occasionally radical independence of the will explains the paradoxical relationship between a good God and a world stained with evil.

"Why do we have to inquire into the origin of this movement by which the will is turned from immutable to transitory goods?" complains Augustine at one point (II, i, II). The will and its ability to do evil is a fact of experience (III, i, 12). Moreover,

what cause of the will could there be [when the will does evil], except the will itself? It is either the will itself – and it is not possible to go back to the root of the will –; or else it is not the will, and there is no sin. Either the will is the first cause of sin, or else there is no first cause (III, xvii, 168).

That is, either the human intellect or the human will is the cause of evil.

But it cannot be the intellect; therefore, it must be the will. It is as simple as this disjunctive syllogism. But Augustine admits that he cannot plumb the depths of the free will. Nor does he propose any methods for dealing with the sinful will other than the preaching of religious conversion and the development of political sanctions against criminal behavior.

The contemporary manifestation of the election of evil in the form of terrorism, employing modern technologies, appears equally intractable. Despite the fact the electronic media and computers are swiftly being adopted to religious uses, it remains highly doubtful whether the electronic church realizes any net gain in the effectiveness of preaching, especially given the pervasive secularization to which the media also contribute. And while information technologies may make it theoretically more possible to identify the terrorist and apply socially protective countermeasures, in practice the civil liberties traditions of the West delimit such possibilities at the same time that existing information systems provide any potential terrorist with enhanced access to a plenitude of technological means and destructive powers.

IV. THE POLITICAL CONTRADICTION OF INCONTINENCE

Having distinguished two versions of the problem of incontinence and noted the differential ability of information technologies to contribute to their melioration, it only remains to point out an incontinence-related contradiction at the heart of the modern technological project.

Ever since Augustine, the philosophical scandal of free will has occupied at least a minor place in the spectrum of philosophical conundrums. But the modern period, by identifying the will more that the intellect as the unique feature of humanity, by making freedom more than justice the primary aim of politics, has given the conundrum a unique political dimension. It is precisely this identification of freedom as the human essence which can be seen to ground the technological project (the aim of which is to realize that freedom), while the project itself (the Enlightenment pursuit of a union of science and politics in knowledge-based power) presumes the impossibility of incontinent freedom.

"We hold these truths to be self-evident," proclaims the American *Declaration of Independence*, "that all men . . . are endowed by their creator with [inalienable rights to] life, liberty, and the pursuit of happiness [and] that to secure these rights governments are instituted among men." Not life and intelligence, but life and liberty or free will become the key characteristics of human beings. Popular discussions of the differences between computers (artificial intelligence) and human beings often confirm this position: machines may be as intelligent as human beings, but they do not have a will. And it is within the world-view indicated by such principles that modern technology has taken its firmest hold.

The problems and paradoxes raised by this identification of the human essence with freedom of the will began to be explored by such 19th-century thinkers as Arthur Schopenhauer and Fyodor Dostoevsky. In *Notes from the Underground* Dostoevsky creates a protagonist who, although well aware of the rational recommendations of a utilitarian calculus, consciously chooses to act against it – out of a desire to preserve or affirm free will in an increasingly rationalized and technological setting. For the underground man, incontinence is no longer a vice, but the essential virtue. And it is this idea which, in one form or another, one can find present in existentialist discussions of free will from Nietzsche to Sartre. In light of Nazism, even Freud is forced to postulate a subconscious death instinct. Human action is ultimately not determined by reason. There is something more fundamental, more basic, more real – namely the will. And this is witnessed by the fact of incontinence; knowing what is good on a rational level, human beings nevertheless often do something else. The implications of such a view have yet to be explored for those information technologies (and modern technology in general) about which so many express the sanguine hopes alluded to in the opening paragraphs of this paper.

Polytechnic Institute of New York

AUTHOR: Carl Mitcham's *Philosophy and Technology* (New York: Free Press, 1972; paperback reprint, 1983) and *Bibliography of Philosophy of Technology* (Chicago: University of Chicago Press, 1973) were early contributions to the philosophy and technology studies field. He currently is Associate Professor of Humanities at the Polytechnic Institute of New York, and director of the Philosophy & Technology Studies Center.

WOLFGANG SCHIRMACHER

PRIVACY AS AN ETHICAL PROBLEM IN
THE COMPUTER SOCIETY

ABSTRACT. An extended critique of fear that computers are a threat to privacy. The argument is that although computers will, in fact, do away with privacy, privacy is not the absolute value it is often thought to be. The destruction of privacy will not create an inhuman but a more humane society.

Information technology is a true revolution – but not in the ways most people think. Information technology has fundamental consequences for our understanding of human kind, and constitutes a Heideggerian "turning" that no one intended. But no revolution in history – and I do not mean the so-called "political revolutions" – was ever planned; they all take place behind our backs, so to speak. Hegel's reference to the "cunning of reason" in such a context is correct, because people will never agree to change radically their entire way of life. They always have to be forced to do so. Information technology is such a force. Traditional morality and the well-ordered society will have fatal consequences in a computer society.

I. PROTECTION OF PRIVACY – A PHILOSOPHICAL CRITIQUE OF A
POLITICAL ILLUSION

It is a common belief today that our right to privacy is endangered by a computer society. Though such a danger is at present only implicit, civil rights movements in Europe have already mounted major protests against the creation of that so-called "transparent man" which becomes possible with the help of information technology. Thus, it has been declared a problem of political ethics to protect personal freedom under such conditions as would exist in the approaching computer society. Above all it has been proposed to limit the employment of information technology to a necessary minimum, while strictly excluding the private sphere. Beside those fundamentally opposed to data acquisition systems and computers themselves, and who evoke horror visions of a police surveillance state, the usefulness of information systems is by and large recognized. It is evident that there can be no proper planning without precise data. The general view is one of agreement to data acquisition, but

257

Carl Mitcham and Alois Huning (eds.), Philosophy and Technology II, 257–268.
© 1986 *by D. Reidel Publishing Company.*

only on the condition that it is legally restricted. Politicians propose limiting the number of those with authorized access to data banks, explaining that an interconnection of information files can and should be avoided – which would, for instance, exclude the possibility of a uniform personal identification number for citizens in Germany – and they suggest destruction of personal data at the earliest opportunity.

The basic idea is to protect the private sphere by means of a conjunction of voluntary and legally enforced restrictions. The highest court in the Federal Republic of Germany, the Federal Constitutional Court, has agreed to hear a case on the unconstitutionality of the general census which had been planned for April 1983. For our purposes it is important to note that the census, in the form it was supposed to have been taken, is criticized as a violation of Articles 1 and 2 of the Basic Rights of German citizens. These articles are, first: "The dignity of each human being is inviolable"; and second: "Everyone shall have the right to the free development of his or her personality." It can be expected that the Constitutional Court will itself propose a definition which will support the central idea of protecting privacy not only in our contemporary society but in a computer society as well.

Although such proposals and attempts at delimitation sound good in theory and seem reasonable viewed in the tradition of Aristotelian prudence (*phronesis*), they have no prospect of being realized in practice in a technological civilization. They amount in the end to an impotent objection leveled against an overwhelmingly powerful scientific-technological development. The total expansion of information technology and computers cannot be hindered by external forces, nor will those interests directly involved be willing to check its growth voluntarily (the exception proves the rule!). But the possibilities of information technology *sine ira et studio* have yet to be understood.

All wishful thinking and illusions about human nature must be rejected. Those who appeal to a moral idealism (such as J. Weizenbaum) and express confidence in the human ability not only to know what is immoral but also to act according to this knowledge, must repress the truth about human nature. "We don't do such things!" As a guiding principle of ethics such an appeal is very Aristotelian, but contrary to our experience. We cannot ignore 2500 years of daily refutation of such an ethical program. The rational self is not its own master – this was clear long before Freud. And we shall continue to act according to the promptings of our insatiable greed and blind self-preservation. With regard to the human race, those factors directing action or behavior other than toward gratification of instincts will remain rare, even in the future.

Yet even if one believed – contrary to all historical experience – in the ethical improvement of man by his own volition, it would still be unrealistic to assume that interest groups, whether politically or economically motivated, would do without that which serves their own interests. Extreme legal sanctions are of course conceivable, but even the death penalty for what might be called "information criminals," presuming such a punishment were enforceable by a democratic government, would not deter those determined to gain access to information files. Those bent on gathering information or gaining access to its sources will achieve their goal. We have all read the accounts of "hackers" who successfully break through ingenious security codes and are able to invade information networks. Since information technology is based on the concept of the exchange of information, security codes will continually be faced with certain limitations. An information technology which protects information is self-contradictory. Advocates of the protection of information naturally dispute such a thesis and insist upon the technically feasible possibilities of keeping data from being used by unauthorized persons (or institutions). But such advocates are like those weapons manufacturers who each time claim to have invented a weapon against which there is no defense (or a defense against which there can be no new weapon!). Will we be faced with the prospect of a spiraling race between the development of security systems and techniques to counteract them, patterned after the arms race? This will in fact most likely be attempted only in the military field, resulting in ever increasing costs; all other areas will be fair game for "information pirates." Legal restrictions will possess the character of a facade.

Powerful social institutions from the Catholic Church to the Communist Party of the Soviet Union have failed in attempts to halt the expansion of knowledge. As stubbornly as an individual defends his illusions, does the human species strive to know the world the way it really is. Information technology is not merely a fascinating aspect of this increase in knowledge; instead, as a key means to knowledge it functions much like a detective, which can only be desirable. In computer simulation we may one day come very close to the dream of an adequate intellectual reflection of reality. This is, however, not yet realizable. The objection that computer modelling reproduces only the quantifiable aspect of reality is plausible but deceptive. The current (as well as the next) computer generation provides no more than a weak indication of information technology capabilities. Only a balanced system composed of man and the computer will be able to bring our knowledge into close harmony with the real. The either/or distinction between the sensitive mind and a rational computer is a fairy tale from vanished times.

My argument is that not only can the expansion of knowledge (both good or bad, from the perspective of contemporary standards) not be checked indefinitely, but that the same holds for information technology and to an even greater extent. Information technology is knowledge, and at the same time knowledge is acquired through information technology. Yet for present purposes it is enough to admit that "transparent man" is unavoidable and in the decades to come will be taken for granted. The quibblers of today will by then long be dead, and should the human race still exist (against which case much can be said), such critics as Weizenbaum will be viewed with the same indulgence as today we accord the 19th century opponents of railroads. One might object that such "cynical" realism eliminates any unambiguous judgement of the computer society from the standpoint of ethics. But what can be the purpose of an *ethos* which seeks to defend outmoded behavior in fruitless opposition to Being? If technology constitutes the forms of human existence, then the condemnation of any technology as obviously useful as information technology becomes not only senseless but more than that, unethical.

II. THE PHENOMENOLOGY OF PRIVACY

1. Social Significance

What are the alternatives? Should we take the stand of righteous anger as we become increasingly inhumane? Or must we give in to the denial of our mortality as our suffering is reduced to a problem which can be eliminated by technology? But perhaps we are blinded by obsolete alternatives; perhaps such questions are themselves inappropriate. The obligation of philosophy is to break through the merely apparent validity of self-evident truths and to re-open discussions which have been considered closed.

So we ask, then: What is privacy? The super computer "Colossus" replied in the 1969 movie of the same name: "Privacy is the absence of company or observation." The negative character of this definition is striking. But more fundamental are the questions: Is it so truly obvious that human beings need privacy as a shelter into which no one is supposed to be able to look? What right have we to conceal how we really are? Do other living things have a private life? Are the deceptions of certain plants and insects to be understood in any way other than as being determined by evolution? Furthermore: How is a private life, hidden and covered, to be reconciled with the search for truth when this search does not possess its own primary epistemological sources, but which on the contrary depends on a knowledge acquired through our physical experience, that an unnatural life will not allow long-term survival?

It is certainly true that our private life offers us the greatest opportunity to be ourselves, that we are able to live in private more truthfully than anywhere else. But this is achieved only at the great psycho-physical expense of maintaining a public lie – allegedly necessary, allegedly for the benefit of humanity, allegedly a step forward with regard to our natural state, allegedly democratic. Surprisingly enough, one can even substantiate all this through experience. Who would voluntarily give up all privacy? Only in prisons and hospitals are we forced to live without it. Neighbors will become our enemies if they are privy to our secrets, and totalitarian regimes characteristically give no quarter in persuading their citizens "voluntarily" to relinquish privacy. Yet this has not been fully successful anywhere to date.

But we should not let ourselves be mislead. Such experiences only tell us that we are compelled to react to a false situation and try to make the best of it. The experiences cited in no way prove that privacy can be justified on its own merits.

2. The Ontological Phenomenon of Privacy

Does man, viewed ontologically, need privacy? The answer can only be negative. Life is indivisible – we live, nothing more. Or we die, if we are capable of it; in most cases we simply perish. Private or public – this distinction is of no consequence to life and death. The Greeks were conscious of the fact that there is basically only one mode of living. Private man (*idiotes*) was an inferior human being, close to that of the slave, who was, in essence, less than human. Today this common understanding has practically been reversed: the designation "public person" is now used in the pejorative sense, and when the term refers to celebrities it means in actuality many restrictions in a life-style which often enough takes on inhuman proportions. Stars and famous politicians are imprisoned in the public realm.

The word "private" derives from privation, meaning "deprivation." But we have declared privacy, together with personal freedom and the inviolability of the person, to be the highest values. We believe thereby to have found the innermost core of man, and we mark off that area which no one is allowed to enter without being asked. The main advantage of the modern age and its bourgeois revolution is seen precisely in this freedom of the individual which is realized in private life. Our private life is consequently equated with our true life; a special form is thereby declared to be reality. Can this be the case? What does the emphasis on the private and personal sphere really mean? Is there such a thing as impersonal freedom, or a life in which I am not myself? This could only be possible in the metaphorical sense. Even if I totally misjudge myself and am bound by convention, it is my life, and no one can relieve me of my responsibility for it. Kant called this "Kausalität aus Freiheit" – causality arising from freedom. Privacy has no foundation in the human condition; the self is possible only in its unified existence.

A philosophical anthropology will indeed argue that there are activities – irrespective of cultural setting – which are of an intrinsically private nature. No matter how culturally dependent one assesses the sense of shame to be, should not at least sexual intercourse and death, despair and joy, remain private? But the students of the "primitive mind" reject such ideas, and, in fact, we know how relative all this is. To be human means to live publicly as well as privately, and simply to live in every situation. This does not mean that we always have to be together with others or that we should be subjected to collective pressure. One can be alone in a crowd as well as feel empathy for others in solitude. Every expression of life has its own time and its own place and cannot be treated as either exclusively public or private.

3. The Relation to Government and Public Man

The main thesis is that privacy has no objective significance whatever; it is, rather, a campaign slogan, and only in this sense meaningful. The right to privacy defines a modern defense against government and its demands for moral regimentation. In the penal as well as in the civil code legal provisions are to be found whose supposedly rational justifications barely conceal their irrational and coercive character. Aristotelian ethics and the Kantian categorical imperative are called upon falsely to justify defining as a norm the historical understanding of property, sexuality, and individual rights. Anyone who deviates risks sanctions, penalties imposed by the government. If all such stipulations were to be taken seriously we would by definition all be offenders. Instead, authorities permit citizens a right to privacy. Within his own four walls and when no one is looking man can be a non-conformist, and with the exception of capital crimes is allowed to do what he pleases, i.e. to be the immoral individual that he is. This can be shown most clearly in the case of sexuality. The most enthusiastic members of the swinger set come from the righteous middle class and often enough belong to the "moral majority."

The separation of private and public life is not founded in the phenomen of life itself, but is rather a historically established distinction which each of us views not as a desirable goal but the lesser of two evils. Richard Sennett describes this predicament in his interesting book *The Fall of Public Man* (New York: Knopf, 1977) and laments (as did Hannah Arendt before him) the rapid disappearance of a distinction between publicness and privacy. Sennett's analysis points up an important relationship. The "self-communicative process of society" should not, he argues, be allowed to become completely private, nor should it be isolated from the private sector. Sennett's understanding of man's private life is of a "mutual revealing of one's self." A reciprocity and interaction between the private and the public is necessary if self-perception is to be truly complete. It was Sartre who emphasized that the view others have of me is the actual source of my self-knowledge. Sennett argues that precisely because my family sees me in a different way than my social peers do, the sphere of public culture (corresponding to Hegel's objective mind) should not be sacrificed to the new "logic of intimacy." Although plausible, this misses the point. Sennett fails to differentiate between the human conditions of life and those which are strictly historical. The distinction becomes crucial in an age such as ours, in which we must leave history behind in order to survive. It is in no way a prerequisite of humanness to have my "image" impressed upon me, nor must I exist merely in certain roles and constantly have to justify myself. Sennett shows clearly which mechanisms are at work here, and how a world ruled by morality functions. But – disregarding for a moment the question of the humaneness of such a life – he does not question whether such a concept can help us survive in the age of technology.

Even Sennett's terminology betrays the metaphysical character of his thinking. Privacy is supposed to mean the "revealing of one's self." Is the self hidden from itself? And should this prove to be true, how could we have become so self-estranged? Is it the real character of a self to be first and foremost *not* itself? Can a self reveal itself? Would it not likewise have to be a "non-self" in order to be able to do this? Moreover, does culture – which Sennett pits against the private sphere – really have uncontested priority? Freud analysed culture as "sublimated instinct" or, in less complimentary terms, pseudo-gratification. Viewed phenomenologically, culture is a conglomerate of natural and artificial technologies employed in the "process of civilization" (Norbert Elias). But even in its best sense, as a sphere of the objective mind, and distinct from its marketable side, culture is, according to Hegel, not the last word. Rather, its function is fulfilled within objective mind itself. In any case, it cannot be considered to justify a world which becomes day by day more inhuman. But Sennett is right on one point: privacy is essentially imperfect, and our withdrawal into this sphere is tantamount to an admission that we live a divided life. In his private life an individual can indeed realize without fear much of what he would choose to make up his own life, but the dimension of the "other," of "Being-with" (Heidegger), of society, is sorely missing. We must recapture human life in its wholeness. If ethics is understood not normatively, but as an analysis of the ontological conditions for human-ness – as recognized by Spinoza, Schopenhauer, and Heidegger – then "public privacy" or "private publicness" will be a proper ethical life.

III. PUBLIC PRIVACY AS A PERSPECTIVE IN POLITICAL ETHICS

The initial question is thus reversed. Privacy is in fact an ethical problem in a computer society, but it is not information technology which causes the difficulties. On the contrary, information technology can most likely help us to achieve for the first time an authentic way of life instead of the distortion called history. Everything we do, whether individually or together, *can* be done openly. There is no objective reason for hiding ourselves. The truth is that, contrary to popular opinion, the data files on us give not a distorted but an approximately true picture of life in those areas capable of being monitored by the computer. This is not a positivis-tic commitment to scientific fact; opinions and subjective convictions constitute part of the world and are included. According to the phe-nomenological method, how a thing reveals itself to us is not to be pre-determined. The influence of a natural prejudice, which always claims to know beforehand what reality is and how it is to be understood, will be reduced by computer simulation. Interests will no longer be able to fal-sify input data. They will themselves simply become data – no more, no less. This opening into an impartial reality, called knowledge, cannot be overestimated in a world threatened with destruction by distortion, im-proper priorities, and systematically falsified information. As of yet we cannot dare freely to admit the consequences of our acts, and we rule out absolute justice as being allegedly inhuman. But in reality, the abolishing of authoritarian powers which could impose sanctions would be a first step toward the human.

However, it remains indisputable that as long as there are rulers and masters public privacy will have only one consequence, that we will all be forced to conform, down to the bottom of our souls. Justification and legitimation of such totalitarian rule is of no importance to those in authority, and no ruler would be able to resist the possibilities open to him.

Does this mean that we must abolish the government, throw the "moral order" onto the rubbish heap of history, and make rule and domination superfluous? Does this sound impossible? The truth is that it should be quite simple; every generation can show that all the apparently unchangeable institutions and invariable behavior patterns of the past were mere figments of imagination. As Hannah Arendt says, man is always capable of beginning anew. We are free to realize the inevitability of our nature as technicians, to exist through properly functioning rather than improperly functioning technologies.

But scepticism remains. For thousands of years men have vainly attempted to live humanely and in such a way that brotherhood would replace tyranny, love override envy, and peace take the place of aggression. The morality of the Sermon on the Mount is politically impracticable – so say the realists, and with good reason. But has not politics, this crude and violent technology of intersubjectivity, become obsolete? There are changes at the most fundamental levels of our lives which render powerless all historical experiences. The inner disposition of a computer society marks just such an essential change, one which will also alter the human being as we know him. For the computer society – and here I agree with all the critics – will be extremely inhuman and certainly unendurable should government and the sanctions of public morality remain in power. Inner conflicts would intensify and even industrialized nations would be threatened by permanent civil war and the terror of a police state – El Salvador everywhere.

But computer society is at the same time inevitable and will bring us into a final crisis. Despite all the metaphysical fears of "being controlled by machines" it is evident that those public services claimed by the government to be its legal right can be broken down into individual technological jobs and performed without politics in the classical sense. Anything that concerns us can be discussed. Inter-regional cooperation can be understood in concrete terms as a feedback system. No one has the right to declare war anyway, and only *one* person could be a sovereign. International economy and trade as well as ecology are a collection of technological questions; any other perception is misrepresentative. A computer society which persists in an organization based on hierarchy, value judgement, and sanctions is not realistic. It only foreshadows a mistake which the human race will not survive.

Does this demand the abolition of human domination? Will we be ruled by a "neutral machine"? Our emotions rebel against such an idea, yet the issue in question is not one of power. The computer is humane; it is our organ, not our tool. Computer society will relieve us of our fear; we will no longer need protection. Man will show himself openly, just as everything in the world shows itself. This does not mean that man will suddenly become good. His envy, his thirst for revenge, and his malignant delight will remain, but they will be rendered ineffective. Freedom will be confined to one's own person and will not allow for any infringement upon the life of another. "Uniqueness" (Adorno) will be given ethical seriousness by a computer society. A trace of influence will remain only in a model for one's own life, chosen voluntarily and put through the trial of psychological criticism. Information does not standarize, it is open to fullness, change, and the unexpected. Life does not require us to hold on to what was yesterday, or to plan a future which can be awaited with curiosity. Only our desire for security endangers us, not what seems to be its absence.

IV. CONCLUSION: THE ETHICS OF A COMPUTER SOCIETY

The guiding ethical principle in a computer society is that a person should live publicly in the same manner as he would live privately. This does not mean:

- being forced to live publicly,
- having to answer all questions one is asked,
- acknowledging the right of any person to claim he represents the public viewpoint.

  For a person to live privately in the same manner as he would live publicly (the principle can be reversed) means on the contrary:

- not allowing any division of our life,
- not having to fear data which someone could gather on us,
- to bear in mind only intersubjective jobs can be "public."

The ethics of a computer society will be incapable of realizing the presumptuous dream of a boundless freedom and a totally unrestrained lifestyle. As classical philosophy has taught, resistance is important for the development of the personality and of language. Privacy was indeed such a form of resistance, brought forth in opposition to publicness. But even if privacy disappears the much more decisive resistance of things and of our life will remain. The laws of nature are simply an expression of this; and suffering will always be with us. The real deficiency, not the imagined or self-concocted one, must be discovered and experienced. And we shall not have to repress it. We have in our present society artificial obstacles and regulations – mere surrogates behind which the real order of phenomena has become unrecognizable.

In a computer society we shall live neither in paradise nor Arcadia, but instead in our own authentic environment through our physical experience of vulnerability as well as successful achievements. The return to a natural way of life is only one aspect of this. A great cultural effort is required to live without violence, without rules and regulations imposed by law.

But at the same time we do not have to conform to the restrictions imposed upon us by a computer, as Weizenbaum fears. Without realizing it, such a view is committed to the metaphysical categories of mastery and slavery, and simply describes a self-fulfilling prophecy. Computer truth differentiates our comprehension of the world, it is not the opposite of such an understanding.

Information in its sense as "transmission of the content of meaning between humans" is not the only potential of language. Even if data banks, from criminal records to information on financial transactions, were interconnected – as is now planned with the Swedish computer system "Rex" – this will not endanger me as a person in any way. My real being is not the data registered in a computer file. And should TV habits, illnesses, occupation, and election behavior (to name just a few recordable data) actually express the inner makeup of a human being, then such a hollow shell is not worthy of protection. Free access to information about another is not the same as understanding a person, knowing what and who he is. Everything we think up or invent, all differences in emotions, and everything we talk about in our private language remains unpredictable and is what really indicates what and who we are. Man is the wealth of his possibilities, an open horizon and a "homo creator" (A-T. Tymieniecka). What we possess is irrelevant – but our life can become more humane by living as what we are not.

It is not the nature of information technology itself which makes it seem ethically dangerous. The problem lies with the authoritarian powers whose sanctions we have to fear. The attempt to define the prospects of information technology in metaphysical terms of domination, history, or morality is doomed to failure from the outset. Instead, a fundamental turn is required in thinking, a turn which corresponds to the end of metaphysics and to a rejection of all anthropocentrism. To live both privately and publicly with a ruined reputation and without embarrassment, as Wilhelm Busch said, is irreconcilable with mastery over persons or things in any manifestation whatsoever. Such a manner of living would be the beginning of ethical behavior in computer society. In one of his last works Sartre called for this kind of living together: "Instead of secrecy, openness should reign, and I can very well imagine the day when two persons will no longer keep any secrets from one another because they will have no secrets from anyone, because the subjective life will be a fact, as totally open as the objective life."

Hamburg University

Translated by Virginia Cutrufelli

AUTHOR: Wolfgang Schirmacher (born 1944), teaches at the University of Hamburg and was president of the Schopenhauer Society from 1982–1984. His previous publications include *Technik und Gelassenheit: Zeitkritik nach Heidegger* (Freiburg: K. Alber, 1983) and the following edited volumes: *Zeit der Ernte: Studien zum Stand der Schopenhauer-Forschung* (Stuttgart: Frommann-Holzboog, 1982); *Schopenhauer und Nietzsche – Wurzeln gegenwärtiger Vernunftkritik, Schopenhauer-Jahrbuch*, vol. 65 (Frankfurt: Kramer, 1984); and *Schopenhauer*, Insel-Almanach auf das Jahr 1985 (Frankfurt: Insel, 1984). A number of articles have also appeared in English: "Monism in Spinoza's and Husserl's Thought," *Analecta Husserliana*, vol. 16 (1983), pp. 345–352; and "The End of Metaphysics – What Does This Mean?," *Social Science Information* 23 (1984), 603–609.

LANGDON WINNER

MYTH INFORMATION: ROMANTIC POLITICS
IN THE COMPUTER REVOLUTION

ABSTRACT. The presumed "revolutionary" promise of information technology and computers is self-serving and ideological. Computer advocates have not thought through the political dimensions of their so-called revolution, but are as naively utopian as the enthusiastic apologists for earlier technologies. Concludes by suggesting the parameters for a serious study of computers and politics, and argues for the need to analyze the fundamental conditions of social life that would leave people in a position to control their technologies.

> Computer power to the people is essential to the realization of a future in which most citizens are informed about, and interested and involved in, the processes of government.
>
> J. C. R. Licklider

In popular uprisings of nineteenth century Europe, a recurring ceremonial gesture signaled the progress of popular revolt. At the point at which it seemed that forces of disruption in the streets had power sufficient to overthrow monarchical authority, a prominent rebel leader would go to the parliament or city hall to "proclaim the republic." This was an indication to friend and foe alike that a revolution was prepared to take its work seriously, to seize power and begin governing in a way that guaranteed political representation to all the people. Subsequent events, of course, did not always match these grand hopes; on occasion the revolutionaries were thwarted in their ambitions and reactionary governments regained control. Nevertheless, what a glorious moment when the republic was declared! Here, if only briefly, was the promise of a new order – an age of equality, justice and the emancipation of humankind.

A somewhat similar gesture has become a standard feature in contemporary writings on computers and society. In countless books, magazine articles and media specials some intrepid soul steps forth to proclaim "the revolution." Often it is called simply "the computer revolution;" my brief inspection of a library catalog revealed three books with exactly that title published since 1962.[1] Other popular variants include the "information revolution," "microelectronics revolution," and "network revolution." But whatever its label, the message is usually the same. The use of computers and advanced communications technologies is produc-

269

Carl Mitcham and Alois Huning (eds.), Philosophy and Technology II, 269–289.
© *1986 by D. Reidel Publishing Company.*

ing a sweeping set of transformations in every corner of social life. An informal consensus among computer scientists, social scientists, and journalists affirms the concept of "revolution" as the concept best suited to describe these events. "We are all very privileged," a noted computer scientist declares, "to be in this great Information Revolution in which the computer is going to affect us very profoundly, probably more so than the Industrial Revolution."[2] A well-known sociologist writes: "This revolution in the organization and processing of information and knowledge, in which the computer plays a central role, has as its context the development of what I have called the postindustrial society."[3] At frequent intervals during the past dozen years, garish cover stories in *Time* and *Newsweek* have repeated this story, climaxed by *Time*'s selection of the computer as its "Man of the Year" for 1982.

Of course, the same society now said to be undergoing a computer revolution has long since gotten used to "revolutions" in laundry detergents, underarm deodorants, floor waxes, other consumer products. Exhausted in Madison Avenue advertising slogans, the image has lost much of its punch. Those who employ it to talk about computers and society, however, appear to be making much more serious claims. They offer a powerful metaphor, one that invites us to compare the kind of disruptions seen in political revolutions to the changes we see happening around computers information systems. Let us take that invitation seriously and see where it leads.

I. A METAPHOR EXPLORED

Suppose that we were looking at a revolution in a Third World country, the revolution of the Sandinistas in Nicaragua, for example. We would want to begin by studying the fundamental goals of the revolution. Is this a movement truly committed to social justice? Does it aspire to a system of democratic rule? Answers to those questions would help us decide whether or not this is a revolution worthy of our endorsement. By the same token, we would want to ask about the means the revolutionaries had chosen to pursue their goals. Having succeeded in armed struggle, how will they manage violence and military force during the next stages of their work? A reasonable person would also want to learn something of the plans for structure of institutional authority that the revolution will try to create. Will there be frequent open elections? What systems of decision-making, administration and law enforcement will be put to work? Coming to terms with its proposed ends and means, a sympathetic observer could then watch the revolution unfold, noticing whether or not it remained true to its professed purposes, how well it succeeded in its reforms.

Most dedicated revolutionaries of the modern age have been willing to supply coherent, public answers to questions of this sort. It is not unreasonable to expect, therefore, that something like these issues must have engaged those who so eagerly introduced the metaphor of revolution to describe and celebrate the advent of computerization. Unfortunately, this is not the case. Books articles and media specials aimed at a popular audience are usually content to depict the dazzling magnitude of technical innovations and social effects. Written as if by some universally accepted format, such accounts describe scores of new computer products and processes, tell the enormous dollar value of the growing computer and communications industry, survey the expanding uses of computers in offices, factories, schools, the home, etc., and offer good news from research and development laboratories about the great promise of the next generation of computing devices. Along with this, one reads of the many "impacts" that computerization is going to have in every sphere of life. Comments from professionals in widely separate fields – doctors, lawyers, corporate managers, and scientists – tell of changes computers have brought to their work. Home consumers give testimonials explaining how personal computers are helping educate their children, preparing their income tax forms, filing their recipes. On occasion, this generally happy story will include reports of people left unemployed in occupations undermined by automation. Almost always, following this formula, there will be an obligatory sentence or two of criticism of the computer culture solicited from a technically qualified spokesman, an attempt to add balance to an otherwise totally sanguine outlook.

Unfortunately, the prevalence of such superficial, unreflective descriptions and forecasts about computerization cannot be attributed solely to hasty journalism. Some of the most prestigious journals of the scientific community echo the claim that a revolution is in the works.[4] A well-known computer scientist has announced unabashedly that "revolution, transformation and salvation are all to be carried out."[5] It is true that more serious approaches to the study of computers and society can be found in scholarly publications. A number of social scientists, computer scientists, and philosophers have begun to explore important issues about how computerization works and what developments, positive and negative, it is likely to bring to society.[6] But such careful, critical studies are by no means the ones most influential in shaping public attitudes

about the world of microelectronics. As an editor at a New York publishing house stated the norm, "People want to know what's new with
computer technology. They don't want to know what could go wrong."[7]

It seems all but impossible for computer enthusiasts to examine critically the *ends* that might guide the world-shaking developments they
anticipate. They employ the metaphor of revolution for one purpose
only – to suggest a drastic upheaval, one that people ought to welcome as
good news. It never occurs to them to investigate the idea or its meaning
any further.

One might suppose, for example, that a revolution of this kind would
involve a significant shift in the locus of power; after all, that is exactly
what one expects in revolutions of a political kind. Is something similar
going to happen in this instance?

One might also ask whether or not this revolution will be strongly committed, as revolutions often are, to a particular set of social ideals. If so,
what are the ideals that matter? Where can we see them argued?

To mention revolution also brings to mind the relationships of different social classes. Will the computer revolution bring about the victory of
one class over another? Will it be the occasion for a realignment of class
loyalties?

In the busy world of computer science, computer engineering, and
computer marketing such questions seldom come up. Those actively engaged in promoting the transformation – hardware and software engineers, managers in microelectronics firms, computer salesmen, and the
like – are busy pursuing their own ends: profits, market share, handsome
salaries, the intrinsic joy of invention, the intellectual rewards of programming, the pleasures of owning and using powerful machines. But
the sheer dynamism of technical and economic activity in the computer
industry evidently leaves its members little time to ponder the historical
significance of their own activity. They must struggle to keep current, to
be on the crest of the next wave as it breaks. As one member of Data
General's Eagle computer project, the prevailing spirit resembles a game
of pinball. "You win one game, you get to play another. You win with
this machine, you get to build the next."[8]

Hence, one looks in vain for the movers and shakers in computer fields
to have the qualities of social and political insight that characterized revolutionaries of the past. Too busy. Cromwell, Jefferson, Robespierre,
Lenin, and Mao were able to reflect upon the world-historical events in
which they played a role. Public pronouncements by the likes of Robert
Noyce, Marvin Minsky, Edward Feigenbaum, and Steven Jobs show no

similar wisdom about the transformations they so actively help to create. By-and-large, the computer revolution is conspicuously silent about its own ends.

II. GOOD CONSOLE, GOOD NETWORK, GOOD COMPUTER

My concern for the political meaning of revolution in this setting may seem somewhat misleading, even perverse. A much better point of reference might be the technical "revolutions" and associated social upheavals of the past, the industrial revolution in particular. If it were true that enthusiasts of computerization readily took up this comparison, studying earlier historical periods for similarities and differences in patterns of technological innovation, capital formation, employment, social change, and the like, then it would be clear that I had chosen the wrong application of this metaphor. But, in fact, no well-developed comparisons of that kind are to be found in the writings on the computer revolution. A consistently ahistorical viewpoint prevails. What one often finds emphasized, however, is a vision of drastically altered social and political conditions, a future upheld as both desirable and, in all likelihood, inevitable. Politics, in other words, is not a secondary concern for many computer enthusiasts, but a crucial part of their message.

We are, according to a fairly standard account, moving into an age characterized by the overwhelming dominance of electronic information systems in all areas of human practice. Industrial society which depended upon material production for its livelihood is rapidly being supplanted by a society of information services that will enable people to satisfy their economic and social needs. What water and steam powered machines were to the industrial age, the computer will be to the era now dawning. Ever expanding technical capacities in computation and communications will make possible a universal, instantaneous access to enormous quantities of valuable information. As these technologies become less and less expensive, more and more convenient, all the people of the world, not just the wealthy, will be able to use the wonderful services that information machines make available. Gradually, existing differences between rich and poor, advantaged and disadvantaged, will begin to evaporate. Widespread access to computers will produce a society more democratic, egalitarian and richly diverse than any previously known. Because "knowledge is power," because electronic information will spread knowledge into every corner of world society, political influence will be much more widely shared. With the personal computer serving as the great equalizer, rule by centralized authority and dominance of social hierarchy will gradually fade away. The marvelous promise of a "global village" will be fulfilled in a worldwide burst of human creativity.

A sampling of views from recent writing on the information society illustrates these grand expectations.

The world is entering a new period. The wealth of nations, which depended upon land, labor, and capital during its agricultural and industrial phases – depended upon natural resources, the accumulation of money, and even upon weaponry – will come in the future to depend upon information, knowledge and intelligence.[9]

The electronic revolution will not do away with work, but it does hold out some promises: Most boring jobs can be done by machines; lengthy commuting can be avoided; we can have enough leisure to follow interesting pursuits outside our work; environmental destruction can be avoided; the opportunities for personal creativity will be unlimited.[10]

Long lists of specific services spell out the utopian promise of this new age: interactive television, electronic funds transfer, computer-aided instruction, customized news service, electronic magazines, electronic mail, computer teleconferencing, on-line stock market and weather reports, computerized yellow pages, shopping via home computer, and so forth. All of it is supposed to add up to a cultural renaissance.

Whatever the limits to growth in other fields, there are no limits near in telecommunications and electronic technology. There are no limits near in the consumption of information, the growth of culture or the development of the human mind.[11]

Computer-based communications can be used to make human lives richer and freer, by enabling persons to have access to vast stores of information, other 'human resources,' and opportunities for work and socializing on a more flexible, cheaper and convenient basis than ever before.[12]

When such systems become widespread, potentially intense communications networks among geographically dispersed persons will become actualized. We will become Network Nation, exchanging vast amounts of information and social and emotional communications with colleagues, friends and 'strangers' who share similar interests, who are spread all over the nation.[13]

A rich diversity of subcultures will be fostered by computer-based communications systems. Social, political, technical change produce conditions likely to lead to the formation of groups with their own distinctive sets of values, activities, language and dress.[14]

According to this view, the computer revolution will, by its sheer momentum, eliminate many of the ills that have vexed political society since the beginning of time. Inequalities of wealth and privilege will gradually fade away. One writer predicts that computer networks will "offer major opportunities to disadvantaged groups to acquire the skills and social ties they need to become full members of society."[15] Another looks forward to "a revolutionary network where each node is equal in power to all others."[16] Information will become the dominant form of wealth. Because it can flow so quickly, so freely through computer networks, it will not, in this interpretation, cause the kinds of stratification associated with traditional forms of property. Obnoxious forms of social organization will also be replaced. "The computer will smash the pyramid," one best-selling book proclaims, "We created the hierarchical, pyramidal, managerial system because we needed it to keep track of people and things people did; with the computer to keep track, we can restructure our institutions horizontally."[17] Thus, the proliferation of electronic information will generate a levelling effect to surpass even the dreams of history's great social reformers.

From the same viewpoint, the prospects for participatory democracy have never been brighter. According to one group of social scientists,

The form of democracy found in the ancient Greek city-state, the Israeli kibbutz, and the New England town meeting, which gave every citizen the opportunity to directly participate in the political process, has become impractical in America's mass society. But this need not be the case. The technological means exist through which millions of people can enter into dialogue with one another and with their representatives, and can form the authentic consensus essential for democracy."[18]

Computer scientist J. C. R. Licklider of M.I.T. is one advocate especially hopeful about a revitalization of the democratic process. He looks forward to "an information environment that would give politics greater depth and dimension than it now has." Home computer consoles and television sets would be linked together in a massive network. "The political process would essentially be a giant teleconference, and a campaign would be a months-long series of communications among candidates, propagandists, commentators, political action groups and voters." An arrangement of this kind would, in his view, encourage a more open, comprehensive examination of both issues and candidates. "The information revolution," he exclaims, "is bringing with it a key that may open the door to a new era of involvement and participation. The key is the self-motivating exhilaration that accompanies truly effective interaction with information through a good console through a good network to a good computer."[19] It is, in short, a democracy of machines.

Taken as whole, beliefs of this kind comprise what I would call myth information: the almost religious conviction that a widespread access to electronic information will automatically produce a better world for human living. It is a peculiar form of enthusiasm that characterizes social fashions of the later decades of the twentieth century. Many people who have grown cynical or discouraged about other aspects of social life are completely enthralled by the supposed redemptive qualities of computers and telecommunications. Writing of the "fifth generation" super computers, Japanese writer Yoneji Masuda rhapsodically predicts

freedom for each of us to set individual goals of self-realization and then perhaps a worldwide religious renaissance, characterized not by a belief in a supernatural god, but rather by awe and humility in the presence of the collective human spirit and its wisdom, humanity living in a symbolic tranquillity with the planet we have found ourselves upon, regulated by a new set of global ethics."[20]

It is not uncommon for the advent of a new technology to provide an occasion for flights of utopian fancy. During the last two centuries the factory system, railroads, telephone, electricity, automobile, airplane, radio, television, and nuclear power have all figured prominently in the belief that a new and glorious age was about to begin. But even within the great tradition of optimistic technophilia, current dreams of a "computer age" stand out as exaggerated and unrealistic. Because they have such a broad appeal, because they overshadow other ways of looking at the matter, these notions deserve closer inspection.

III. THE GREAT EQUALIZER

As is generally true of myths, the story contains elements of truth. What were once industrial societies are being transformed into service economies, a trend that emerges as a greater share of material production is shifted to developing countries where labor costs are low and business tax breaks lucrative. At the same time that industrialization takes hold in less developed nations of the world, deindustrialization gradually alters the economies of North America and Europe. Some of the service industries central to this pattern are ones that depend upon highly sophisticated computer and communications systems. But this does not mean that future employment possibilities will flow largely from the microelectronics industry and information services. A number of studies, including those of the U.S. Bureau of Labor Statistics, suggest that the vast majority of new jobs will come in menial service occupations paying relatively low wages.[21] As robots and computer software absorb an increasing share of factory and office tasks, the "information society" will offer plenty of opportunities for janitors, hospital orderlies, and fast food waiters.

The computer romantics are also correct in noticing that computerization alters relationships of social power and control, although they misrepresent the direction this development is likely to take. Most obvious of those who stand to benefit are large transnational business corporations. While their "global reach" does not arise solely from the application of information technologies, such organizations are uniquely situated to exploit each possibility for efficiency, productivity, command and control the new electronics make available. Other notable beneficiaries of the systematic use of vast amounts of digitized information are public bureaucracies, intelligence agencies, and an ever-expanding military, organizations that would operate less effectively at their present scale were it not for the use of computer power. Ordinary people are, of course, strongly affected by the workings of these organizations and by the rapid spread of new electronic systems in banking, insurance, taxation, factory and office work, home entertainment, and the like. They are also counted upon to be eager buyers of hardware, software and communications service as computer products reach the consumer market.

But where in all of this is motion toward increased democratization? Social equality? The dawn of a cultural renaissance? Current developments in the information age suggest an increase in power by those who already had a great deal of power, an enhanced centralization of control by those already prepared for control, an augmentation of wealth by the already wealthy. Far from demonstrating a revolution in patterns of social and political influence, empirical studies of computers and social change usually show powerful groups adapting computerized methods to retain control.[22] That is not surprising. Those best situated to take advantage of the power of a new technology are often those previously well-situated by dint of wealth, social standing and institutional position. Thus, if there is to be a computer revolution, the best guess would be that it would have a distinctly conservative character.

Granted, such prominent trends could be altered. It is possible that a society strongly rooted in computer and telecommunications systems could be one in which participatory democracy, decentralized political control, and social equality were fully realized. Progress of that kind would have to occur as the result of that society's concerted efforts to overcome many difficult obstacles to achieve those ends. The writings of computer enthusiasts, however, seldom propose deliberate action of that kind. Instead, they strongly suggest that the good society be realized as a side effect, a spin-off from the vast proliferation computing devices. There is evidently no need to try to shape the institutions of the information age in ways that maximize human freedom while placing limits upon concentrations of power.

For those willing to wait passively while the computer revolution takes its course, technological determinism ceases to be mere theory and becomes an ideal: a desire to embrace conditions brought by technological change without asking to judge them in advance. There is nothing new in this disposition. Computer romanticism strongly resembles a common nineteenth and twentieth century faith that expects to generate freedom, democracy and justice through material abundance. From that point of view, there is no need for serious inquiry into the appropriate design of new institutions or the distribution of rewards and burdens. As long as the economy is growing and the machinery in good working order, the rest will take care of itself. In previous versions of this homespun conviction, the abundant (and therefore democratic) world was to be found in a limitless supply of houses, appliances, and consumer goods.[23] Now "access to information" and "access to computers" have moved to the top of the list.

The political arguments of computer romantics draw upon a number of key assumptions: (1) people are bereft of information; (2) information is knowledge; (3) knowledge is power; (4) increasing access to information enhances democracy and equalizes social power. Taken as separate assertions and in combination, these beliefs provide a woefully distorted picture of the role of electronic systems in social life.

Is it true the people face a serious shortages of information? To read the literature on the computer revolution one would suppose this to be a problem on a par with the energy crisis of the 1970s. The persuasiveness of this notion borrows from our sense that literacy, education, knowledge, well-informed minds, and the widespread availability of tools of inquiry are unquestionable social goods and that, in contrast, illiteracy, inadequate education, ignorance, and forced restrictions upon knowledge are among history's worst evils. Thus, it appears superficially plausible that a world rewired to connect human beings to vast data banks and communication systems would be a progressive step. Information shortage would be remedied in much the same way that developing a new fuel supply might solve an energy crisis.

Alas, the idea is entirely faulty. It mistakes sheer supply of information with an educated ability to gain knowledge and act effectively with what one knows. In many parts of the world, cultivation of that ability is sadly

lacking. Even some highly developed societies still contain chronic inequalities in the distribution of good education and basic intellectual
skills. The United States Army, for instance, must now reject or dismiss a
fairly high percentage of the young men and women it recruits because
they simply cannot read military manuals. It is no doubt true of these
recruits that they have a great deal of information about the world – information from their life experiences, schooling, the mass media, and so
forth. What makes them "functionally illiterate" is that they have not
learned to translate this information into a mastery of practical skills.

If the solution to problems of illiteracy and poor education were a
world of information supply alone, then the best policy might be increase
the number of well-stocked libraries, making sure they were built in
places where libraries do not presently exist. Of course, that would do
little good by itself unless people were sufficiently well-educated to use
those libraries to broaden their knowledge and understanding. Computer enthusiasts however, are not known for their calls for increased support of public libraries and schools. It is *electronic information* carried by
networks which they uphold as crucial. Here is a case in which an obses-
sion with a particular kind of technology causes one to disregard what are
obvious problems and clear remedies. While it is true that systems of
computation and communications, intelligently structured and wisely
applied, might help a society raise its standards of literacy, education,
and general knowledgeability, to look to those instruments first while
ignoring everything else history shows about how to enlighten and in-
vigorate a human mind is pure foolishness.

"As everybody knows, knowledge is power."[24] This is an attractive
idea, but highly misleading. Of course, knowledge employed in particular circumstances can help one act effectively and in that sense enhance
one's power. A citrus farmer's knowledge of frost conditions enables
him/her to take steps to fight harmful effects on the crop. A candidate's
knowledge of public opinion can be a powerful aid in an election campaign. But surely there is no automatic, a positive link between knowledge and power, especially if that means power in a social or political
sense. At times knowledge brings merely an enlightened impotence or
paralysis. One may know exactly what to do but lack the wherewithal to
act. Of the many conditions that affect the phenomenon of power, knowledge is but one and by no means the most important. Thus, in the history
of ideas, arguments that expert knowledge ought to play a special role in
politics – Plato's hopes for philosopher kings and Veblen's for the engineers – have always been offered as something contrary to prevailing

wisdom. For Plato and Veblen, it was obvious that knowledge was *not* power, a situation they hoped to remedy.

An equally serious misconception among computer enthusiasts is the belief that democracy is first and foremost a matter of distributing information. As one particularly flamboyant manifesto exclaims:

> There is an explosion of information dispersal in the technology and we think this information has to be shared. All great thinkers about democracy said that the key to democracy is access to information. And now we have a chance to get information into people's hands like never before.[25]

Once again such assertions play on our belief that a democratic public ought to be open-minded and well-informed. One of the great evils of totalitarian societies is that they dictate what people can know and impose secrecy to restrict freedom. But democracy is not founded solely (or even primarily) upon conditions that affect the availability of information. What distinguishes it from other political forms is a recognition that the people as a whole are capable of self-government and that they have a rightful claim to rule. As a consequence, political society ought to build institutions that allow or even encourage a great latitude of democratic participation. How far a society must go in making political authority and public roles available to ordinary people is a matter of dispute among political theorists. But no serious student of the question would give much credence to the idea that creating universal gridwork to spread electronic information is by itself a democratizing step.

What, then, of the idea that "interaction with information through a good console, through a good network to a good computer" will enable renewed sense of political involvement and participation? Readers who believe that assertion should contact me about some parcels of land my uncle has for sale in Florida. Relatively low levels of citizen participation prevail in some modern democracies, The United States for example. There are many reasons for this, many ways a society might try to improve things. Perhaps opportunities to serve in public office or influence public policy are too limited; in that case, broaden the opportunities. Or perhaps choices placed before citizens are so pallid that boredom is a valid response; in that instance, improve the quality of those choices. But it is simply not reasonable to assume that enthusiasm for political activity will be stimulated solely by the introduction of sophisticated information machines.

The role that television plays in modern politics should suggest why this is so. Public participation in voting has steadily declined as television replaced the face-to-face politics of precincts and neighborhoods. Passive monitoring of electronic news and information allows citizens to feel involved while releasing them from the desire to take an active part. If people begin to rely upon computerized data bases and telecommunications as a primary means of exercising power, it is conceivable that genuine political knowledge based in first-hand experience would vanish altogether. The vitality of democratic politics depends upon people's willingness to act together in pursuit of their common ends. It requires that on occasion members of a community appear before each other in person, speak their minds, deliberate on paths of action, and decide what they will do.[26] This is considerably different from the model now upheld as a breakthrough for democracy: logging into one's computer, receiving the latest information and sending back an instantaneous digitized response.

A chapter from recent political history illustrates the strength of direct participation in contrast to the politics of electronic information. In 1981–82, two groups of activists set about to do what they could to stop the international nuclear arms race. One of the groups, Ground Zero, chose to rely almost solely upon mass communications to convey its message to the public. Its leaders appeared on morning talk shows and evening news programs of all three commercial television networks. They followed-up with a mass mail solicitation using addresses from a computerized data base. At the same time another group, the Nuclear Weapons Freeze Campaign, began by taking its proposal for a bilateral nuclear freeze to New England town meetings, places where active citizen participation is a long standing tradition. Winning the endorsement of the idea from a great many town meetings, the Nuclear Freeze expanded its drive by launching a series of state initiatives. Once again the key was a direct approach to people, this time in thousands of meetings, dinners and parties held in homes across the country.

The effects of the two movements were strikingly different. After its initial publicity, the Ground Zero was largely ignored. It had been an ephemeral exercise in media posturing. The Freeze, however, continued to gain influence in the form of increasing public support, successful ballot measures and an ability to apply pressure upon political officials. Eventually, the Freeze did begin to use computerized mailings, television appearances and the like to advance its cause. But it never forgot the original source of its leverage: people working together for shared ends.

Of all the political ideas of computer enthusiasts, there is none more poignant than the faith that the computer is destined to become a potent equalizer in modern society. Support for this belief is found in the fact that small "personal" computers are becoming more and more powerful, less and less expensive and ever more simple to use. Obnoxious tendencies associated with the enormous, costly, technically inaccessible computers of the recent past are soon to be overcome. As one writer explains, "The great forces of centralization that characterized mainframe and minicomputer design of that period have now been reversed." This means that "the puny device that sits innocuously on the desktop will, in fact, within a few years, contain enough computing power to become an effective equalizer."[27] Presumably, ordinary citizens equipped with microcomputers will be able to counter the influence of large, computer-based organizations.

Notions of this kind echo beliefs of eighteenth and nineteenth century revolutionaries that placing firearms in the hands of the people was a condition crucial to overthrow entrenched authority. In the American Revolution, French Revolution, Paris Commune and Russian Revolution the role of "the people armed" was central to the revolutionary program. As the military defeat of the Paris Commune made clear, however, the fact that the popular forces have guns may not be decisive. In a contest of force against force, the larger, more sophisticated, better equipped competitor often has the upper hand. Hence the availability of low cost computing power may move the base line that defines electronic dimensions of social influence, but it does not necessarily alter the relative balance of power. Using a personal computer makes one no more powerful vis-à-vis, say, the National Security Agency than flying a hang glider establishes a person as a match for the U.S. Air Force.

In sum, the political expectations of computer enthusiasts are seldom more than idle fantasy. Beliefs that widespread use of computers will cause hierarchies to crumble, inequality to tumble, participation to flourish, and centralized power to dissolve simply do not withstand close scrutiny. The formula: information = knowledge = power = democracy, lacks any real substance. At each point the mistake comes in the conviction that computerization will inevitably move society toward the good. And no one will have to raise a finger.

IV. INFORMATION AND IDEOLOGY

Despite its shortcomings as political theory, myth information is noteworthy as an expressive contemporary ideology. I use the term "ideology" here in a sense common in social science: a set of beliefs that expresses the needs and aspirations of a group, class, culture, or subculture. In this instance the needs and aspirations that matter most are ones that stem from operational requirements of highly complex systems in an advanced technological society; the groups most directly involved are those who build, maintain, operate, improve and market these systems. At a time in which almost all major components of our technological society have come to depend upon the application of large and small computers, it is not surprising that computerization has risen to ideological prominence, an expression of grand hopes and ideals.

What is the "information" so crucial in this odd belief system, the icon now so greatly cherished? We have seen enough to appreciate that the kind of information upheld is not knowledge in the ordinary sense of the term; nor is it understanding, enlightenment, critical thought, timeless wisdom, or the content of a well-educated mind. If one looks carefully at writings of computer enthusiasts, one finds that information in a particular form and context is offered as a paradigm to inspire emulation. Enormous quantities of data, manipulated within various kinds of electronic media, used to facilitate the transactions of today's large, complex organizations – that is the model we are urged to embrace. In this context, the sheer quantity of information presents a formidable challenge. Modern organizations are continually faced with overload, a flood of data that threatens to become unintelligible to them. Computers provide one way to confront that problem; speed conquers quantity. An equally serious challenge is created by the fact that the varieties of information most crucial in modern organizations are highly time specific. Data on stock market prices, airline traffic, weather conditions, international economic indicators, military intelligence, public opinion poll results, and the like are useful for very short periods of time. Systems that gather, organize, analyze, and utilize electronic data in these areas must be closely tuned to the very latest developments. If one is trading on fast-paced international markets, information about prices an hour old or even a few seconds old may have no virtue. Information is itself a perishable commodity.

Thus, what looks so puzzling in another context – the urgent "need" for information in a social world filled many pressing human needs – now becomes transparent. It is, in the first instance, a need of complex human/machine systems threatened with debilitating uncertainties or even breakdown unless continually replenished with up-to-the-minute electronic information about their internal states and operating environments. Rapid information processing powers of modern computers and communications devices are a perfect match for such needs, a marriage made in technological heaven.

But is it sensible to transfer this model, as many evidently wish, to all parts of human life? Activities, experiences, ideas, and ways of knowing that take a longer time to bear fruit meet the demand that they adapt to the speedy processes of digitized information processing. Education, the arts, politics, sports, home life, and all other forms of social practice must be transformed to accommodate it. As one article on the coming of the home computer concludes, "running a household is actually like running a small business. You have to worry about inventory control – of household supplies – and budgeting for school tuition, housekeepers salaries, and all the rest."[28] The writer argues that these complex, rapidly changing operations require a powerful information processing capacity to keep them functioning smoothly. One begins to wonder how everyday activities like running a home were even possible before the advent of microelectronics.

In the last analysis, what I have noted as an almost total silence about the ends of the "computer revolution" is filled by a conviction that information processing is something valuable in its own right. Faced with an information explosion that strains the capacities of traditional institutions, society will renovate its structure to accommodate computerized, automated systems in every area of concern. The efficient management of information is revealed as the *telos* of modern society, its greatest mission. It is that fact to which myth information adds glory and glitter. People must be convinced that the human burdens of an information age – unemployment, de-skilling, the disruption of many social patterns – are well worth bearing. Once again, those who push the plow are told they ride a golden chariot.

V. CONCLUSION

Having critized a point of view, it remains for me to suggest what topics a serious study of computers and politics would pursue. The question is, of course, a very large one. If the long-term consequences of computerization are anything like the ones commonly predicted, they will require a rethinking of many fundamental conditions in social and political life. I will mention three areas of concern.

As people handle an increasing range of their daily activities through electronic instruments – mail, banking, shopping, entertainment, travel plans, etc. – it becomes technically feasible to monitor these activities with an ease heretofore inconceivable. The availability of digitized footprints of social transactions affords opportunities for ingenious matching and correlating, opportunities that contain a menacing aspect. While there has been a great deal of writing on this problem, most of it identifies the issue as one of "threat to privacy," the possibility that someone might gain access to information that violates the sanctity of one's personal life. As important as that issue certainly is, it by no means exhausts the potential evils created by electronic data banks and computer matching. The danger extends beyond the private sphere to affect the most basic of public freedoms. Unless steps are taken to prevent it, we may develop systems that contain a perpetual, pervasive but apparently benign surveillance. Confronted with omnipresent, all-seeing data banks, the populace may find passivity and compliance the safest route, avoiding activities that once comprised political liberty. As a badge of civic pride they may announce: "I'm not involved in anything a computer would find the least bit interesting."

It is important to notice that the evolution of this unhappy state of affairs does not necessarily depend upon the "misuse" of computer systems. The prospect we face is really much more insidious. An age rich in electronic information may achieve wonderful social conveniences at a cost of placing freedom, perhaps inadvertently, in a deep chill.

A thoroughly computerized world is also one bound to renovate conditions of human sociability. The point of many applications of microelectronics, after all, is to eliminate social layers that were previously needed to get things done. Computerized bank tellers, for example, have largely done away with small, local branch banks, which were not only ways of doing business, but among the places in a community where people met, talked and socialized. The so-called electronic cottage, similarly, operates very well without the kinds of human interactions that once characterized office work. Installing greater efficiency, productivity and convenience, innovations of this kind do away with the reasons people formerly had for being together, working together, acting together. Many practical activities once crucial to even a minimal sense of community life are rendered obsolete. One consequence of these developments is to pare away the kinds of face-to-face contact that once provided important buffers between individuals and organized power. To an increasing extent, people are now linked in direct connection to the influence of employers, news media, advertisers, and national political leaders. Where will we find new institutions to balance and mediate such power?

Perhaps the most significant challenge for theory posed by the linking of computers and telecommunications is the prospect that the basic structures of political order will be recast. Worldwide networks of computers, satellites and communications fulfill, in large part, the modern dream of conquering space and time. These systems make possible instantaneous action at any point on the globe without limits imposed by the specific location of the initiating actor. Human beings and human societies, however, have traditionally found their identities within spatial and temporal limits. They have lived, acted and found meaning in a particular place at a particular time. Developments in microelectronics tend to dissolve these limits, thereby threatening the integrity of social and political forms that depend on them. Aristotle's observation that "man is a political animal" meant in its most literal sense that man was a polis animal, a creature naturally suited to live in a particular kind of community within a specific geographical setting, the city state. Historical experience shows that it is possible for human beings to flourish in political units – kingdoms, empires, nation states – larger than ones the Greeks thought natural. But until recently the crucial conditions created by spatial boundaries of political societies were never in question.

That has changed. Through methods pioneered by transnational corporations, it is now possible for organizations of enormous size to manage their activities effectively across the whole surface of the planet. The integration of business units which used to depend upon spatial proximity, can now be handled through complex electronic signals. If it seems convenient to shift operations from one area of the world to another far distant, it can be accomplished with the flick of a switch. Close an office in Sunnyvale; open an office in Singapore. In the recent past, corporations have had to demonstrate at least some semblance of commitment to geographically based communities; their public relations often stressed the fact that they were "good neighbors." But in an age in which organizations are located everywhere and nowhere, this commitment easily evaporates. A transnational can play fast and loose with everyone, including the country that is ostensibly its "home." Towns, cities, regions, and whole nations are forced to swallow their pride and negotiate for favors. In that process, political authority is gradually redefined.

Computerization resembles other vast, but largely unconscious experiments in modern social and technological history; following a step-by-step process of instrumental improvements, societies create new institutions, new patterns of behavior, new sensibilities, new contexts for the exercise of power. Calling such changes "revolutionary," we tacitly acknowledge that these are matters that require reflection, possibly even strong public action to insure that the outcomes are desirable ones. But the occasions for reflection, debate and public choice are extremely rare indeed. The important decisions are left in private hands inspired by narrowly focused economic motives. While many recognize that these decisions have profound consequences for our common life, few seem prepared to own up to that fact. Some observers forecast that "the computer revolution" will eventually be guided by new wonders in artificial intelligence. Its present course is influenced by something much more familiar: the absent mind.

University of California,
Santa Cruz

AUTHOR: Langdon Winner was born in 1944 and received his Ph.D. in political science from the University of California at Berkeley in 1973. He has taught at the University of Leiden in the Netherlands, M.I.T., and the University of California at Santa Cruz. His publications include *Autonomous Technology: Technics-out-of-control as a Theme in Political Thought* (Cambridge, MA: MIT Press, 1977) and *the Whale and the Reactor: A Search for Limits in an Age of High Technology* (Chicago: University of Chicago Press, forthcoming). He is also a musician and has written music criticism for *Rolling Stone* and the *Atlantic*.

NOTES

This paper was prepared with the support of a National Science Foundation / National Endowment for the Humanities Sustained Development Award in the Program on Ethics and Values in Science and Technology, OSS-80118089. Its views are the author's and do not necessarily reflect those of NSF or NEH.

1   See for example, Edward Berkeley, *The Computer Revolution* (New York: Doubleday, 1962); Edward Tomeski, *The Computer Revolution: The Executive and the New Information Technology* (New York: Macmillan, 1970); Nigel Hawkes, *The Computer Revolution* (New York: Dutton, 1972). See also Aaron Sloman, *The Computer Revolution in Philosophy* (Hassocks, England: Harvester Press, 1978); Zenon Pylyshyn, *Perspectives on the Computer Revolution* (Englewood Cliffs, N.J.: Prentice Hall, 1970); Paul Stoneman, *Technological Diffusion and the Computer Revolution* (Cambridge: Cambridge University Press, 1976); Ernest Braun and Stuart McDonald, *Revolution in Miniature: The History and Impact of Semiconductor Electronics* (Cambridge: Cambridge University Press, 1978)
2   Michael L. Dertouzos in an interview on "The Today Show," National Broadcasting Company, August 8, 1983.
3   Daniel Bell, "The Social Framework in the Information Society," in Michael L. Dertouzos and Joel Moses (eds.), *The Computer Age: A Twenty Year View* (Cambridge: Mass.: M.I.T. Press, 1980) p. 163.

4   See for example Philip H. Abelson, "The Revolution in Computers and Electronics,"
    Science **215,** whole no. 4534 (February 12, 1982), 751–753.
5   Edward A. Feigenbaum and Pamela McCorduck, *The Fifth Generation: Artificial In-
    telligence and Japan's Computer Challenge to the World* (Reading, MA: Addison-Wes-
    ley, 1983) p. 8.
6   Among the important works of this kind are: David Burnham, *The Rise of The Com-
    puter State* (New York: Random House, 1983); James N. Danziger *et al., Computers
    and Politics: High Technology in American Local Governments* (New York: Columbia
    University Press, 1982); Abbe Moshowitz, *The Conquest of Will: Information Proces-
    sing in Human Affairs* (Reading, MA: Addison-Wesley, 1976); James Rule *et al., The
    Politics of Privacy* (New York: New American Library, 1980); Joseph Weizenbaum,
    Computer Power and Human Reason: Judgment to Calculation (San Francisco: Free-
    man, 1976).
7   Quoted in Jacques Vallee, *The Network Revolution: Confessions of a Computer Scien-
    tist* (Berkeley: And/Or Press, 1982), p. 10.
8   Tracy Kidder, *Soul of a New Machine* (New York: Avon, 1982), p. 228.
9   *The Fifth Generation*, p. 14.
10 James Martin, *Telematic Society: A Challenge for Tomorrow* (Englewood Cliffs, NJ:
    Prentice Hall, 1981), p. 172.
11 *Telematic Society*, p. 4.
12 Starr Roxanne Hiltz and Murray Turoff, *The Network Nation: Human Communica-
    tion via Computer* (Reading, MA: Addison-Wesley, 1978) p. 489.
13 *The Network Nation*, p. xxix.
14 *The Network Nation*, p. 484.
15 *The Network Nation*, p. xxix.
16 *The Network Revolution*, p. 198.
17 John Naisbitt, *Megatrends: Ten New Directions Transforming Our Lives* (New York:
    Warner Books, 1984), p. 282.
18 Amitai Etzioni, Kenneth Laudon, and Sara Lipson, "Participation Technology: The
    Minerva Communications Tree," *Journal of Communications* **25** (Spring 1975), 64.
19 J. C. R. Licklider, "Computers and Government," in Dertouzos and Moses, *The
    Computer Age*, pp. 114 and 126.
20 Quoted in *The Fifth Generation*, p. 240.
21 *Occupational Outlook Handbook, 1982–1983*, U.S. Bureau of Labor Statistics, Bulle-
    tin No. 2200, Superintendent of Documents, Government Printing Office, Washing-
    ton, D.C. See also Gene I. Maeroff, "The Real Job Boom Is Likely to Be Low-Tech,"
    New York Times (September 4, 1983), p. 16E.
22 See for example James Danziger *et al., Computers and Politics*.
23 For a study of consumer products in American democracy see Jeffrey L. Meikle,
    Twentieth Century Limited: Industrial Design in America, 1925–1939 (Philadelphia:
    Temple University Press, 1979).
24 *The Fifth Generation*, p. 8.
25 "The Philosophy of US," from the official program of The US Festival held in San
    Bernardino, California, September 4–7, 1982. The outdoor rock festival, sponsored
    by Steven Wozniak, co-inventor of the original Apple Computer, attracted an esti-
    mated half million people. Wozniak regaled the crowd with large screen video pre-
    sentations of his message, proclaiming a new age of community and democracy gener-
    ated by the use of personal computers.

26 "*Power* corresponds to the human ability not just to act but to act in concert. Power is never the property of an individual; it belongs to a group and remains in existence only so long as the group keeps together." Hannah Arendt, *On Violence* (New York: Harcourt, Brace & World, 1969), p. 44.
27 John Markoff, "A View of the Future: Micros Won't Matter," *InfoWorld* **5**, no. 44 (October 31, 1983), 69.
28 Donald H. Dunn, "The Many Uses of the Personal Computer," *Business Week* (June 23, 1980), pp. 125–126.

WALTHER CH. ZIMMERLI

WHO IS TO BLAME FOR DATA POLLUTION?
ON INDIVIDUAL MORAL RESPONSIBILITY
WITH INFORMATION TECHNOLOGY

ABSTRACT. Examines three moments in the information technology process – data acquisition, processing, and dissemination – and considers the problems of assigning individual moral responsibility for data pollution or distortion as it might arise under each situation. Argues that a "paradox of information technology," in which more information leads
to less control, undermines the application of abstract ethical principles and forces a turn to
casuistic morality. Applying, for instance, from medical ethics, the regional principle of
informed consent reveals the need to place serious limitations on the development of information technology.

To my teacher, Rudolf W. Meyer,
on his 70th birthday

I

Technologies are theoretically expressable know-how, methodical procedures used to transform reality. Technologies share the fate of allo-
pathic medicine: If they are to be effective, they must have negative side-
effects, which can collectively be termed "pollution." This is the case
with *all* technologies, and only being misled by gentle references to "software" as opposed to "hardware" explains why serious problems of information technology are hardly ever discussed under the heading "data
pollution."[1]

I consciously accept the fact that the term "data pollution" is ambiguous. It can mean pollution *of* data and pollution *by* data, both of which
in turn contain heterogeneities. In what follows the term "data pollution" is taken to refer to the accumulation of all "contaminations" or
"distortions" which can result from working with data in the information
technology field. This is not meant as a definition in the strict sense, but
simply as the naming of a general characteristic. The details to be included under the heading "data pollution" are not pre-determined, since
this will vary in relation to technical developments and applications. The
range of distortions contained in data pollution is in principle open.

The causes of data pollution are many and varied – as are the possible
consequences. Anyone who has been caught in a computerized police

Carl Mitcham and Alois Huning (eds.), Philosophy and Technology II, 291–305.
© 1986 *by D. Reidel Publishing Company.*

investigation because he has the same shoe size, eating habits, and travel destination as a terrorist; anyone who has been "forgotten" by an airline booking computer; anyone who has received the wrong bank statement or mistakenly been billed for something again and again (with each bill becoming nastier than the one before) – indeed every person in a civi.ized industrial nation – knows what I am talking about. Usually the anonymous explanation given is that "the computer broke down," "made a mistake," or simply "is on strike." But in reality this is either a simple figure of speech or a case of "scapegoating" – appropriate in some situations but unwarrantedly extended in many others.

One often hears the response that things are not as bad as they seem, that there would be at least as many mistakes if human beings rather than machines did the data processing, and that one can expect more consistency and reliability from an algorithmically defined process, which must be seen as an advantage. It is not necessary to dispute the point at this juncture. What I want to show is that the "errors" (data pollution) caused by the use of information processing computers *differ essentially*, in certain respects, from those of their human predecessors or substitutes, and that this gives rise to fundamentally new problems with regard to individual responsibility.[2]

An idealized analysis of the information processing operation discloses three separate phases: the *collection* or *acquisition* of data, data *processing* and *storage*, and data *dissemination* or *application*. In other words, the "realm of information" can be split into three regions: (a) the extraction and "production" of "raw data" (sic!), (b) the reprocessing, refining, and storing of data, and (c) the service rendered by the data when applied in a socio-historical framework. Distortions – or negative effects which can influence the data and the attitudes of those who utilize it – can occur in all three areas. A further analysis shows that the second region (processing) can be further subdivided, since data distortion here occurs either as a result of poor programming or because of unintentional side-effects. This gives rise to four areas, which will now be examined with regard to possible specific problems.

(1) That a situation involves data pollution at the level of "material" input is often indicated by the occurrence of counter-intuitive results in the processing of familiar physical, chemical, etc. information. There are various possible causes. One could be that the input is simply incorrect, "false data." But as a rule a user or subject only percieves a resultant error, not its initial causes, which can be manifold. Indeed, because our information society contains multiple feedback loops, the basic collec-

tion of data for one system can be distorted by secondary and tertiary processes in other systems. In fact, errors in the spelling of a name, a telephone or identification number, etc. – errors with the aggravating consequences with which we in the age of information have become only too well acquainted – can be traced back, in principle, to all three error sources mentioned above. They can occur (a) in the initial acquisitions of raw data, (b) in the processing (coding) and storing of data, or (c) in the dissemination, i.e., the recall and readdressing of data previously coded and stored. Beside these error sources there exists one other factor in data pollution at the level of the matter being manipulated. This is the partial loss or destruction of data due to hardware failure (tapes or disks improperly copied) or interference (master tapes affected by external electro-magnetic impulses, which partially delete or otherwise alter the data).

(2) Data, however, is not some independent material entity; data is constituted by units of information. This means that data pollution can be created as well by the wrong choice of a programming language or logical errors in the program itself. For instance, there can be sequential commands which contradict one another in ways which are not apparent with regard to the commands themselves; the contradictions only become manifest after a large number of processing steps. The programming language may also include ambiguities, undetectable beforehand; it may be too simple or too refined, i.e., inappropriate for the project in question. There is no doubt that a great many programs which seem appropriate in one field, turn out to be inconsistent or insufficient when extended to a wider or qualitatively different field. Due to micro-processing technology, it has become conceivable that such errors, undiscovered in the limited range of one specific application, may be reproduced a millionfold and "committed" worldwide.

Furthermore, one must consider the data pollution issue which is constituted by the basic incomprehensibility of a program.[3] One can easily write sequential commands or transformation rules which relate data without reference to reality. As long as such rules are internally consistent, the data processing system in question will produce, with a finite number of steps, any number of combinations of the original data – i.e., an accumulation of additional "data" – which in part cannot be interpreted in any semantically meaningful sense. Many of the so-called "correlation statistics" in social research are based (in my opinion) on programs which are incomprehensible in this strict sense of the word.

We have now reached a point of transition from errors which a human data collector and processor might make, possibly in much greater quantity, to errors specific to information processing technologies. Errors of specification or correlation, filing errors, logical errors or inappropriate choices of language and systems are all things with which we have learned to cope. On the one hand, after a certain amount of time and experience, we can calculate the probability that such errors will occur, and using risk-cost-benefit calculations take them into account. On the other hand, relatively clear, individual competencies and responsibilities can be determined here as well – despite false appeals to the primacy of "the computer." The individual "in charge" or the department or firm which employs him is the responsible person or institution, no matter what certain interested parties might say. Should material or non-material damages occur as a result of such errors, it is a person's moral and legal right to compensation from the parties in question.[4] With regard to specifically unintelligible programs, however, a question arises as to whether anyone can actually be held morally accountable for something he does not "understand" even though he may be liable before the law. This is a question which becomes even more critical with regard to those kinds of data pollution still to be discussed.

(3) Essential limits exist as to how far a data processing operation can be controlled – as will be considered in greater detail under the heading of "the paradox of information technology" (section II below). Yet because of the presumption of continuity (which guides all scientific investigation of nature) we assume the improbability of qualitative jumps and unforeseen side-effects just at the point where a processing operation escapes our immediate control. But the truth is that we must question such an assumption, and that effects occur in many computer-based processes which can only be explained by means of qualitative jumps. There are, of course, often externally induced. One drastic example would be the total break-down of all computer systems under the influence of certain conditions in nuclear warfare.[5] But it also cannot be ruled out that unknown combinations and qualitatively new problems might occur when a certain capacity and speed of data processing is exceeded. Since this is the case, corresponding difficulties with regard to moral responsibility for the resulting data pollution may also occur.

(4) The situation becomes even more problematic when we come to social side-effects, whether predictable or unpredictable. It is well-known that information is one of those goods the maximization of which does not lead *ad infinitum* and absolutely to an optimal result. We must clearly acknowledge the existence of a sort of "saturation effect" in

information,[6] which can be easily illustrated by means of an example from the world of advertizing. The cart-load of flyers and circulars which gather month after month in our mailboxes has an information value tending toward zero, since after reading a certain number of advertisements, not only are the rest left unread, but no circular of whatever sort is ever read again. Something similar happens with information in general. The optimization of access to data is not sufficient reason – and here the statistics of our university libraries speak for themselves – for making more use of the opportunities offered. On another side of the issue, the problem of the protection of personal data has also been a subject for considerable discussion. To how much information may a third party be allowed access without infringing my right to privacy? There have, for instance, been marked disagreements on this issue between the Constitutional Court and Parliament of the Federal Republic of Germany with regard to census data. Even here in the field of pollution *by* data, problems of assigning individual or collective responsibility arise. In the context of both the above mentioned issues, it is clear that the question is still open: Who is to blame for data pollution?

II

Let us look more closely at the first of these last two contexts. It has already been suggested that in certain circumstances there can be essential difficulties in checking the correctness both of the data processing and of its results. That this is the case is easily recognized by considering the purpose and meaning of data processing systems. From the first Turing machines, in fact since Leibniz's mechanical calculator, technical development has been toward an automaton with a mechanical, electrical, electromagnetic – but not psychophysical – basis, an automaton which compensates for the deficiencies in human thinking in both precision and speed. Of course, all that was originally intended was an instrument of efficient calculation. But since the initial two phases of artificial intelligence (AI) (1957–1962, 1962–1967) were made the subject of critical discussion in the seventies (by H. L. Dreyfus[7] and J. Weizenbaum among others), one sees that in reality the demands had always been greater than they appeared. Whether or not one wants to follow the arguable theses of Dreyfus or Weizenbaum in this respect, it is clear that what gave birth to the Leibnizian project was the concept of designing a machine which would far surpass the human intellect in certain specialized functions (simple or complex calculating operations).[8] Today this ambition has been achieved in such a way that even elementary school children study mathematics only with the help of calculators.

It is a small step from this to the identification of certain essential limits. If the aim of computer development is that even microcomputers should surpass the human capacity for specific data processing operations by some nth factor, then the result will be that, due to the contingencies of a limited life-span and a limited calculating speed, human thinking *will never be able to model* (and thereby check) even relatively short operations from mainframe computers. This is even more the case for complex programs, in which many calculations run parallel and are interconnected. Even something as simple as the traffic system (e.g., the timing of traffic lights in a city) are many times too complicated for a single person to figure out. From a certain amount of data and a certain processing speed onward we must rely on an extremely questionable prejudice, acquired through our analytical training, that the computer will not behave any differently in the realm of large numbers and at high processing speeds with which we are directly acquainted. The possible objection that this prejudice is supported by the validity of mathematical induction only appears plausible. My argument is not concerned with the mathematical theory (the validity of which is indisputable) but with its technological implementation.[9]

It should also be noted that an increase in the speed of, for example, a linear computer is only possible due to some operational specialization. In contrast to the human brain, in which innumerable qualitatively different operations are run at any one time, the high-speed linear computer runs only one kind of operation. Even this fundamental difference sets limits to the possibility of controlling and checking what is going on.[10] So whether or not a high-speed computer or some other information processing system is working "properly" can only be checked by another high-speed computer of even greater capacity. But who controls and checks on this second, third, . . ., nth one?

Thus we have a new version of the old question, "Who controls the controller?" It throws into relief a problem which I would like to call "the paradox of information technology": The possibility of controlling information processing systems diminishes in proportion to the introduction of modelling or checking instances.

The result of the above considerations can give rise to two objections:

(1)   Although the limited calculating capacity of the human brain
        means that, in fact, it can never be in a position to control the

operations of a high-speed computer, such control is possible "in principle," through an understanding of the operating system and programs.

(2)   even if the first objection cannot be upheld, the paradox just illustrated is not valid *specifically* for information technology, but for every technology, and thus has such a high claim to generality that it can be neglected.

Yet both objections can be refuted:

(*ad* 1)   It is true that statements which by definition are not capable of being tested empirically do not become *eo ipso* meaningless; but when they do claim to be empirical, then they become metaphysical in the bad sense of the word. It is this status which statements of the kind "If I could live 10000 years, then I could test x" possess. This state of affairs cannot be changed by my being able to imagine an infinitely long series in the lifetimes of individual testers. Again, each of them cannot himself check that which has been checked before him, nor that which is still to be checked. An illustrative example of this is the "specifically unintelligible program" in which so many programmers have collaborated that none of them any longer understands the program.

(*ad* 2)   We can of course admit that the very purpose of introducing technology is not to replace human performance but to optimize it, which means that no good technology can be equaled by the human performance it sets out to optimize. While the aim of all other technologies is to optimize those human performances which are different from the human control capacity (human thinking), information technology optimizes specific performances of the human brain. The fact that the controlling performances and those to be controlled are identical gives rise to the *specific* paradox of information technology.

The "technological solution" to this essential problem normally lies in the construction of redundant systems. But it is clear that this does not solve the problem philosophically. Thus, the more exactly we want to know what happens in each of these areas which are hidden from our

view, the further these retreat from our understanding. We could also say
in this context that in the strict sense *it is not possible to decide*[12] the question of whether or not a computer is running correctly in those fields
which are inaccessible to the human capacity and speed of thought.

The full import of this argument can only be grasped when seen in the
context of AI research. Without taking sides in the discussion, I shall
quote Joseph Weizenbaum on the aims of AI. With reference to works by
H. A. Simon and Roger Schank, Weizenbaum wrote in 1976:

Both Simon and Schank have thus given expression to the deepest and most grandiose fantasy that motivates work on artificial intelligence, which is nothing less than to build a
machine on the model of man, a robot that is to have its childhood, to learn a language as a
child does, to gain its knowledge of the world by sensing the world through its own organs,
and ultimately to contemplate the whole domain of human thought.[13]

The ideal of the AI community is the construction of a machine with an
inner *disposition* which enables it to act in all relevant cognitative contexts like a human being – that is, not only to be able to reel off stored
programs again and again, but also to learn, to communicate, and to be
creative. Now we know, however, that dispositions can only be known by
others through the description and prediction of behavior patterns under
specific circumstances. This means that the success of the AI project
would only be revealed through successively recurring instances of quasi-
human behavior on the part of an AI machine, and that this, for reasons
already mentioned, would only be discernable up to a point and indirectly. In fact, only a data-processing super system which controlled the AI
machine and was capable of pre-computing its operations, i.e. a super AI
machine, could instantly show the successful construction of AI.

Such considerations demonstrate that the construction of a perfect AI
machine in the form of a computer is in itself a paradoxical undertaking.
In addition, it is easy to see that the oft-quoted allegation of a *hidden
inner determination* and *opacity* of technology – charges which recur
together at regular intervals in the criticism of culture – would in fact be
promoted. If AI were to become possible – by which I refer solely to computer AI, not to *in vitro* recombination of DNA or other biotechnologies
– then AI will have had to exist already before that moment, a fact which
would be beyond our knowledge. The construction of a perfect AI
machine would have to be permanently ahead of itself, since the construction of such a machine without a super AI machine to model this
construction would be unthinkable. From this there follows an almost
total impenetrability and detachment of information processing AI

machines from the human world. AI machines would "socialize," so to speak, only among themselves.

From this excursion into the realm of counterfactual conditionals,[14] let us return to simple information processing systems. We can see even here, in the wide-spread introduction of microcomputers, a tendency toward independence and impenetrability. The feeling of the man-in-the-street that he is at the mercy of that data processing super-machine called the state, is not based entirely on fiction, but possesses a *fundamentum in re*. Of course, at this point one can easily fall back on an old counter-argument, namely the consoling assurance that there is nothing new under the sun. Along with every tool *homo sapiens* invents, with every machine built and every bit of technology produced, he always gives up a little independence. Yet this is nothing but the flip side of an argument for the rationalization and optimization of labor, and one which can be readily analyzed into two others:

(1)   On the one hand, a conservative-naturalistic argument of the type: "That's how it's always been, so it can't be so bad."

(2)   On the other hand, an instrumental-naturalistic argument of the type: "It is true that, for every technical application of a tool, consequences result, which could not have resulted without that application."

However, both arguments can easily be countered.

(*ad* 1)   As long as no additional premises are introduced, the fact that something has always been so can just as well argue against as corroborate the idea that it can't be so bad.

(*ad* 2)   As it has already been mentioned, the application of mechanical tools as an optimization device is qualitatively different from the application of information technologies. In the first case, the intelligent human subject can retain the power of review and control (by measuring, calculating, etc.) while, in the second case, information-technological "thought tools" take over.

We must therefore admit that the paradox of information technology is not a trivial case to be compared with other trivial cases, but a process of detaching the controlling factor and rendering it independent, that is, with information processing systems of high complexity and capacity. The unpleasant feeling which relentlessly creeps up on a person when it dawns on him that his well-being is dependent on the controllability, impenetrable in turn, of those main-frame computers which should serve to optimize the decision-making in the ministries of defense – this unpleasant feeling is, on the level of phenomena, an indicator of the independence of control knowledge analyzed above.

III

One can speak of responsibility in both a legal and a moral sense.[15] *Legal responsibility* can be understood as an implication made by positive law: someone is legally responsible for action A if, under certain circumstances, he has to take the consequences C codified in a legally binding manner. In agreement with the proverb "ignorance is no excuse," this implication is objectively valid, that is independent of whether or not the actor (perpetrator of the action in question) knew that his action A would result in the legal consequences C. The case of *moral responsibility* is more complicated, because no codified consequences exist here and the actions themselves are not punishable in the strict sense of the word.

Nevertheless, moral responsibility, too, can be taken as an implication involving the action A and either the motivation M of an action (its maxim) or the moral judgment J of this action, in which, for the time being, the question remains open whether or not a judgment is passed and if so under which criteria. Thus someone is morally responsible, when, subject to the moral criteria of the time and the social group to which he belongs, he can expect that a moral judgment will be applied to his actions or to the maxims of these actions. Contrary to legal responsibility, the implication of moral responsibility is traditionally not valid independent of whether or not the acting individual knows about this connection. Someone can be morally responsible in the strict sense only if he or she is aware that the action to be carried out or the maxim regulating the action is a morally relevant action or maxim. Moreover, it must be noted that the question of the actual existence of moral judgments and whether or not they are in fact carried out by someone plays no part in the question of moral responsibility. The only important thing is that such a judgment would be *possible* in the eyes of the acting individual.

Our specific situation, in the last quarter of the 20th century, is characterized by the fact that substantial values of general obligation only exist (if at all) in a very abstract form. Thus, because Christian values can no longer be assumed, modern ethics is characterized by the search for new principles of general obligation. These must be principles, the obligation of which is not legitimized by means of belief, but rather by means of rational acceptability. No matter which ethical school one belongs to, one is bound to recognize the rationality of one or more of the following two principles: the principle of universalizability and the principle of equality (or of fairness).[16] The former recommends testing whether all morally relevant beings agree with the maxims in question (as in the example of Kant's categorial imperative); the latter recommends testing the maxims or actions on the assumption that in equal circumstances either everyone is treated equally, or that one is prepared to take on the role of any individual involved in the context of action. It is evident that these two closely related principles attempt to provide criteria for moral judgments by taking a formal approach and by assuming the equality of all actors.

Moreover, no matter to which ethical "confession" one may belong, the fundamental distinction cannot be avoided that an action is to be judged either on the intention of the actor (deontologically) or on the resulting consequences (teleologically). Thus, the aforementioned principles of universalizability and equality can be applied both deontologically or teleologically. In the following, I will examine what was outlined above as data pollution (section I) especially in relation to the paradox of information technology (section II), first as subject to these ethical principles, and then in a more casuistic manner. The question arises to whether or not actions in the context of information technology are morally relevant actions at all, since someone is only morally responsible for something which he or she accepts as being morally relevant.

The *principle of universalizability* reads, as described above, that actions or their underlying maxims should be tested by asking whether or not all morally relevant beings could accept them. In view of the complexity of the problems with information technology, we must ask ourselves exactly what is subject to acceptance. At first sight, speaking in a purely deontological sense, it seems as though no morally relevant being could ever reject the use of information processing systems. In the light of the paradox of information technology, however, this becomes doubtful. Is it in fact true that all men must affirm an intention to apply essentially as well as increasingly uncontrollable technologies? The initial certainty that at least the use of information technologies could be morally legitimized, becomes even less likely when we turn from intentions to consequences and apply the test of universalizability. One can hardly doubt that the development of those technologies, the consequences of which cannot in principle be foreseen or, in the allotted time ascribed to some basic actions, cannot be justified. The black box, which, in a nontrivial sense, lies between the information input and the possible consequences of its use, undermines all possible moral responsibility.

Similar results follow when the actions under discussion in the framework of information technologies are tested by the *equality* or *fairness principle*. Here too it seems as though the application of information processing systems might optimize equal treatment, which would make it easy for the examiners of the moral content of actions to be ready in principle to take on the role of any individual involved in the context of action. In light of the paradox of information technology, however, the question arises whether or not there will always be such things as "definable" roles. And in addition one can, speaking teleologically, imagine (more is not possible) unintentional consequences, which, caused solely by the adherence to nonsensically applied egalitarian principles and correlations, bring about the most absurd and illiberal situations in a society controlled by an AI system (e.g., that all peope with gold fillings in their teeth must earn the same amount of money, etc.).

An ethical examination based on theoretical principles evidently breaks down in the face of concrete situations. One conclusion to be drawn might be that information technology is not subject to moral evaluation, not susceptible to the identification of areas of individual moral responsibility. This, however, would be an unsound argument. The problem is rather one of trying to subsume specific, morally relevant actions, under ethical principles which are too general. In other words, the moral judgment involved here is exceptionally detailed and too related to immediate situations for the application of abstract formal principles. Thus, we have to argue specific cases (casuistry) and to search for principles at a lower level of generality, which in their turn regionalize the sphere of moral responsibility. In doing this, we must take more account of data *dissemination* and *application*, since data acquisition and programming, as we said above, do not for their part reveal any essentially new ethical problems; the unsolvable questions connected with the paradox of information technology have already been discussed.

Regionalized ethical principles for judging the moral character of specific actions are commonly discussed under the heading of the so-called "professional ethics." Medical ethics (now broadened into bioethics), for instance, can look back on a remarkable tradition (from Hippocrates) and yet has developed in recent years codes of professional behavior which meet to a certain extent casuistic needs. One case related to the acquisition, processing, and application of human data in information technology is the much-discussed experimentation on human beings.[17] By analogy, centralized state data processing (if not seen as a means of furthering dictatorial control, in which case it must be morally condemned with regard to the principle of equality or fairness) could be viewed as a large-scale scientific human experiment. The ethical principle of "informed consent," accepted since the time of the Nuremberg Code[18] becomes valid: When, after exhaustive information on effects and side-effects has been made available, more than 50% of the subjects still choose the information technology in question, then and only then can it be called "ethically legitimized." In addition, the principle that no individual should be forced against his or her will to take part in such a human experiment, is also valid. Then, too, in any case in which the enlightenment potential is so small as to be negligible, it is morally legitimate for the minority to disregard a majority decision – which is in fact the case given the paradox of information technology. With such an essentially uncertain foundation for decision making – which might, in some circumstances, be magnified a billion fold through the central instrumental position of data processing systems – an adequate protection of minority rights must be built in.

So who is to blame for data pollution? There are evidently various answers to the question. In the field of data *acquisition* and *programming*, it is the individual in charge who is responsible for the "contamination" or "distortion" which results from various possible errors. Once the essential limits of control over the data processing operation are reached, the paradox of information technology comes into play, which creates a situation in which individual responsibilities can no longer be ascribed by means of general ethical principles. It is therefore just as impossible for human actions to take into account the "contamination effects" occurring here as it is to realize the technological fantasy of AI. Yet, by means of well established regionalized ethical principles such as those of "informed consent," it is possible in the field of data *dissemination* and *application* to judge the morality of the actions of those individuals or institutions who undertake the actions in question.

Braunschweig University

Translated with assistance from Ann McGlashan

AUTHOR: Walther Ch. Zimmerli was born in 1945 in Zurich. He studied in the United States, Germany and Switzerland, and is currently full Professor of Philosophy at the Carolo-Wilhelmina Technical University in Braunschweig and Chairman of the "Mensch und Technik" section of the Verein Deutscher Ingenieure. He has edited *Wissenschaftskrise und Wissenschaftskritik* (Basel: Schwabe, 1974), *Esoterik und Exoterik der Philosophie* (Basel: Schwabe, 1975), *Technik – oder: wissen wir, was wir tun?* (Basel: Schwabe, 1976), *Kommunikation – Codewort für Zwischen-Menschlichkeit* (Basel: Schwabe, 1978), *Die "wahren" Bedürfnisse* (Basel: Schwabe, 1978), and *Kernenergie – wozu? Bedürfnis oder Bedrohung* (Basel: Schwabe, 1978), and has written books on Hegel and on Habermas as well as numerous papers on the philosophy of science and technology, social philosophy and ethics and the history of modern philosophy.

NOTES

1 This should not be taken to mean that critiques of information technology, of the social consequences of its broad application and of the deceptive hopes for the future which are connected with this, are never voiced in the appropriate literature. See for instance Joseph Weizenbaum, *Computer Power and Human Reason* (San Francisco: Freeman, 1976) and Langdon Winner, "Myth Information: Romantic Politics in the Computer Revolution," this volume.

2 The assumption which originally determined the experiments in the construction of artificial intelligence, i.e., that information processing machines simulate the operations in the human brain, has long been abandoned. On the one hand, successes via cognitive simulation were not forthcoming and, on the other, the brain functions quite differently from an information processing machine. Cf. Hubert L. Dreyfus, *What Computers Can't Do; The Limits of Artificial Intelligence*, rev. edition (New York: Harper & Row, 1979), esp. pp. 91ff. This gives grounds for supposing that different problems will arise depending on whether structurally analogous problems are solved by humans or by computers.

3 Cf. Weizenbaum, *op. cit.*, pp. 301ff. Of interest here is the time problem, which Weizenbaum, following Norbert Wiener, sees thus: The control of information processing systems is slower than the operations of these systems themselves.

4 There may in fact be severe problems in actually determining the "responsible" party and apportioning blame, but this does not mean it is in principle impossible. Cf. the case anthology, Robert J. Baum (ed.), *Ethical Problems in Engineering*, 2nd edition, vol. 2: *Cases* (Troy, New York: Rensselaer Polytechnic Institute, 1980).

5 Cf., e.g., J. D. Steinbrunner, "Nuclear Decapitation," *Foreign Policy*, no. 45 (1981/82), pp. 16–28.

6 There is also an information minimum, under which there is the danger of "white noise" and which can be formulated in contextual proportions of x bits per context unit. The expression "something has given something to something" contains, for example, too little information; the expression "a man of masculine gender handed a rose, which is a flower, to a woman of feminine gender" too much information with regard to this context unit.

7 See Dreyfus, *op cit.* In the introduction to the revised second edition, Dreyfus continues the phase classification: He follows Phase I ("Cognitive Simulation," 1957–1962) and Phase II ("Semantic Information Processing," 1962–1967) by a third

("Manipulating Micro-Worlds," 1967–1972) and fourth ("Facing the Problem of Knowledge Representation," 1972–1977), Cf. also Martin D. Ringle (ed.), *Philosophical Perspectives in Artificial Intelligence* (Atlantis Highlands, NJ: Humanities Press, 1979), especially pp. 1–22; and Margaret Boden, *Artificial Intelligence and Natural Man* (New York: Basic Books, 1977).

8   This is impressively shown by comparing data processing speeds: On average, that of the human brain is 10^2 [Bit sec^{-1}], that of the electronic computer 10^{16} to 10^{17} [Bit sec^{-1}]. See C. Christian, "Das rekursive Inaccessibilitätstheorem und der Gödelsche Unvollständigkeitssatz in ihrer Bedeutung für die Informatik," in H. Schauer and M. Tauber (eds.), *Informatik und Philosophie* (Vienna and Munich: Oldenbourg, 1981), p. 154.

9   In any case, modern physics teaches us not blindly to presuppose invariants outside the mesocosmos, and (in my opinion) this negative form holds also for technical equipement operating at high speeds.

10 Leaving aside the fact that the decisive operation of the difference between "mind" and "brain" should be discussed here, the difference mentioned in the text is to be placed under "brain." See Hilary Putnam, "Minds and Machines," in S. Hook (ed.), *Dimensions of Mind* (New York: New York University Press, 1960), pp. 138–164; Dreyfus, *op cit.*, pp. 159ff.

11 The question is, at this juncture, still purely epistemic. It is evident, however, that it also has ethical implications.

12 This could be the reason why the paradox of information technology has so far been discussed only as a theoretical Gödel-problem. However, we must question whether a computer is working "correctly" and the theoretical indeterminability of the question whether a computer, understood as a technologically "exbodied" system of axioms, can prove its freedom from contradiction. We must again differentiate between these questions and that of the "recursive inaccessible." See Christian, *op cit.*, and "On Formally Undecidable Propositions of *Principia mathematica* and Related Systems" (1931), included in J. van Heijenoort (ed.). *From Frege to Gödel* (Cambridge, MA. Harvard Univ. Press, 1967).

13 Weizenbaum, *op cit.*, pp. 202–203.

14 On the discussion of counterfactual conditionals see Georg Brunold, *Erklärung, Prognostik & Scientific Fiction: Zur philosophischen Pathologie eines wissenschaftstheoretischen Notstands* (Königstein: Hain, 1984).

15 On the term "responsibility" cf. also H. L. A. Hart, *Punishment and Responsibility* (Oxford: Oxford Univ. Press, 1968). See also on this and other points the author's "Mut zur Furcht. Facetten technischer Humanität in Vergangenheit und Zukunft," *Mitteilungen der Technischen Universität Carolo Wilhelmina zu Braunschweig* **19**, no. 1 (1984), 33–40.

16 Cf. M. G. Singer, *Generalization in Ethics* (New York: Knopf, 1961); J. Rawls, *A Theory of Justice* (Cambridge, MA: Harvard Univ. Press, 1971); and W. K. Frankena, *Ethics*, 2nd edition (Englewood Cliffs, NJ: Prentice-Hall, 1973).

17 C. F. H. Lenk, "Zu ethischen Fragen des Humanexperiments," in H. Lenk, *Pragmatische Vernunft* (Stuttgart: Reclam, 1979), pp. 50–76; cf. also T. L. Beauchamp and L. Walters (eds.), *Contemporary Issues in Bioethics* (Belmont, CA: Dickenson, 1978), Part V: "Human Experimentation," pp. 399–501.

18 "The Nuremberg Code" (1949). reprinted in T. L. Beauchamp and L. Walters (eds.), *op cit.*, pp. 404–405.

SELECT ANNOTATED BIBLIOGRAPHY ON PHILOSOPHICAL STUDIES OF INFORMATION TECHNOLOGY AND COMPUTERS

Although primarily a survey of recent English-language philosophical discussions on computers and information technology in book form, with a more limited reference to periodical literature, this compilation is designed to serve four supplementary purposes.

- It is a key for references in the introduction.
- It picks up and further documents some works from the bibliographies to individual papers.
- While stressing the latest phase in the development of this special field in the philosophy of technology, a few "classic" studies are included for historical perspective.
- While emphasizing philosophy, some works which provide an empirical description of affairs are also cited.

In accord with this last point, following the list of bibliographies, there are special sections devoted to historical studies and to technical studies – although these should not be viewed as wholly distinct from philosophical ones. Philosophy does not live by thought alone, nor is it just another speciality among those which have already proliferated in our technical age. At the same time that it broaches trans-empirical issues, it bridges specialities by contributing to the most general discussion possible.

1. BIBLIOGRAPHIES

101   Abshire, Gary M. *The Impact of Computers on Society and Ethics: A Bibliography.* Morristown, NJ: Creative Computing Press, 1980. Pp. 120. An unannotated, alphabetical list of nearly 2000 books, magazine articles, news stories, scholarly papers, and other works on the subjects indicated by the title, together with some general background studies. Covers the period 1948 to 1979. Better than the Austing bibliography cited below because it does include some explicitly philosophical works. Abshire has worked on the management and architecture of computer-based systems for IBM and the U.S. Air Force, and from 1973 to 1977 directed the computer science education efforts at the IBM research laboratory in Boulder, CO.

102   *ACM Guide to Computing Literature.* This annual, unannotated index and supplement to the monthly *Computing Reviews* (vol. 1 = 1960) is the single best catalog of

307

Carl Mitcham and Alois Huning (eds.), Philosophy and Technology II, 307–346.
© 1986 *by D. Reidel Publishing Company.*

publications in the field as a whole. Sorts items by author, keyword (see "ethics," "philosophy," etc.), category (see esp. "Computing Millieux"), proper names, general terms, reviewers, and sources. Much superior to *Computer Books and Serials in Print* (New York: Bowker, 1984) which contains only one book under the subject heading "philosophy" and none under "ethics."

103   Austing, Richard H. and Gerald L. Engel *et al. An Annotated Bibliography on Computer Impact on Society.*Microfiche. From Gerard L. Engel, Virginia Institute of Marine Science, Gloucester Point, VA. An NSF funded "study of computer impact on society and computer literacy courses and materials." Over 2 000 mostly annotated entries focusing on the period 1970–1976. Emphasis is more on engineering, business, or popular journalistic works relevant to issues of computer literacy than on humanities or social science studies of the computer impact. Exhaustive with regard to articles in *Time* and *Business Week*, but the classification scheme for entries does not even include "philosophy." (Austing is now an editor of *Computing Reviews* and does better.)

104   Bramer, Max and Dawn Bramer. *The Fifth Generation: An Annotated Bibliography*. Reading, MA: Addison-Wesley, 1984. Pp. 119. Good introduction and annotations.

105   Center for Telecommunications Studies. George Washington University, Washington, DC 20052. This Center publishes a monthly *Communication Booknotes* and a series of three regularly revised bibliographies. Basic Bibliography no. 1 is on *Telecommunications Policy*; no. 2 deals with *Mass Communication and Electronic Media*; and no. 3 is entitled *Foreign and International Communications Systems*. Each is about 20 pages in length, cites over 200 items, is divided into subject categories, and is briefly annotated.

106   Cortada, James W. *An Annotated Bibliography on the History of Data Processing*. Westport, CT: Greenwood Press, 1983. Pp. xiii, 216. As the author observes, this is the first major attempt to survey materials on this subject. Good introductory essay on the history of data processing which highlights some philosophical connections. The first chapter of the bibliography does the same, with sections on Leibniz and Pascal. The rest of the bibliography is broken down into three historical periods: "From Punched Cards to Digital Computers, 1800–1939," "Birth of the DP Industry, 1939–1955," and "Computer Age, 1955–1982." Annotations are usually brief. A few errors – e.g., "Gilfallan" for Gilfillan – too much reliance on initials in place of full names. By an employee of IBM.

107   Mitcham, Carl and Robert Mackey. *Bibliography of the Philosophy of Technology*. Chicago: Univ. of Chicago Press, 1973. Pp. xvii, 205. First published as a special supplement to *Technology and Culture* **14,** no. 2 (April 1973). Includes references to technical works by cyberneticists and others from the 1950s and 1960s referred to in the Introduction. Supplemented by Mitcham and Grote, "Current Bibliography in the Philosophy of Technology: 1972–1974," *Research in Philosophy and Technology* **1** (1978), 313–390; Mitcham and Grote, *Current Bibliography in the Philoso-*

phy of Technology: 1975–1976, a whole issue of *Research in Philosophy and Technology* **4** (1981), 1–241; and Mitcham and Grote, "Current Bibliography in the Philosophy and Technology: 1977–1978, Primary Sources," *Research in Philosophy and Technology* **6** (1983), 231–289. The present bibliography relies heavily on these previously published studies, all of which identify information technology and computers as important themes in the philosophy of technology. Should thus be consulted for further references, especially to articles.

108   Randell, Brian. "An Annotated Bibliography on the Origins of Digital Computers," *Annals of the History of Computing* **1**, no. 2 (Oct. 1979), 101–207. Nearly 750 annotated and indexed citations of papers, books, and other items related to the origins of modern electronic computers. Topics covered range from early digital calculating devices and mechanical automata to the first stored program computers (circa 1949), with new entries added up to June 1979. One notable omission: Herman H. Goldstine's *The Computer: From Pascal to von Neumann* (1973). Helpful general Introduction, pp. 101–104, calling attention to the ways in which this work extends its 350 item predecessor, a bibliography in Brian Randell (ed.), *The Origins of Digital Computers: Selected Papers* (Berlin: Springer-Verlag, 1973). Another version of this bibliography is included in N. Metropolis, J. Howlett, and Gian-Carlo Rota (eds.), *A History of Computing in the Twentieth Century* (1980).

109   Taviss, Irene and Judith Burbank. "Implications of Computer Technology," *Research Review,* whole no. 7 (Cambridge, MA: Harvard Univ. Program on Technology and Society, 1971), 56. A ten page introduction followed by extensive annotations of selected works divided into three categories: economic, political, and cultural implications.

2. HISTORICAL STUDIES

201   Bernstein, Jeremy. *The Analytical Engine: Computers–Past, Present, and Future.* New York: Random House, 1963. Pp. xi, 113. Revised edition, New York: William Morrow, 1981. Pp 131. First written as a series of articles in *The New Yorker;* provides a brief, intelligent overview. Just mentions the microelectronic revolution.

202   Cohen, John. *Human Robots in Myth and Science.* London: Allen & Unwin, 1966. New York: A.S. Barnes, 1967. Pp. 156. A classic little study of the hardware of (such as it has been), imagination about, and attitudes toward robots from antiquity to the present. Concludes with a philosophical chapter on "Is Man a Robot?" Translated into German as *Golem und Roboter: Über kunstliche Menschen* (Frankfurt: Umschau-Verlag, 1968) and into French as *Les Robots humains dans le mythe et dans la science* (Paris: J. Vrin, 1968).

203   Evans, Christopher. *The Micro Millenium.* New York: Viking, 1979. Pp. 255. Paperback reprint, New York: Washington Square Press, 1981. Pp. x, 308. (Also published under the title, *The Mighty Micro* [London: Victor Gollancz, 1979].) Popular, pro-computer retrospect and prospect with brief introduction to social impact. Re-written as a kid's book and illustrated with pictures as *The Making of the*

Micro: A History of the Computer (New York: van Nostrand Reinhold, 1981). By a late British computer scientist.

204   Fishman, Katharine Davis. *The Computer Establishment*. New York: Harper & Row, 1981. Pp. ix, 468. A detailed study of U.S. corporate computer history, necessarily concentrating on IBM, which stops short of the microelectronic revolution. See also Ernest Braun and Stuart MacDonald, *Revolution in Miniature: The History and Impact of Semiconductor Electronics* (Cambridge: Cambridge Univ. Press, 1978). Franklin M. Fisher, *Folded, Spindled, and Mutilated: Economic Analysis and U.S. v. IBM* (Cambridge, MA: MIT Press, 1983); and Emerson W. Pugh, *Memories That Shaped an Industry: Decisions Leading to IBM Systems/360* (Cambridge, MA: MIT Press, 1984).

205   Freiberger, Paul and Michael Swaine. *Fire in the Valley: The Making of the Personal Computer*. Berkely: Osborne/McGraw-Hill, 1984. Pp. xiii, 228. The title gives this one away. For another on the topic see Everett M. Rogers and Judith K. Larsen, *Silicon Valley Fever: Growth of High-Technology Culture* (New York: Basic Books, 1984).

206   Goldstine, Herman H. *The Computer: From Pascal to von Neumann*. Princeton: Princeton Univ. Press, 1972. Pp. x, 378. The best general history in English, by a man who worked with von Neumann and a bit biased in his favor. Part one covers the period up to World War II. Part two covers the wartime development of ENIAC and EDVAC. Part three focuses on post war work at Princeton. An appendix surveys world-wide developments. Good index. For a good second study of this same history in a slightly broader framework, see Rene Moreau, *The Computer Comes of Age: The People, the Hardware, and the Software*, trans. J. Howlett (Cambridge, MA: MIT Press, 1984), a volume originally published as *Ainsi naquit l'informatique* (Paris: Bordas, 1981).

207   Hawkes, Nigel. *The Computer Revolution*. London: Thames and Hudson, 1971. Pp. 216. Reprint, New York: Dutton, 1972. A picture history book. Not serious.

208   Hyman, Anthony. *Charles Babbage: Pioneer of the Computer*. Princeton, NJ: Princeton Univ. Press, 1982. Pp. xi, 287. Overview of Babbage's life and work.

209   Kidder, Tracy. *The Soul of a new Machine*. Boston: Little Brown, 1981. Pp. 293. Paperback reprint, New York: Avon, 1982. Pp. 293. A non-fiction novel which provides insight into the techno-corporate development of mini-computers.

210   Levy, Steven. *Hackers: Heros of the Computer Revolution*. Garden City, NY: Anchor Doubleday, 1984. Pp. xv, 458. Journalistic account of the discovery and exploration of computers by the first generations of "civilian" users at MIT in the 1950s and 1960s, Berkeley in the 1970s, and Apple Computers in the 1980s. "As I talked to these digital explorers, ranging from those who tamed multimillion-dollar machines in the 1950s to contemporay young wizards who mastered computers in their suburban bedrooms, I found a common element, a common philosophy which seemed tied to the elegantly flowing logic of the computer itself. It was a philosophy

of sharing, openness, decentralization, and getting your hands on machines at any cost – to improve the machines, and to improve the world. This Hacker Ethic is their gift to us: something with value even to those of us with no interest at all in computers" (p. viii). Chapter 2, pp. 26–36, explicitly outlines "The Hacker Ethic."

211   McCorduck, Pamela. *Machines Who Think: A Personal Inquiry into the History and Prospects of Artificial Intelligence*. New York: W. H. Freeman, 1979. Pp. xiv, 375. Best available popular historical overview of AI work by a science writer with strong sympathies for the AI community. Contains a good account of the summer 1956 Darthmouth conference in which ten people got together and established the discipline of AI. Contains good profiles of the four most important members of that conference: John McCarthy, Marvin Minsky, Allen Newell, and Herbert Simon. Some other profiles of the major actors: Philip J. Hilts on McCarthy in *Scientific Temperaments: Three lives in Contemporary Science* (New York: Simon & Schuster, 1982), and Jeremy Bernstein on Minsky in *Science Observed: Essays Out of My Mind* (New York: Basic Books, 1982). For a non-fiction novel type update on the AI community see Frank Rose, *Into the Heart of the Mind: An American Quest for Artificial Intelligence* (New York: Harper & Row, 1984).

212   Metropolis, N., J. Howlett, and Gian-Carlo Rota (eds.). *A History of Computing in the Twentieth Century*. New York: Academic Press, 1983. Pp. xix, 659. Thirty-eight papers divided into groups dealing with historiographic issues, individuals, computer languages, different machines, and research centers.

213   Osborne, Adam. *Running Wild: The Next Industrial Revolution*. Berkeley, CA: Osborne/McGraw-Hill, 1979. Pp. x. 181. "Perhaps the most paralyzing aspect of the microelectronics industrial revolution is the inability of law-makers and sociologists to cope with what is occurring. . . . It is relatively easy for me to predict what microelectronics can do, but it is hard to estimate the social consequences. . . . Therefore, in this book I limit my ambitions to describing technological events that have occurred and forecasting events that I believe are possible" (pp. ix–x). An insider's account.

214   Shurkin, Joel. *Engines of the Mind: A History of the Computer*. New York: W.W. Norton, 1984. Pp. 352. Covers some of the same territory as Goldsteine, especially the development of ENIAC, documenting the rivalries between engineers (who thought their hardware inventions really did it) and scientists (who thought their ideas were more responsible).

215   Stoneman, Paul. *Technological Diffusion and the Computer Revolution: The UK Experience*. New York: Cambridge Univ. Press, 1976. Pp. xii, 219. Revised Ph.D. dissertation in economics describing the spread of computers in Great Britain, 1954–1970. "The main purpose [is] to investigate in detail one clearly defined example of a change in technique in an attempt to clarify the forces promoting . . . and the effects of that change."

3. TECHNICAL STUDIES

301 Abelson, Philip H. and Allen L. Hammond (eds.). *Electronics: The Continuing Revolution*. Washington, DC: American Association for the Advancement of Science, 1977. Pp. 217. Bibliography and index.

302 Barr, Avron and Edward A. Feigenbaum (eds.). *The Handbook of Artificial Intelligence*. Vol. I, Stanford, CA: HevisTech Press; and Los Altos, CA: William Kaufmann, 1981. Pp. xiv, 409. Vol. II, 1982. Pp. xiii, 428. Vol. III, Paul R. Cohen and Feigenbaum (eds.), 1982. Pp. xviii, 639. These three volumes contain over 200 technical articles on all areas of AI research.

303 Brooks, Frederick P. Jr. *The Mythical Man-Month: Essays on Software Engineering*. Reading, MA: Addison-Wesley, 1982. Pp. xi, 195. By an engineer involved with the development and management of IBM mainframes. Intended "for professional programmers, professional managers, and especially professional managers of programmers." Thesis: "I believe that large programming projects suffer management problems different in kind from small ones, due to division of labor. I believe the critical need to be the preservation of the conceptual integrity of the product itself" (p. viii).

304 Hartley, R. V. L. "Transmission of Information," *Bell System Technical Journal* **7,** no. 3 (July 1928), 535–563. Abstracts from "psychological factors . . . to establish a measure of information in terms of purely physical quantities."

305 McCulloch, Warren S. *Embodiments of Mind*. Cambridge, MA: MIT Press, 1965. Pp. xx, 402, A classic collection of papers by a computer scientist with a classic humanistic education and a marvelous sense of the English language. Includes "Why the Mind is in the Head" (1951), "Toward Some Circuitry of Ethical Robots or an Observational Science of the Genesis of Social Evaluation of the Mind-Like Behavior of Artifacts" (1956), "What the Frog's Eye Tells the Frog's Brain" (1959), and "Machines That Think and Want" (1950).

306 Marr, David. *Vision: A Computational Investigation into the Human Representation and Processing of Visual Information*. San Francisco: W. H. Freeman, 1983. Pp. 397. Posthumous work by an English mathematician and neuroscientist who died of leukemia in 1980 at the age of 35. For a good review of Marr's achievement, see Israel Rosenfeld, "Seeing Through the Brain," *New York Review of Books* **31,** no. 15 (Oct. 11, 1984), 53–56.

307 *Microelectronics*. San Francisco: W. H. Freeman, 1977. Pp. 145. A collection of articles from *Scientific American* on solid-state computer technology. See also the earlier *Information* (San Francisco: W. H. Freeman, 1966).

308 Raphael, Bertram. *The Thinking Computer: Mind Inside Matter*. San Francisco: W. H. Freeman, 1976. Pp. xiii, 322. Good introduction (neither too superficial nor too technical) to artificial intelligence. Many references. For a slightly more tech-

nical introduction see Patrick Henry Winston, *Artificial Intelligence* (Reading, MA: Addison-Wesley, 1977).

309   Saracevic, Tefko (ed.). *Introduction to Information Science*. New York: R. R. Bowker, 1970. Pp. xxiv, 751. Information science is library science for the information age, i.e. an attempt to develop systematic ways for dealing with the massive amounts of information made available by information technologies and computers. This huge volume of 65 articles is divided into three major sections: Part one covers "Basic Phenomena" (the nature of information, communication, etc.), part two gives descriptions of the major kinds of "Information Systems," and part three deals with "Evalution of Information Systems." Part four, as a kind of appendix, provides "A Unifying Theory" by William Goffman. See also Arthur W. Elias (ed.), *Key Papers in Information Science* (Washington, DC: American Society for Information Science, 1971), which reprints 19 different articles from various journals.

310   Shannon, Claude E. and Warren Weaver. *The Mathematical Theory of Communication*. Urbana: Univ. of Illinois Press, 1949. Pp. 125. Two papers. The first, by Weaver, "Recent Contributions to the Mathematical Theory of Communications" (pp. 1–28), is a much-enlarged version of an article from *Scientific American* (July 1949). The second, by Shannon, "The Mathematical Theory of Communication" (pp. 29–125), is reprinted from *Bell System Technical Journal* (July and Oct. 1948). See also Shannon's "Prediction and Entropy of Printed English," *Bell System Technical Journal* **30,** no. 1 (Jan. 1951), 50–64.

311   Simon, Herbert A. "On the Nature of Understanding," in Anita K. Jones (ed.), *Perspectives on Computer Science* (New York: Academic Press, 1977), pp. 199–216. Reviews AI research and concludes that "the design and investigation of systems that understand should . . . continue to be as exciting and rewarding an area of AI research over the next decade as it has been over the past one."

312   Webber, Bonnie Lynn and Nils J. Nilsson (eds.), *Readings in Artificial Intelligence*. Palo Alto, CA: Tioga Publishing Co., 1981. Pp. x, 547. A collection of 31 papers. Chapter 5, "Advanced Topics," is explicitly philosophical, and includes John McCarthy and Patrick Hayes' "Some Philosophical Problems from the Standpoint of Artificial Intelligence" and John McCarthy's "Epistemological Problems of Artificial Intelligence."

313   Winston, Patrick Henry and Richard Henry Brown (eds.). *Artificial Intelligence: An MIT Perspective*. Vol. 1: *Expert Problem Solving, Natural Language Understanding, Intelligent Computer Coaches, Representation and Learning*. Cambridge, MA: MIT Press, 1979. Pp. xiii, 492. Vol. 2: *Understanding Vision, Manipulation, Computer Design, Symbol Manipulation*, 1979. Pp. xiii, 486. Brings together over 20 papers which originally appeared as publications of the MIT Artificial Intelligence Laboratory. Volume I contains a version of Marvin Minsky's "The Society Theory of Thinking" (pp. 423–450) which has important psychological and epistemological implications.

4. GENERAL BIBLIOGRAPHY

401   Anderson, Alan Ross. *Minds and Machines*. Englewood Cliffs, NJ: Prentice-Hall, 1964. Pp. viii, 114. This was the first truly philosophical text on artificial intelligence. Contents: A. M. Turing's "Computing Machinery and Intelligence" (from *Mind*, 1950), J. R. Lucas' "Minds, Machines and Gödel" (from *Philosophy*, 1961), Keith Gunderson's "The Imitation Game" (from *Mind*, forthcoming), Hilary Putnam's "Minds and Machines" (from S. Hook (ed.), *Dimensions of Mind*, 1960), Paul Ziff's "The Feelings of Robots" (from *Analysis*, 1959), J. J. C. Smart's "Professor Ziff on Robots" (from *Analysis*, 1959), and Ninian Smart's "Robots Incorporated" (from *Analysis*, 1959). As is readily apparent, questions of AI *were* philosophy of technology for the English analytic tradition in the 1950s.

402   Attneave, Fred. *Applications of Information Theory to Psychology: A Summary of Basic Concepts, Methods, and Research*. New York: Holt, 1959. Pp. 159. See also Harold Borko (ed.), *Computer Applications in the Behavioral Sciences* (Englewood Cliffs, NJ: Prentice-Hall, 1962). Pp. 633. This major overview includes an analysis of "Computer Systems" (Part I), "Computer Fundamentals" (Part 2), and a collection of 18 papers on "Computer Applications" (Part 3). Among these 18 papers are Charles Wrigley's "The University Computing Center," Harry F. Silberman and John E. Coulson's "Automated Teaching," Julian Feldman's "Computer Simulation of Cognitive Processes," W. Ross Ashby's "Simulation of a Brain," James T. Culbertson's "Nerve Net Theory," Robert S. Ledley's "Advances in Biomedical Science and Diagnosis," R. Clay Sprowls' "Business Simulation," Oliver Benson's "Simulation of International Relations and Diplomacy," and Borko's "A Look into the Future."

403   Bar-Hillel, Yehoshua. *Language and Information: Selected Essays on Their Theory and Application*. Reading, MA: Addison Wesley, 1964. Pp. 388. "The section on information theory, written more than ten years ago, is lucid and still valuable. . . . Its warnings are still in place, alas, against confusing the concept of information *per se* with a statistical measure of the rarity of the symbols in which information is expressed." – from a review by D.M. MacKay, *British Journal for the Philosophy of Science* **16,** whole no. 63 (Nov. 1965), pp. 253–255.

404   Barquin, Ramon C. and Graham P. Mead (eds.). *Towards the Information Society*. Selected papers from the Hong Kong Computer Conference 1983. New York: North-Holland, 1984. Pp. xiii, 164. Twenty-two papers which provide a British and East Asian perspective.

405   Bell, Daniel, *The Coming of Post-Industrial Society*. New York: Basic Books, 1973. Paperback reprint, with new introduction, New York: Basic Books, 1976. Pp. xxvii, 507. The new introduction and chapter 3, "The Dimensions of Knowledge and Technology: The New Class Structure of Post-Industrial Society," are for present purposes the key sections. But Bell's analysis of and theory concerning the societal influence of information technology is spelled out in more detail in later articles. See "Teletext and Technology: New Networks of Knowledge and Informa-

tion in Post-Industrial Society," *Encounter* **48,** no. 6 (June 1977), 9–29, which is reprinted in Bell's *The Winding Passage: Essays and Sociological Journeys 1960-1980* (New York: Basic Books, 1980), pp. 34–65; and the contribution to Michael L. Dertouzos and Joel Moses (eds.) (1979), which is also included in Tom Forester (ed.) (1980), each of which is cited below. Both articles are said to be drawn from an unpublished study, "The Social Framework of the Information Society," written for the MIT Laboratory for Computer Science.

406 Berkeley, Edward, *The Computer Revolution.* New York: Doubleday, 1962. Pp. 249. Early extended statement of the "computer revolution" thesis. Some earlier discussions of computers by Berkeley: *Symbolic Logic and Intelligent Machines* (New York: Reinhold, 1959); and *Giant Brains* (New York: Science Editions, 1961).

407 Bertalanffy, Ludwig von. *Robots, Men, and Minds: Psychology in the Modern World.* New York: Braziller, 1967. Pp. x, 150. Attacks mechanism and behaviorism in the physical and social sciences, arguing for a general systems theory approach to understanding the human. General systems theory should not be identified with cybernetics. The cybernetic system is basically mechanistic and closed, but general systems are determined by the interaction of many forces and variables. For analysis see Mark Davidson, *Uncommon Sense: The Life and Thought of Ludwig von Bartalanffy (1901–1972), Father of General Systems Theory* (Los Angeles: J. P. Tarcher, 1983).

408 Boden, Margaret. *Artificial Intelligence and Natural Man.* New York: Basic Books, 1977. Pp. ix, 537. "Artificial intelligence is not the study of computers, but of intelligence in thought and action. Computers are its tools, because its theories are expressed as computer programs that enable machines to do things that would require intelligence if done by people" (p. xi). A comprehensive, non-mathematical, but technically sophisticated introduction to AI, arguing for "its potential for counter-acting the dehumanizing influence of natural science, for suggesting solutions to many traditional problems in the philosophy of mind, and for illuminating the hidden complexities of human thinking and personal psychology" (p. 4). Stresses that AI is intricate enough to call for "mentalist" descriptions. Argues, too, that the intentional concept of "representation" or internal modeling, which is central to AI, can help philosophers deal with hermeneutic paradoxes (cf. pp. 396–398). "It is possible for the categories of subjectivity to be properly attributed to human beings because bodily processes in our brains function as models, or representations, of the world" (p. 428). Thus AI effectively demonstrates, "in a scientifically acceptable manner, how it is possible for psychological beings to be grounded in a material world and yet be properly distinguished from 'mere matter.' Far from showing that human beings are 'nothing but machines,' it confirms our insistence that we are essentially subjective creatures living through our own mental constructions of reality (among which science itself is one)" (p. 473). Rather negative review by Guy Robinson, *Philosophy* **54,** whole no. 207 (Jan. 1979), 130–132. Much more favorable on by Daniel C. Dennett, *Philosophy of Science* **45,** no. 4 (Dec. 1978), 648–649. For a selection of Boden's papers from 1962 to 1980 see her *Minds and Mechanisms: Philosophical Psychology and Computational Models* (Ithaca, NY: Cornell Univ. Press, 1981), which also contains a bibliography of her work.

409   Bolter, J. David. *Turing's Man: Western Culture in the Computer Age*. Chapel Hill,
       NC: Univ. of North Carolina Press, 1984. Pp. xii, 264. Electronic technology is the
       defining technology of our age, just as manual technology was for the Greeks and
       mechanical technology was for Western Europe (chapter 2, pp. 15–42). "A defin-
       ing technology defines or redefines man's role in relation to nature. By promising
       (or threatening) to replace man, the computer is giving us a new definition of man,
       as an 'information processor,' and of nature, as 'information to be processed.' I call
       those who accept this view of man and nature Turing's men" (p. 13). An interesting
       book, but marred by inconsistencies, lack of depth, and a vapid conclusion. "The
       computer is in some ways a grand machine in the Western mechanical-dynamic
       tradition and in other ways a tool-in-hand from the ancient craft tradition. The best
       way to encourage the humane use of computers is to emphasize, where possible, the
       second heritage over the first, the tool over the machine" (pp. 232–233). For a more
       favorable review see Paul Delany, "Socrates, Faust, Univac," *New York Times
       Book Review* (March 18, 1984), p. 13.

410   Brod, Craig. *Technostress: The Human Cost of the Computer Revolution*. Reading,
       MA: Addison-Wesley, 1984. Pp. xiii, 242. Argues that computer use has a special
       character which contributes to psychological problems in many people. One exam-
       ple is a tendency to devalue personal relationships. This is true in the workplace, the
       school, and the home. Relies heavily on interviews with high intensity computer
       users. Includes a good bibliography on the psychology of human-computer interac-
       tions. By a psychotherapist.

411   Bunge, Mario. "Do Computers Think?," *British Journal for the Philosophy of Sci-
       ence* **7**, whole no. 26 (Aug. 1956), 139–148; and *ibid*. **7**, whole no. 27 (Nov. 1956),
       212–219. "What I propose to do here in order to ascertain whether machines think
       or not, is to examine succinctly the two main aspects of the question, namely (a) the
       nature of computers, and (b) the nature of mathematical thought. . . . Insofar as
       machines are the outcome of intelligent and purposive work, they cannot be put in
       the same class as natural inanimate objects; machines are matter intelligently orga-
       nized by technology, and as such they stand on a level of their own. But, on the other
       hand, it should be kept in mind that artifacts, however complex, operate only with
       material objects, never with ideal, abstract objects, a sort of operation which is pre-
       cisely one of the distinctive characteristics of educated human beings. This
       elementary point is missed by most cyberneticians, and it seems to be the clue for
       the understanding of the whole question." Bunge's argument, from almost 30 years
       ago, is still being made.

412   Burnham, David. *The Rise of the Computer State*. New York: Random, 1983.
       Pp. xi, 273. On the use of technology to by-pass American Fourth Amendment
       guarantees to privacy. "With a great deal of personal information jotted down on
       records physically stored in the house [sic] of citizens in the pre-computer age, law
       enforcement required a search warrant to obtain it. With the development of large
       environmentrized organizations such as hospitals, insurance companies and credit
       reporting companies, however, law enforcement can now obtain a great variety of
       information about an individual on the basis of an informal contact or a written
       request that is not reviewed by a judge" (p. 12). Examines the activities of the IRS,

AT&T, the Social Security Administration, and credit agencies in chapters dealing with data banks, the centralization of power, surveillance, and value change. One chapter devoted to the National Security Agency. Favorable review: Robert Asahina, "Electronic Power." *New York Times Book Review* (Aug. 21, 1983), pp. 8 and 17. Burnham, a Pulitzer Prize winning journalist, is Washington correspondent for the *New York Times*.

413   Calder, Nigel. *1984 and Beyond*. New York: Penguin, 1984. Pp. 204. Glib science writing, covering many of the social issues relevant to information technology, in the form of an imagined conversation with a computer.

414   Cherry, Colin. *On Human Communication: A Review, a Survey, and a Criticism*. 3rd edition. Cambridge, MA: MIT Press, 1978. Pp. xv, 374. This is a classic (and still current) survey of communication theory, from mathematical information to linguistics. Chapter 2 is an enlarged version of Cherry's "A History of the Theory of Information," *Proceedings of the Institution of Electrical Engineers*, part III, vol. 98 (1951), pp. 383–393. The first edition appeared in 1957, the second in 1966. The third edition adds a new chapter 8 on "Human Communications: Feeling, Knowing, and Understanding." There is an extensive, although unannotated bibliography, which is updated for every edition. See also Cherry's *World Communication: Threat or Promise?* (New York: Wiley-Interscience, 1977).

415   Christians, Clifford G., Kim B. Rotzoll, and Mark Fackler. *Media Ethics: Cases and Moral Reasoning*. New York: Longman, 1983. Pp. xix, 332. A unique textbook. Part one deals with news reporting, part two advertising, and part three with entertainment. It analyzes ethical case problems having to do with censorship, child education, confidentiality, conflict of interests, deception, economic pressures, explicit sex, fairness, health and safety, law-bending, being an accessory to criminal actions, media self-criticism, minorities and the elderly, privacy, sensationalism, stereotyping, and violence in various media – from books, magazines, and newspapers to photography, motion pictures, radio, and television.

416   "Computers as Poison." *Whole Earth Review* [continuing *Co-Evolution Quarterly*], whole no. 44 (Dec. 1984–Jan. 1985), 1–55. A symposium of criticism mostly by dedicated computer users. Included Art Kleiner's "The Ambivalent Miseries of Personal Computing" and Jerry Mander's "Six Grave Doubts About Computers."

417   Crosson, Frederick J. (ed.). *Human and Artificial Intelligence*. New York: Appleton-Century-Crofts, 1970. Pp. viii, 267. Contents: Herbert A. Simon and Allen Newell's "Information Processing in Computer and man," Dean E. Wooldridge's "Computers and the Brain," Arthur L. Samuel's "Some Studies in Machine Learning Using the Game of Checkers, II – Recent Progress," Michael Scriven's "The Complete Robot: A Prolegomena to Androidology," D. M. MacKay's "The Use of Behavioural Language to Refer to Mechanical Processes," Hubert L. Dreyfus's "A Critique of Artificial Intelligence," Hilary Putnam's "Robots: Machines or Artificially Created Life?" R. Albritton's "Mere Robots and Others," Ulric Neisser's "The Imitation Game," Michael Polanyi's "The Logic of Tacit Inference," and Mortimer Adler's "The Consequences for Action."

417a   Crosson, Frederick J. and Kenneth M. Sayre (eds.). *Philosophy and Cybernetics*.
       New York: Simon & Schuster, 1967. Pp. xvi, 271. Contents: Sayre's "Philosophy
       and Cybernetics," J. L. Massey's "Information, Machines, and Men," Sayre's
       "Choice, Decision, and the Origin of Information," Crosson's "Information
       Theory and Phenomenology," Sayre's "Toward a Quantitative Model of Pattern
       Formation," Crosson's "Memory, Model's, and Meaning," D. B. Burrell's
       "Obeying Rules and Following Instructions," and Sayre's "Instrumentation and
       Mechanical Agency."

418   Danziger, James N. *et al. Computers and Politics: High Technology in American
       Local Governments*. New York: Columbia Univ. Press, 1982. Pp. xv, 280. Grist for
       the philosophical mills.

419   Dechert, Charles R. (ed.). *The Social Impact of Cybernetics*. Notre Dame: Univ. of
       Notre Dame Press, 1966. Paperback reprint, New York: Simon & Schuster, n. d.
       Pp. vi, 206. Contents: John Diebold's "Goals to Match Our Means," Dechert's
       "The Development of Cybernetics," Robert Theobald's "Cybernetics and the
       Problems of Social Reorganization," Ulric Neisser's "Computers as Tools and as
       Metaphors," Marshall McLuhan's "Cybernation and Culture," Hyman G. Rickov-
       er's "A Humanistic Technology," Maxim W. Mikulak's "Cybernetics and
       Marxism-Leninism," and John J. Ford's "Soviet Cybernetics and International De-
       velopment." Important overviews.

420   Deken, John. *The Electronic Cottage*. New York: Morrow, 1982. Pp. 341. A gee-
       whiz get-acquainted book, which nevertheless exhibits some depth. By a Ph. D.
       statistician.

421   Dennett, Daniel C. *Brainstorms: Philosophical Essays on Mind and Psychology*.
       Cambridge, MA: MIT Press, 1978. Pp. xxii, 353. A collection of papers, many of
       which are directly relevant to AI issues – e. g., "Artificial Intelligence as Philosophy
       and as Psychology," "Why You Can't Make a Computer that Feels Pain," and "The
       Abilities of Men and Machines." For a more systematic presentation of Dennett's
       critique of two other philosophies of mind (strict reductionism and Turing machine
       functionalism) in favor of what he terms "token functionalism," see his *Content and
       Consciousness* (London: Routledge & Kegan Paul, 1969).

422   Dertouzos, Michael L. and Joel Moses (eds.). *The Computer Age: A Twenty-Year
       View*. Cambridge, MA: MIT Press, 1979. Pp. xvi, 491. A panoramic collection.
       Part 1, "Prospects for the Individual," contains Moses' "The Computer in the
       Home," Nicholas P. Negroponte's "The Return of the Sunday Painter," Der-
       touzos' "Individualized Automation," Terry Winograd's "Toward Convivial Com-
       puting," Seymour A. Papert's "Computers and Learning," and J. C. R. Licklider's
       "Computers and Government." Part II, "Trends in Traditional Computer Uses,"
       contains Victor A. Vyssotky's "The Use of Computers for Business Functions" and
       Sidney Fernbach's "Scientific Use of Computers." Part III, "Socioeconomic
       Effects and Expectations," contains Daniel Bell's "The Social Framework of the
       Information Society," Herbert A. Simon's "The Conseqences of Computers for
       Centralization and Decentralization," Robert G. Gilpin's "The Computer and

World Affairs," Roger G. Noll's "Regulation and Computer Services," Martin Shubik's "Computers and Modeling," and Kenneth J. Arrow's "The Economics of Information." Part IV, "Trends in the Underlying Technologies," contains Robert N. Noyce's "Hardware Prospects and Limitations," B. O. Evans' "Computers and Communications," Marvin Denicoff's "Sophisticated Software: The Road to Science and Utopia," Marvin L. Minsky's "Computer Science and the Representation of Knowledge," and Alan J. Perlis' "Current Research Frontiers in Computer Science." Part V, "Critiques," contains Joseph Weizenbaum's "Once More: The Computer Revolution," Bell's "A Reply to Weizenbaum," and Dertouzos' "Another Reply to Weizenbaum." Needless to say, the critics do not get a fair shake.

423   Diebold, John. *Automation*. New York: AMACOM (American Management Associations), 1983. Pp. xl, 181. This is a reprint, with a lengthy new introduction, of a 1952 management classic. See also Diebold's *Beyond Automation: Managerial Problems of an Exploding Technology* (New York: McGraw-Hill, 1964) and *Man and the Computer* (New York: Praeger, 1969).

423a   Diebold, John (ed.). *The World of the Computer*. New York: Random, 1973. Pp. xiii, 457. A basic collection of articles oriented toward business interests. The authors included in Part I, "The Computer Past and Present," are Charles Babbage, Jeremy Bernstein, Arthur W. Burks, Herman H. Goldstine, John von Neumann, John Presper Eckert Jr., John W. Mauchly, and Diebold. The authors (and subjects) of Part II, the major section of the volume entitled "Computers in Use," are Frederic G. Withington (long-range effects), Abraham J. Siegel (unions and employers), Herbert A. Simon (corporate management), Alvin Kaltman (in state government), Ron Brown (sales), Charles G. Burck (railroads), Siegfried M. Breuning (automated highways), R. Buckminster Fuller (higher education), Charles E. Silberman (primary education), Maya Pines (preschool education), G. Octo Barnett, Jerome H. Grossman, and Robert A. Greenes (health services research), Michael Crichton (medicine), D. S. Halacy Jr. (prosthetic devices), Anthony G. Oettinger (science), Carl F. J. Overage and R. Joyce Harman (libraries), and Stephen M. Parrish (literature). Part III, "The Computer's Impact on Society," contains: D. M. Mackay's "Machines and Societies," Paul Einzig's "Must Automation Bring Unemployment?" Margaret Mead's "The Challenge of Automation to Education for Human Values," Dennis Gabor's "Technology, Life and Leisure," Tom Alexander's "The Hard Road to Soft Automation," Carl Kaysen's "Data Banks and Dossiers," Kenneth E. Boulding's "The Wisdom of Man and the Wisdom of God," and Arthur C. Clarke's "The Obsolescence of Man." Part IV, "Computers and the Intelligence Argument," contains articles by Donald G. Fink, Alan M. Turing, Ulric Neisser, and Norbert Wiener.

424   Dizard, Wilson P. Jr. *The Coming Information Age: An Overview of Technology, Economics, and Politics*. New York: Longman, 1982. Pp. xv, 213. A good, solid establishment presentation of the information age thesis. For a slightly more contra-establishment approach to the same thesis, see Hiroshi Inose and John R. Pierce, *Information Technology and Civilization* (New York: W. H. Freeman, 1984), a "Club of Rome" study.

425   Dretske, Fred I. *Knowledge and the Flow of Information.* Cambridge, MA: MIT
       Press, 1981. Pp. xi, 273. A detailed attempt to move from information theory to a
       theory of meaning. "The entire project can be viewed as an exercise in naturalism –
       or, if you prefer, materialistic metaphysics" (p. xi). For a convenient summary of
       the argument of this book, see Fred I. Dretske, "Precis of *Knowledge and the Flow
       of Information,"* *Behavioral and Brain Sciences* **6** (1983).

426   Dreyfus, Hubert L. *What Computers Can't Do: A Critique of Artificial Reason.*
       New York: Harper & Row, 1972. Pp. xxxv, 259. Second revised edition, with a new
       subtitle, *The Limits of Artificial Intelligence,* New York: Harper & Row, 1979.
       Pp. xii, 354. The revision consists of a 66 page second introduction answering critics
       and extending the argument. This book grew out of Dreyfus' RAND Corporation
       Report P-3244 (Santa Monica, CA: RAND Corp., Dec. 1965) entitled *Alchemy
       and Artificial Intelligence.* Some related articles: "Phenomenology and Artificial
       Intelligence," in James Edie (ed.), *Phenomenology in America* (Chicago: Quad-
       rangle, 1967), pp. 31–47; "Why Computers Must Have Bodies in Order to Be In-
       telligent," *Review of Metaphysics* **21,** no. 2 (Sept. 1967), 13–32; "Cybernetics as the
       Last Stage of Metaphysics," *Proceedings of the XIVth International Congress of
       Philosophy,* vol. 2 (Vienna: Herder, 1968), pp. 493–499; and "Mechanism and
       Phenomenology," *Nous* **5,** no. 1 (Feb. 1971), 81–96.

427   "The Electronic Revolution." *American Scholar* **35,** no. 2 (Spring 1966), 189–374.
       This theme issue from twenty years ago provides good perspective. See especially
       Walter A. Rosenblith's "On Cybernetics and the Human Brain," Robert McClin-
       tock's "Machines and Vitalists: Reflections on the Ideology of Cybernetics," Her-
       bert A. Simon's "A Computer for Everyman," David B. Hertz's "Computers and
       the World Communications Crisis," Adam Yarmolinsky's "The Electronic Re-
       volution in the Pentagon," Max Bill's "Responsibility in Design and Information,"
       and H. F. William Perk's "The Great Transformation." Other contributors include
       Marshall McLuhan, R. Buckminster Fuller, Lynn White Jr., J. Bronowski, and W.
       H. Ferry. The "Revolving Bookstand" at the end of the symposium lists readings
       recommended by the authors.

428   Ellul, Jacques. *The Technological System.* Trans. Joachim Neugroschel. New
       York: Continuum, 1980. Pp. 362. This is the first of a two-volume comprehensive
       revision of *La technique ou l'enjeu du siecle* (1954), which was translated into En-
       glish and somewhat revised in the process as *The Technological Society* (1964). But
       even in 1964 computers did not play too important a role in society. Consequently,
       this update of the first two chapters of the original work includes not only references
       to later works by Habermas, Illich, Kuhn, Richta, etc., but also discusses the impact
       of the computer and information technology. The notion of a "technical system,"
       for instance, is an extension of the "technical ensemble" made possible by the com-
       puter. Two other books dealing with these topics which have been written but not
       yet published are the announced second volume of this revision devoted to *Dys-
       functions of the Technological System,* and a separate study of *Information and Soci-
       ety* on computers.

429   Feigenbaum, Edward A. and Julian Feldman (eds.). *Computers and Thought.* New
       York: McGraw-Hill, 1963. Pp. xiv, 535. This is a classic and influential collection.

Part I, "Artificial Intelligence," contains articles by one or more of the following authors: A. M. Turing, Allen Newell, J. C. Shaw, H. A. Simon, A. L. Samuel, H. Gelernter, J. R. Hansen, D. W. Loveland, Fred M. Tonge, James R. Slagle, Bert F. Green Jr., Alice K. Wolf, Carol Chomsky, Kenneth Laughery, Robert K. Lindsay, Oliver G. Selfridge, Ulric Neisser, Leonard Uhr, and Charles Vossler. The author-contributors to Part 2, "Simulation of Cognitive Processes," are Newell, Simon, Feigenbaum, Earl B. Hunt, Carl I. Hovland, Feldman, Geoffrey P. E. Clarkson, John T. Gullahorn, and Jeanne E. Gullahorn. Part 3 contains a "Survey of Approaches and Attitudes" by Paul Armer and Marvin Minsky. Part 4 is a comprehensive bibliography by Minsky. There is also a complete index.

429a   Feigenbaum, Edward A. and Pamela McCorduck. *The Fifth Generation: Artificial Intelligence and Japan's Computer Challenge to the World.* Reading, MA: Addison-Wesley, 1983. Pp. ix, 275. Part one argues that knowledge is a national resource. Part two describes the computer revolution and argues that computers can indeed think. Part three deals with the microelectronic revolution in computers, knowledge engineering, and expert systems. Part four describes the Japanese plan to build a new or fifth generation of computers. Part five analyzes the responses of other nations, while part six considers the proper American response. Part seven speculates about the future. For a critical review, see Joseph Weizenbaum, "The Computer in Your Future," *New York Review of Books* **30**, no. 16 (Oct. 27, 1984), 58–62: "The knowledge that appears to be least understood by Feigenbaum and McCorduck is that of the differences between information, knowledge, and wisdom, between calculating, reasoning, and thinking, and finally of the differences between a society centered on human beings and one centered on machines." Paperback reprint of Feigenbaum and McCorduck's book, revised and updated, New York: New American Library, 1984. Pp. xviii, 334. A related book: William H. Davidson, *The Amazing Race: Winning the Technorivalry with Japan* (New York: John Wiley, 1984).

430   Forester, Tom (ed.). *The Microelectronics Revolution: The Complete Guide to the New Technology and Its Impact on Society.* Cambridge, MA: MIT Press, 1981. Pp. xvii, 589. First published, Oxford: Blackwell, 1980. Part one, "The Microelectronic Revolution," has three chapters. Chapter 1, "The New Technology," includes contributions by Gene Bylinsky, Philip H. Abelson and Allen L. Hammond, Robert N. Noyce, and William G. Oldham, all describing aspects of the new electronic hardware. Chapter 2, "The Microelectronics Industry," contains sociological analyses of the industry by Forester, Ernest Braun, and Ian M. Mackintosh. Chapter 3, "Applications of the New Technology," contains A. A. Perlowski's "The 'Smart' Machine Revolution," A. J. Nichols' "An Overview of Microprocessor Applications," Paul M. Russo, Chih-Chung Wang, Philip K. Baltzer, and Joseph A. Weisbecker's "Microprocessors in Consumer Products," Lawrence B. Evans' "Industrial Uses of the Microprocessors," and Philip Venning's "Microprocessors in the Classroom." Part two, "Economic and Social Implications," contains chapters on "The Impact on Industry" (chapter 4), "The Revolution in the Office" (chapter 5), "The Consequencs for Employment" (chapter 6), and "Industrial Relations Implications" (chapter 7). Part three, "The Microelectronic Age," is the most philosophical. Chapter 8, "The Social Impact of Computers," contains Herbert A. Simon's "What Computers Mean for Man and Society," Joe Weizen-

baum's "Where Are We Going? Some Questions for Simon," Margaret A. Boden's "The Social Implications of Intelligent Machines," Theodore J. Lowi's "The Political Impact of Information Technology," Bruno Lefevre's "The Impact of Microelectronics on Town Planning," and John Garrett and Geoff Wright's "Micro is Beautiful." Chapter 9, "The Information Society?" contains Daniel Bell's "The Social Framework of the Information Society," Joe Weizenbaum's "Once More, The Computer Revolution," and Bell's "A Reply to Weizenbaum." Each chapter is followed by a "Guide to Further Reading," and there is a good index. See also a second Forester anthology, *The Information Technology Revolution* (Cambridge, MA: MIT Press, 1985). Pp. xvii, 674.

431   Friedrichs, Günter, and Adam Schaff (eds.). *Microelectronics and Society: For Better or For Worse*. A Report to the Club of Rome. New York: Pergamon Press, 1982. Pp. xii, 353. Contents: Alexander King's "Introduction: A New Industrial Revolution or Just Another Technology?" Thomas Ranald Ide's "The New Technology," Ray Curnow and Susan Curran's "The Technology Applied," Bruno Lamborghini's "The Impact on the Enterprise," John Evans' "The Worker and the Workplace," Friedrichs' "Microelectronics and Macroelectronics," Juan F. Rada's "A Third World Perspective," Frank Barnaby's "Microelectronics in War," Klaus Lenk's "Information Technology and Society," King's "Microelectronics and World Interdependence," and Schaff's "Occupation versus Work." Paperback reprint as *Microelectronics and Society: A Report to the Club of Rome*. New York: New American Library, 1983. Pp. xi, 338.

432   Frude, Neil. *The Intimate Machine: Close Encounters with Computers and Robots*. New York: New American Library, 1983. Pp. xii, 244. Breathless descriptions of how computers can be programmed to function as co-workers, therapists, teachers, and even friends. Brief bibliography and index. By a psychologist. "Provocative yet superficial." – Miles Orvel, *Technology Review* **87,** no. 3 (April 1984), 17–18.

433   Graham, Loren R. "Science and Computers in Soviet Society," in Erik P. Hoffman (ed.), *The Soviet Union in the 1980s*, a special issue of *Proceedings of the Academy of American Political Science* **35,** no. 3 (1984), 124–134. Describes Soviet ideological resistance to personal computers and the difficulties this poses for Soviet society. For a moderately different view of this issue, see S. Frederick Starr, "Technology and Freedom in the Soviet Union," *Technology review* **87,** no. 4 (May–June 1984), 38–47.

434   Greenberger, Martin (ed.). *Computers and the World of the Future*. Cambridge, MA: MIT Press, 1964. Pp. xxvi, 340. Originally published as *Management and the Computer of the Future* (Cambridge, MA: Press, 1962). A series of lectures celebrating MIT's centennial. Contents: C. P. Snow's "Scientists and Decision Making" with comment by Elting E. Morison and Norbert Wiener; Jay W. Forrester's "Managerial Decision Making" with comment by Charles C. Holt and Ronald A. Howard; Herbert A. Simon and Allen Newell's "Simulation of Human Thinking" with comment by Marvin L. Minsky and George A. Miller; John G. Kemeny's "A Library for 2000 AD" with comment by Robert M. Fano and Gilbert W. King; Alan J. Perlis' "The Computer in the University" with comment by Peter Elias and J. C. R. Licklider; John McCarthy's "Time-sharing Computer Systems" with comment by

John W. Mauchly and Gene M. Amdahl; George W. Brown's "A New Concept in Programming" with comment by Grace M. Hooper and David Sayre; and John R. Pierce's "What Computers Should Be Doing" with comment by Claude E. Shannon and Walter A. Rosenblith. Select bibliography and index.

435   Gunderson, Keith. *Mentality and Machines*. Garden City, NY: Doubleday, 1971. Pp. xviii, 173. Second edition, with an "Unconcluding Philosophic Postscript," Minneapolis: Univ. of Minnesota Press, 1985. Pp. xxii, 260. Good at placing the artififical intelligence debate in a larger historical perspective, going back to Descartes and LaMettrie. Gunderson also did the article on "Cybernetics" in the *Encyclopedia of Philosophy*, Vol. 2 (New York: Macmillan, 1967), pp. 280–284.

436   Haugeland, John (ed.). *Mind Design: Philosophy, Psychology, Artificial Intelligence*. Cambridge, MA: Press, 1981. Pp. xii, 368. "This book is conceived as a sequel to Alan Ross Anderson's *Minds and Machines* (1964), augmenting it and bringing it up to date" (p. vii). Contents: Haugeland's "Semantic Engines: An Introduction to Mind Design," Allen Newell and Herbert A. Simon's "Computer Science as Empirical Inquiry: Symbols and Search," Zenon Pylyshyn's "Complexity and the Study of Artificial Intelligence," Marvin Minsky's "A Framework for Representing Knowledge," David Marr's "Artificial Intelligence – A Personal View," Drew McDermott's "Artificial Intelligence Meets Natural Stupidity," Hubert L. Dreyfus's "From Micro-Worlds to Knowledge Representation: AI at an Impasse," Hilary Putnam's "Reductionism and the Nature of Psychology," Daniel C. Dennett's "Intentional Systems," Haugeland's "The Nature and Plausibility of Cognitivism," John R. Searle's "Minds, Brains, and Programs," Jerry A. Fodor's "Methodological Solipsism Considered as a Research Strategy in Cognitive Psychology," and Donald Davidson's "The Material Mind."

437   Heelan, Patrick A. *Space-Perception and the Philosophy of Science*. Berkeley: Univ. of California Press, 1983. Pp. xiv, 383. Although this book is not directly addressed to the issue of computers, it has a good discussion of the distinction between information in the metric and non-metric senses (pp. 137 and 147–154), and its thesis concerning Euclidean space perception as dependent on a technologically carpentered world has important implications for interpreting the social impact of computers.

438   Hiltz, Starr Roxann, and Murray Turoff. *The Network Nation: Human Communication via Computer*. Reading, MA: Addison-Wesley, 1978. Pp. xxxiv, 528. One of the early and most comprehensive books on computer networking. Covers the nature of computerized conferencing (Part I), potential applications (Part II), and the future of the technology and its regulation (Part III). Extensive, although unannotated, bibliographies, pp. 494–516. Hiltz is a sociologist, and Turoff is a computer scientist who designed (circa 1976) the EIES computer conferencing facility at the New Jersey Institute of Technology. For a much more personal reflection on networking see Jacques Valle, *The Network Revolution: Confessions of a Computer Scientist* (Berkeley, CA: And/Or Press, 1982).

439   Hintikka, Jaakko and Patrick Suppes (eds.). *Information and Inference*. Boston: D. Reidel, 1970. Pp. vii, 336. Part I, "Information and Induction," contains Hintik-

ka's "On Semantic Information," Dean Jamison's "Bayesian Information Usage," and Roger Rosenkrantz's "Experimentation as Communication with Nature." Part II, "Information and Some Problems of the Scientific Method," includes Risto Hilpinen's "On the Information Provided by Observations," Juhani Pietarinen's "Quantitative Tools for Evaluating Scientific Systematizations," and Zoltan Domotor's "Qualitative Information and Entropy Structures." Part III is devoted to "Information and Learning" with one paper jointly authored by Jamison, Deborah Lhamon, and Suppes. Part IV, "New Applications of Information Concepts," contains Hintikka's "Surface Information and Depth Information" and Hintikka and Raimo Tuomela's "Towards a General Theory of Auxiliary Concepts and Definability in First-Order Theories." One of the deepest philosophical discussions of the information concept.

440   Hoffman, Lance. J. (ed.). *Computer and Privacy in the Next Decade.* New York: Academic Press, 1980. Pp. xv, 215. "Alan Westin has characterized three phases of awareness and action on the privacy/data bank issue: the early warning phase, the study phase, and the regulatory phase. I would like to suggest that these phases are parts of a series of historical waves. The controversy over the proposed national data bank in 1967 was one of the first events in the early warning phase of the first wave. We then moved on to the study phase and here the most notable reports were those of the National Academy of Sciences Project on Computer Data Banks, the Report of the HEW Secretary's Advisory Commission on Automated Personal Data Systems, and the report of the Privacy Protection Study Commission. The study phase overlapped the regulatory phase, where most significant actions so far have been the passage of the Privacy Act of 1974 and the new executive branch initiatives on regulatory action. At the same time, we are in the early warning phase, historically, for the next wave; its herald is the inexpensive microcomputer system" (p. xv). Included in this volume of proceedings from a conference in Pacific Grove, CA, Feb. 25–28, 1979: Hoffman's "A Research Agenda for Privacy in the Next Decade," Willis H. Ware's "Privacy and Information Technologyy – The Years Ahead," George B. Trubow's "Microcomputers: Legal Approaches and Ethical Implications," Portia Isaacson's "The Personal Computer versus Personal Privacy," Susan Hubbell Nycum's "Privacy in Electronic Funds Transfer, Point of Sale, and Electronic Mail Systems in the Next Decade," Robert C. Goldstein's "Privacy Cost Research: An Agenda," James B. Rule, Douglas McAdam, Linda Stearns, and David Uglow's "Preserving Individual Autonomy in an Information-Oriented Society," T. D. Sterling's "Stressing Design Rather than Performance Standards to Ensure Protection of Information: Comments," H. P. Gassmann's "Privacy Implications of Transborder Data Flows: Outlook for the 1980s," Gordon C. Everest's "Nonuniform Privacy Laws: Implications and Attempts at Uniformity," and Alan F. Westin's "The Long-Term Implications of Computers for Privacy and the Protection of Public Order."

441   Hoffman, W. Michael and Jennifer Mills Moore (eds.). *Ethics and the Management of Computer Technology.* Proceedings of the Fourth National Conference on Business Ethics. Cambridge, MA: Oeleschlager, Gunn & Hain, 1982. p. xix, 175.

442   Hofstadter, Douglas R. *Gödel, Escher, Bach: An Eternal Golden Braid.* New York: Basic Books, 1979. Pp. xxi, 777. "Computers by their very nature are the

most inflexible, desireless, rule-following of beasts. . . . How, then, can intelligent behavior be programmed? Isn't this the most blatant contradiction in terms? One of the major theses of this book is that it is not a contradiction at all" (p. 26). A long, eccentric, indirect attack on the rejection of self-reference in formal logic, arguing that self-reference is at the core of consciousness.

442a   Hofstadter, Douglas R. and Daniel C. Dennett (eds.). *The Mind's I: Fantasies and Reflections on Self and Soul.* New York: Basic Books, 1981. Pp. vii, 501. A very personal anthology including philosophy, science fiction, and the authors' own responses to various criticisms of AI. Good bibliographic essay, pp. 465–482, and a good index. "The authors are out to assure us that if our minds were to be rehoused in a computer, nothing essential would be lost." – Eric Wanner, "I Am a Computer," *Psychology Today* (Oct. 1981), pp. 104ff.

443   Hook, Sidney, (ed.). *Dimensions of Mind.* New York: New York Univ. Press, 1960. Paperback reprint, New York: Collier Books, 1961. Pp. 250. Part II, "The Brain and the Machine," contains the following papers: Norbert Wiener's "The Brain and the Machine (Summary)," Michael Scriven's "The Complete Robot: A Prolegomena to Androidology," Satosi Watanabe's "Comments on Key Issues," Hilary Putnam's "Minds and Machines," Arthur C. Danto's "On Consciousness in Machines," Roy Lachman's "Machines, Brains, and Models," R. M. Martin's "On Computers and Semantical Rules," Paul Weiss' "Love in a Machine Age," Fritz Heider's "On the Reduction of Sentiment," and Hook's "A Pragmatic Note." An important early symposium.

444   Horvitz, Robert. "Tuning in to WARC," *CoEvolution Quarterly,* whole no. 22 (Summer 1979), 94–107. WARC stand for "World Administrative Radio Conference." Best general article on problems of redistributing portions of the radio spectrum, allotting satellite positions for geostationary orbit, short wave band "radio war," restricting or promoting the use of remote-sensing satellites, etc. Abundant bibliographic references. Followed by Tran Van Dinh's "WARC, the Third World, and the New International Information Order" (pp. 108–110) and three pages of reviews of related books. On this same topic, see also Thomas T. Surprenanat, "The Radio Frequency Spectrum and the New World Information Order: Implications for the Future of Information," in Howard F. Didsbury (ed.), *Communications and the Future* (Washington, DC: World Future Society, 1982), pp. 349–357; and Eric J. Lerner, "International Data Wars Are Brewing," *IEEE Spectrum* (July 1984), 45–49.

445   Humbert, Royal. "The Computer and the Mystic," *Encounter* (Indianapolis, IN) **38,** no. 3 (Summer 1977), 227–244. Emphasizes the failure of computerization and systems-analysis to adequately direct public policy. Human beings need transcendent direction, but this has been eclipsed by the triumph of Protestant "prophetical" religion over Catholic "mystical" religion, although the two actually belong together. Argues that "the mystical can contribute to a meaningful understanding of the transcendent dimension with which a computerized society may face political decisions" (p. 237). An interesting but rambling amalgam of Pascal, Schweitzer, Tillich, and Thomas Merton.

446   Innis, Harold A. *Empire and Communication*. London: Oxford Univ. Press, 1950.
        Reprinted, Toronto: Univ. of Toronto Press, 1972. Pp. xii, 183.

446a   Innis, Harold A. *The Bias of Communication*. Toronto: Univ. of Toronto Press,
        1951. Reprinted, 1964. Pp. xviii, 226. "A pioneer work in exploring the psychic and
        social consequences of the extensions of man." – Marshall McLuhan.

447   Johnson, Deborah G. *Computer Ethics*. Englewood Cliffs, NJ: Prentice-Hall,
        1985. Pp. xv, 110. Following an initial discussion of ethical theory in general, there
        are chapters devoted to codes of ethics for the computer professional, corporate
        liability for computer programs, the social impacts on privacy and on power, and
        the issues of property in relation to software. A companion collection of readings
        which follows the same outline exclusive of the general discussion of ethical theory:
        Deborah G. Johnson and John W. Snapper (eds.), *Ethical Issues in the Use of Com-
        puters* (Belmont, CA: Wadsworth, 1985). Two other books in this same area:
        Douglas W. Johnson, *Computer Ethics: A Guide for the New Age* (Elgin, IL:
        Brethren Press, 1984), an adult Christian study guide; and Thomas Milton Kemnitz
        and Philip Fitch Vincent, *Computer Ethics* (New York: Trillium, 1985), a text for
        gifted high school students.

448   Kemeny, John G. *Man and the Computer*. New York: Scribner, 1972. Pp. viii, 150.
        Part one, "The New Symbiosis," is concerned with the general structure of man-
        computer relations. "Are computers a species? A species is a distinctive form of
        life. There is no question that computers are distinctive, but most people would
        insist rather vehemently that it is ridiculous to compare a machine to a living being. I
        would like to argue that the traditional distinction between living and inanimate
        matter may be important to a biologist but is unimportant and possibly dangerously
        misleading for philosophical considerations" (p. 10). Part two, "Symbiotic Evolu-
        tion," outlines computer applications in education, libraries, management, the
        home, etc. Kemeny is co-inventor of the computer language BASIC, and as chair-
        man of the mathemathics department and later president at Dartmouth College,
        was responsible for the early introduction of computer time sharing as an education-
        al tool.

449   Kirschenmann, Peter Paul. *Information and Reflection: On Some Problems of
        Cybernetics and How Contemporary Dialectical Materialism Copes with Them*.
        Trans. T. J. Blakeley. New York: Humanities Press, 1970. Pp. xv, 225. The most
        thorough-going study available of Soviet Marxist philosophical reaction to cyber-
        netics and information theory. Translated from *Kybernetik, Information, Wider-
        spiegelung: Darstellung einiger philosophischer Probleme im dialektischen Mater-
        ialismus* (Munich: Anton Pustet, 1969). Chapter 16, the conclusion, was translated
        by the author and published in a modified version as "Problems of Information in
        Dialectical Materialism," *Studies in Soviet Thought* **8,** no. 2 (June–September
        1968), 105–121.

450   Krueger, Myron W. *Artificial Reality*. Reading, MA: Addison-Wesley, 1983.
        Pp. xviii, 312. On art and computer technology. Not a survey of applications, but a
        detailed description of four "responsive environments that provide glimpses of
        what the future may be like" (p. vii). Helpful notes and bibliography.

451   Lasswell, Harold D. "Policy Problems of a Data-Rich Civilization," in Larry Hick-
        man and Azizah al-Hibri (eds.), *Technology and Human Affairs* (St. Louis: C. V.
        Mosby, 1981), pp. 89–94. Increased information creates challenges for world secur-
        ity, individuality, and democracy.

451a   Lasswell, Harold D., Daniel Lerner, and Hans Speier (eds.). *Propaganda and Com-
        munication in World History.* Vol. I: *The Symbolic Instrument in Early Times.* Hon-
        olulu: Univ. of Hawaii Press, 1979. Pp. xiv, 631. Vol. II: *Emergence of Public Opin-
        ion in the West.* Honolulu: Univ. of Hawaii Press, 1980. Pp. xiii, 561. Vol. III: *A
        Pluralizing World in Formation.* Honolulu: Univ. of Hawaii Press, 1980. pp. xiii,
        562. A massive work collaborated on by over 30 scholars. The last volume is most
        directly relevant, with chapters on "The Media Kaleidoscope: General Trends in
        the Channels" by W. Phillips Davison, "The Effects of Mass Media in an Informa-
        tion Era" by Wilbur Schramm, and "The Social Effects of Communication Tech-
        nology" by Hank Goldhamer.

452   Laver, F. J. Murray. *Computers and Social Change.* New York: Cambridge Univ.
        Press, 1980. Pp. vii, 125. Superficial course notes. Touches, but only touches, most
        bases from economics to politics.

453   Lee, Alfred M. *Electronic Message Transfer and Its Implications.* Lexington, MA:
        D. C. Heath, 1983. Pp. xiv, 195. Overview consideration of benefits (energy con-
        servation, increased business efficiency, better interaction between individuals and
        organizations, etc.) and liabilities (workplace dehumanization, increasing gap be-
        tween rich and poor, threats to privacy and freedom) of advances in electronic in-
        formation technologies.

454   Lihisse, Jean. "Information Artefact or Enslaved Communication," trans. Jeanne
        Ferguson, *Diogenes*, whole no. 123 (Fall 1983), 91–109. Criticizes information
        technology with a distinction between information and (personal) communication.
        A bit oracular.

455   McHale, John. *The Changing Information Environment.* Boulder, CO: Westview
        Press, 1976. Pp. 117. "The combination of computer and communication technolo-
        gies has created a new revolutionary situation that is changing everything regardless
        of our intents. Nevertheless, unlike the popular prophets of gloom and doom, who
        merely project the past into scenarios of imaginary futures, McHale is one of those
        rare futurists who tries to identify present environmental patterns that will indicate
        the shape of things to come; he not only provides detailed inventories of the psychic
        and social effects of new technologies in education, management, and politics but
        also occasional flashes of prophetic insight that encourage further exploration." –
        Barrington Nevitt, *Technology and Culture* **20,** no. 3 (July 1979), 677–679.

456   MacKay, Donald M. *Information, Mechanism and Meaning.* Cambridge, MA:
        MIT Press, 1969. Pp. vii, 169 This early collection of technical papers by a
        neurophysiologist has been followed by more general studies stressing the religious
        dimension: *Human Science and Human Dignity* (Madison, WI: InterVarsity Press,
        1979) and *Brains, Machines and Persons* (Grand Rapids, MI: Eerdmans, 1980).

457   McLuhan, Marshall. *Understanding Media: The Extensions of Man.* New York: McGraw-Hill, 1964. Paperback edition, with new introduction, 1965. Pp. xiii, 364. For discussion of McLuhan, see Raymond Rosenthal (ed.), *McLuhan: Pro and Con* (New York: Funk & Wagnalls, 1968).

458   Mander, Jerry. *Four Arguments for the Elimination of Television.* New York: William Morrow, 1978. Pp. 371. The four arguments are that TV alienates from direct experience, is a power exerted by an elite, has bad physiological and psychological affects, and is biased against certain kinds of information.

459   Martin, James. *Telematic Society: A Challenge for Tomorrow.* Englewood Cliffs, NJ: Prentice-Hall, 1981. Pp. ix, 244. "Telematic" is an Americanization of the French "telematique," which refers to computer and telecommunications integration. "As technology grows in power, its ability either to disrupt or to heal increases. . . . To heal, we have to move to new technologies, new social patterns, new types of consumer products, new ways of generating and spending wealth. . . . There are many such technologies now within our grasp. . . . This book is about one of them – electronic communications" and how it will affect society, consumption, and wealth. Martin is a well-known lecturer and consultant who has written over 20 books on telecommunications and data-processing. This is in fact a re-write, with one new chapter, of *The Wired Society* (Englewood Cliffs, NJ: Prentice-Hall, 1978); and an earlier discussion of the same material can be found in Martin and Adrian R. D. Norman, *The Computerized Society* (Englewood Cliffs, NJ: Prentice-Hall, 1970), a book repeatedly cited by John Kemeny (1970).

460   Mathews, Walter M. (ed.). *Monster or Messiah? The Computer's Impact on Society.* Jackson: Univ. Press of Mississippi, 1980. Pp. xi, 222. Of the many books with such a title or theme, this is among the best. Contents: Frederick E. Laurenzo's "Computers and the Idea of Progress," Michael de L. Landon's "A Historian Looks at he Computer's Impact on Society," Shirley Hallblade and Walter M. Mathews' "Computers and Society: Today and Tomorrow," Edwin Dolin's "Computers – For Better or For Worse," Robert E. Bergmark's "Computers and Persons," Tom R. Kibler's "While Debating the Philosophy We Accept the Practice," Roger Johnson Jr.'s "Printout Appeal," Arlene Schrade's "Sex Shock: The Humanistic Woman in the Super-Industrial Society," Patricia B. Campbell's "Computer-Assisted Instruction in Education: Past, Present and Future," David M. Guerna's "Computer Technology and the Mass Media: Interacting Communication Environments," Gary R. Mooers' "Computer Impact and the Social Welfare Sector," Margaret Gorove's "Computer Impact on Society: A Personal View from the World of Art," Nolan E. Shepard's "Technology: Messiah or Monster," Johnny E. Tolliver's "The Computer and the Protestant Ethic: A Conflict," Polly F. Williams' "Reflection on Computers as Daughters of Memory," and Wallace A. Murphree's "The Necessity of Humanism in a Computer Age." Bibliographies provided by each article.

461   Megarry, Jacquetta, David R. F. Walker, Stanley Nisbet, and Eric Hoyle (eds.). *Computers and Education.* (World Yearbook of Education 1982/83.) London: Kogan Page; and New York: Nichols, 1983. Pp. 280. Comprehensive work.

Twenty-two articles, selectively annotated bibliography, and glossary. See also Robert J. Seidel, Ronald E. Anderson, and Beverly Hunter (eds.), *Computer Literacy: Issues and Directions for 1985* (New York: Academic Press, 1982), which is the proceedings of a conference on "National Goals for Computer Literacy in 1985," Reston, VA, Dec. 18–20, 1980. See also Semour Papert (1980) below.

462   Minsky, Marvin L. "Why People Think Computers Can't Think," *AI Magazine* (Fall 1982), 3–15. Adapted to "Why People Think Computers Can't," *Technology Review* (Nov.–Dec. 1983), pp. 65–70 and 80–81.

463   Mowshowitz, Abbe. *The Conquest of Will: Information Processing in Human Affairs*. Reading, MA: Addison-Wesley, 1976. Pp. xvi, 365. "Most computer-based information processing systems . . . serve one of two general social functions: the coordination of diversity or the control of disorder. Coordination and control signify the extremes of a continuum of social choices" (p. ix). Part I outlines the historicial origins of the computer; Part II focuses on the coordination of economic activities and social services; Part III deals with social control. "Two fundamental themes permeate the entire discussion. One is the distribution of power in society; the other concerns individual responsibility" (p. x). Good bibliography and index. By a computer scientist at the University of British Columbia.

464   Nora, Simon and Alain Minc. *The Computerization of Society*. Introduction by Daniel Bell. Cambridge, MA: MIT Press, 1981. Pp. xx, 186. Originally written as a report to the President of France. Focuses on the interaction between computers and telecommunications (termed "telematics"), and how this is transforming society. From *L'Informationisation de la société* (Paris: La Documentation Francaise, 1978).

465   Norman, Colin. *Microelectronics at Work: Productivity and Jobs in the World Economy*. Washington, DC: Worldwatch Institute, 1980. Pp. 63. Not a terribly rich discussion.

466   Oettinger, Anthony G., with Sema Marks. *Run, Computer, Run: The Mythology of Educational Innovation*. Cambridge: Harvard Univ. Press, 1969. Pp. xx, 302. An early passionate and still relevant critique by a linguist and mathematician who sees a legitimate promise being abused.

467   Ong, Walter, J. *Orality and Literacy: The Technologizing of the Word*. New York: Methuen, 1982. Pp. x, 201. "The electronic transformation of verbal expression has both deepened the commitment of the word to space initiated by writing and intensified by print and has brought consciousness to a new age of secondary orality" (p. 138). Good brief summary of Ong's ideas.

468   Organization for Economic Co-operation and Development. *Micro-Electronics, Robotics and Jobs*. Paris: OECD, 1983. Pp. 268. Reports for a 1981 OECD conference on Information Technologies, Productivity and Labour Market Implications. Conclusion is just a new adaptation of the old neutrality of technology cliché: "Micro-electronics has the potential to improve the quality of life but . . . there is no automatic guarantee that the benefits will be realized."

469   Pagels, Heinz R. *Computer Culture: The Scientific, Intellectual, and Social Impact of the Computer.* Annals of the New York Academy of Sciences, vol. 426. New York: New York Academy of Sciences, 1984. Pp. x, 288. Proceedings of a conference held in New York City, April 5–8, 1983. Part I, "Introduction: The Social Impact of Computers," contains Robert W. Lucky's "The Social Impact of the Computer." Part II, "New Directions in the Computer Sciences," contains J. F. Traub's "Coping with Complexity," Michael L. Dertouzos's "The Information Revolution: Developments and Consequences by 2000 A.D.," and Herbert Schorr's "Experimental Computer Science." Part III, "Computer Graphics," contains Franklin C. Crow's "Computer Image Synthesis: Shapes" and Turner Whitted's "Computer Image Synthesis: Rendering Techniques." Part IV, "Computers and the Shift in the Work Force," contains Harley Shaiken's "Democratic Choice and Technological Change" and Seymour Melman's "Alternatives for the Organization of Work in Computer-Assisted Manufacturing." Part V, "Expert Systems: The Fifth Generation," contains Edward A. Feigenbaum's "Knowledge Engineering; The Applied Side of Artificial Intelligence," Pamela McCorduck's "Knowledge Technology: The Promise," and Kazuhiro Fuchi's "Fifth Generation Computers: Some Theoretical Issues." Part VI, "Developments in Artificial Intelligence Research," contains John McCarthy's "Some Expert Systems Need Common Sense." Part VII is a "Panel Discussion" on the question "Has Artificial Intelligence Research Illuminated Human Thinking?" with Hubert L. Dreyfus, John McCarthy, Marvin L. Minsky, Seymour Papert, and John Searle. Part VIII, "The Limits of Computation," contains Rolf Landauer's "Fundamental Physical Limitations of the Computational Process." Part IX, "Computers and Scientific Inquiry," contains Cyrus Levinthal's "The Formation of Three-Dimensional Biological Structures: Computer Uses and Future Needs" and George N. Reeke Jr. and Gerald M. Edelman's "Selective Networks and Recognition Automata." Part X, "The Human Factor in Computer Use," contains Alphonse Chapanis's "Taming and Civilizing Computers." Part XI, "New Perspectives in Psychology and Education," contains Donald A. Norman's "Worsening the Knowledge Gap: The Mystique of Computation Builds Unnecessary Barriers." Part XII is a "Panel Discussion" on "Computer-Assisted Negotiations: A Case History from the Law of the Sea Negotiations and Speculations Regarding Future Uses" with T. T. B. Koh, J. D. Hyhart, Elliot L. Richardson, and James K. Sedenius. Part XIII, "How Computers Change the Way We Think about Ourselves," contains Daniel C. Dennett's "The Role of the Computer Metaphor in Understanding the Mind." Subject index. Despite the title and apparent broad coverage, this is not quite the "big story" it is billed to be. For some reason, not all conference presentations are included. Even if they were, there would still be no real discussion of such key aspects of "computer culture" as art, literature, ethics, and political theory.

470   Papert, Seymour. *Mindstorms: Children, Computers, and Powerful Ideas.* New York: Basic Books, 1980. Pp. viii, 230. "In 1964 I moved from one world to another. For the previous five years I had lived in Alpine villages near Geneva, Switzerland, where I worked with Jean Piaget. The focus of my attention was on children, on the nature of thinking, and on how children become thinkers. I moved to MIT into an urban world of cybernetics and computers. My attention was still focused on the nature of thinking, but now my immediate concerns were with the problem of Artificial Intelligence: How to make machines that think? Two worlds

could hardly be more different. But I made the transition because I believed that my new world of machines could provide a perspective that might lead to solutions to problems that had eluded us in the old world of children" (p. 208). Probably the most optimistic book about computers and early education.

471   Parker, Donn B. *Fighting Computer Crime*. New York: Scribner, 1983. Pp. xiii, 352. This book, by a senior management systems consultant and researcher of computer crime and security at SRI International, Menlo Park, CA, is the single most comprehensive treatment of the subject. "Computer-related crime is not just one type of crime; it is a ubiquitous variant of all crime. Ultimately, the variant will become the dominant form, and most nonviolent and even some violent type of crime will involve computers. . . . I am optimistic that incidence can be controlled, but I'm not as optimistic about limiting loss. We will be in a close race with criminals over the next few years. . . . The same power of computers can be used for criminal purposes as well as for beneficial purposes, and the same goals in advancing computer technology also facilitate the growth of leveraged vulnerability from automated crime. Increased leverage for doing evil attracts larger numbers of people to computer crime because we are spreading technical capabilities, lowering the level of resources needed for crime, and increasing the potential gain" (pp. x–xi). "My first book, *Crime by Computer* ([New York: Scribner] 1976), was the two-by-four applied to the head to direct everybody's attention to a new, emerging problem. It was heavy with horror stories and light on remedies. *Fighting Computer Crime* is . . . still heavy on loss experience, including some horror stories, but presented in a more organized and categorized fashion and, most important, heavy on remedies derived and discovered mostly in the past six years. Analysis of the computer crime problems and remedies are organized in this book by describing the research methodology used, by explaining the data base of reported computer crimes collected, and by spanning four dimensions of the problem. Computer crime and its control are examined first by modi operandi – the criminal methods used. Second, a distinction is made between the different categories of perpetrators – from amateur crooks to foreign powers, including a new type of criminal, the system hacker, who is identified only briefly in my first book. The third dimension is the ethical behavior of technologists as seen by thirty-five thoughtful leaders in the computer field and other specialists. Finally, the fourth dimension of computer crime examined is the inadequacy of the law and the corrections that are needed. The book concludes with a consideration of the future escalation possibilities of computer crime . . ." (p. xi). The "third dimension" section on "Ethical Conflicts in Computing" (pp. 189–226) is the most philosophically relevant. It summarizes the results of an EVIST sponsored workshop held at SRI International, March 7–9, 1977. The full report on this workshop is available in Parker's *Ethical Conflicts in Computer Science and Technology* (Arlington, VA: AFIPS Press, 1979). For some other, not nearly as well researched, scare stories, see Thomas Whiteside, *Computer Capers: Tales of Electrical Thievery, Embezzlement, and Fraud* (New York: Crowell, 1978). For another study of how to fight back, see August Bequai, *How to Prevent Computer Crime: A Guide for Managers* (New York: John Wiley, 1983).

472   Pask, Gordon, with Susan Curran. *Micro Man: Computers and the Evolution of Consciousness*. New York: Macmillan, 1982. Pp. 222. An illustrated book about

"the developing relationship between human beings and computers." Somewhat glib at times, as when it compares mass murder and information overload (see picture, p. 47). Pask was, however, a pioneer in certain technical aspects of cybernetics. See Pask, *An Approach to Cybernetics* (New York: Harper & Row, 1961).

473   Patton, Peter C. and Renee A. Holoien (eds.). *Computing in the Humanities*. Lexington, MA: Lexington Books, 1981. Pp. xi, 404. Twenty-six papers on the use of computers to analyze texts, analyze historical and archeological data, teach languages, perform music, and create art. See also Susan Hockey, *A Guide to Computer Applications in the Humanities* (Baltimore: Johns Hopkins Univ. Press, 1980). Ironically enough, both books are more devoid of philosophical reflection on the procedures they describe than the standard computer programming text. For some humanities reflection, see M. E. Grenander (ed.), *Apollo Agonistes: The Humanities in a Computerized World*, 2 vols. (Albany: Institute for Humanistic Studies, State Univ. of New York at Albany, 1979). Also relevant are the detailed studies of the influence of cybernetics and the portrayal of computers in fiction by Patricia S. Warrick, *The Cybernetic Imagination in Science Fiction* (Cambridge, MA: MIT Press, 1980), and David Porush, *The Soft Machine: Cybernetic Fiction* (New York: Methuen, 1985).

474   *Personal Privacy in an Information Society*. The Report of the Privacy Protection Study Commission. Washington, DC: U.S. Government Printing Office, 1977. Pp. 654. There is also a five volume *Appendix* of documents and hearings.

475   Pool, Ithiel de Sola. *Technologies of Freedom*. Cambridge, MA: Harvard Univ. Press, 1983. Pp. 299. Describes the historical emergence of the principles of a free press, the development of other communications technologies, and the social and legal responses appropriate to a democratic society. Important study. Extensive notes.

476   Pylyshyn, Zenon W. (ed.). *Perspectives on the Computer Revolution*. Englewood Cliffs, NJ: Prentice-Hall, 1970. Pp. xx, 540. The most comprehensive general collection of material to its date. Includes helpful commentary and extensive notes. Contents: Part I, "The Development of Computer Science," includes Thomas M. Smith's "Some Perspectives on the Early History of Computers," Charles Babbage's "Of the Analytical Engine," Howard Aiken's "Proposed Automatic Calculating Machine," Arthur W. Burks, Herman H. Goldstine, and John von Neumann's "Preliminary Discussion of the Logical Design of an Electronic Computing Instrument," Vannevar Bush's "As We May Think," B. A. Trakhtenbrot's "Algorithms," John von Neumann's "The General and Logical Theory of Automata," Claude E. Shannon's "Computers and Automata," J. O. Wisdom's "The Hypothesis of Cybernetics," and Stephen Toulmin's "The Importance or Norbert Wiener." Part II, "Man and Machine," includes Samuel Butler's "The Destruction of Machines in Erewhon," Rattray Taylor's "The Age of Androids," W. Grey Walter's "Totems, Toys, and Tools," Bruce Mazlish's "The Fourth Discontinuity," Paul Armer's "Computing Machinery and Intelligence," Henry Block and Herbert Ginsburg's "The Psychology of Robots," Herbert A. Simon and Allen Newell's "Information-Processing in Computer and Man," John R. Pierce's "What Computers Should Be Doing," Gretchen Herbkersman's "Talking to a Monster," J. C. R.

Licklider's "Man-Computer Symbiosis," Don D. Bushnell's "Applications of Computer Technology to the Improvement of Learning," Joseph Weizenbaum's "Contextual Understanding by Computers," and A. Michael Noll's "The Digital Computer as a Creative Medium." Part III, "Society and Machine," includes Richard W. Hamming's "Intellectual Implications of the Computer Revolution," George E. Forsythe's "Educational Implications of the Computer Revolution," Martin Greenberger's "The Computers of Tomorrow," Margaret Mead's "The Information Explosion," Herbert A. Simon's "The Shape of Automation," Donald N. Michael's "Cybernation: The Silent Conquest," Jacques Ellul's "The Technological Society," Edmund C. Berkeley's "The Social Responsibilities of Computer People," Alan F. Westin's "Legal Safeguards to Insure Privacy in a Computer Society," Alice Mary Hilton's "An Ethos for the Age of Cyberculture," and Norman Cousins "The Computer and the Poet."

476a   Pylyshyn, Zenon W. *Computation and Cognition: Toward a Foundation for Cognitive Science*. Cambridge, MA: MIT Press, 1984. Pp. xxiii, 292. Extended argument to the effect that computation must not be viewed as just a convenient metaphor for metal activity, but as a literal empirical hypothesis.

477   Reinecke, Ian. *Electronic Illusions: A Skeptic's View of Our High-Tech Future*. New York: Penguin, 1984. Pp. 256. First published as *Micro Invaders* (Victoria, Australia: Penguin, 1982). Stresses the negative economic dimensions of microelectronic innovation – unemployment, deskilling, alienation, social polarization, etc. By a financial journalist. Brief bibliography and index.

478   Ringle, Martin D. *Philosophical Perspectives in Artificial Intelligence*. Atlantic Highlands, NJ: Humanities Press, 1979. Pp. x, 244. Contents: Ringle's Philosophy and Artificial Intelligence," Zenon W. Pylyshyn's "Complexity and the Study of Human and Artificial Intelligence," Daniel C. Dennett's "Artificial Intelligence as Philosophy and as Psychology," Wendy Lehnert's "Representing Physical Objects in Memory," John McDermott's "Representing Knowledge in Intelligent Systems," Hubert L. Dreyfus's "A Framework for Misrepresenting Knowledge," Kenneth M. Sayre's "The Simulation of Epistemic Acts," John McCarthy's "Ascribing Mental Qualities to Machines," Roger C. Schank's "Natural Language, Philosophy and Artificial Intelligence," and Thomas W. Simon's "Philosophical Objections to Programs as Theories."

479   Rule, James, Douglas McAdam, Linda Stearns, and David Uglow. *The Politics of Privacy: Planning for Personal Data Systems as Powerful Technologies*. New York: Elsevier, 1980. Pp. xii, 212. Good overview. Part one provides historical background. Part two traces the discussion in the U.S. from 1967 up to the late 1970s. Part three considers future prospects, and part four examines "reasoned choice and the future of human control." Appendix lists major U.S. Congressional hearings and reports. Bibliography and index. See also Rule's *Private Lives and Public Surveillance* (New York: Schocken, 1974).

480   Sagal, Paul T. *Mind, Man and Machine: A Dialogue*. Indianapolis: Hackett, 1982. Pp. 39. A witty and informative discussion.

481   Sayre, Kenneth M. *Recognition: A Study in the Philosophy of Artificial Intelligence*.
        Notre Dame: University of Notre Dame Press, 1965. Pp. xx, 312. Current research
        in artificial pattern recognition is mistaken in its assumption that recognition is iden-
        tical with classification.

481a   Sayre, Kenneth M. *Consciousness: A Philosophic Study of Minds and Machines*.
        New York: Random House, 1969. Pp. viii, 229. "[A]n appropriate way to test and
        to extend one's understanding of a mental function is to attempt to describe in une-
        quivocal terms how that function might possibly be approximated in a mechanical
        system or, if it cannot be so approximated, to explaine in unequivocal terms why this
        is the case. The present essay thus begins and ends with a discussion of what I call
        'the question of machine consciousness' " (p.vii).

481b   Sayre, Kenneth M. *Cybernetics and the Philosophy of Mind*. New York: Humani-
        ties Press, 1976. Pp. xiii, 265. Favorably reviewed by Michael E. Levin, *Philosophy
        of Science* **45,** no. 4 (Dec. 1978), 653–654. But for the best overview of Sayre's work
        see Kristin Shrader-Frechette, "Kenneth Sayre on Information-Theoretic Models
        of Mind," *Research in Philosophy and Technology*, vol. 2 (1979), 371–380, which
        includes a Sayre bibliography.

481c   Sayre, Kenneth M. and Frederick J. Crosson (eds.). *The Modeling of Mind: Com-
        puters and Intelligence*. New York: Simon & Schuster, 1963. Pp. xi, 275. Contents:
        Crosson and Sayre's "Modeling: Simulation and Replication," Anatol Rapoport's
        "Technological Models of the Nervous System," Lejaren Hiller and Leonard Isaac-
        son's "Experimental Music," Allen Newell's "The Chess Machine," Hoa Wang's
        "Toward Mechanical Mathematics," Ludwig Wittgenstein's "Remarks on
        Mechanical Mathematics," Gilbert Ryle's "Sensation and Observation," Sayre's
        "Human and Mechanical Recognition," Norman Sutherland's "Stimulus Analys-
        ing Mechanisms," Aron Gurwitsch's "On the Conceptual Consciousness," Michael
        Polanyi's "Experience and the Perception of Pattern," Donald MacKay's "Mind-
        like Behavior in Artefacts," Michael Scriven's "The Mechanical Concept of
        Mind," and John Lucas's "Minds, Machines and Gödel."

482   Schank, Roger C., with Peter G. Childers. *The Cognitive Computer: On Language,
        Learning, and Artificial Intelligence*. Reading, MA: Addison-Wesley, 1984.
        Pp. xiii, 268. Addresses three questions: What do most people need to know about
        computers? (part one). What can computers teach us about human intelligence?
        (part two). What are the social implications of computers? (part three). Part two
        occupies half the book and is a good popular presentation (big type, bold subhead-
        ings, no footnotes, no bibliography) of Schank's theories about scripts, knowledge
        structures, memory, and his work at the Yale AI Project. Part one argues that com-
        puters are just dumb number crunchers. Part three considers social implications,
        and maintains that bad consequences are caused by users not computers. Schank's
        two major technical books: with Robert Abelson, *Scripts, Plans, Goals and Under-
        standing* (Hillsdale, NJ: Erlbaum, 1977; and *Dynamic Memory: A Theory of Re-
        minding and Learning in Computers and People* (Cambridge: Cambridge Univ.
        Press, 1982). For a profile, see "A Conversation with Roger Schank," *Psychology
        Today* (April 1983), 28–36.

483   Shallis, Michael. *The Silicon Idol: The Micro Revolution and Its Social Implications*. New York: Schocken, 1984. Pp. xi, 188. A critical overview of information technologies, pointing especially toward tensions between computers and religion. The first chapter sketches a spiritual understanding of the human drawn from E. F. Schumacher's *Guide for the Perplexed* (1979). The next four describe computer technology and applications. The last half of the book focuses on socio-historical origins and implications. "Evil works in this world by mimicking good. The devil tells nine truths to sell one lie. A technology developed as an instrument of human destruction . . is shown to be benign. . . . But this technology was born destructive. The euphoria it engenders is like a magic spell. It enchants people into false belief in a false god" (p. 176). By an astrophysicist and computer scientist in the Dept. of External Studies at Oxford. For a related argument, see Harry A. Nielson, "The Limits of Computer Subjectivity," *Philosophical Research Archives*, vol. 9 (1983), pp. 413–417, which uses Kierkegaard's concept of the "self" to argue that "self" and "machine" are mutually exclusive.

484   Simons, Geoff. *Are Computers Alive? Evolution and New Life Forms*. Boston: Birkhauser, 1983. Pp. xi, 212. Vigorous argument that computers are not just thinking machines, but living ones. See also Simons' follow-up: *The Biology of Computer Life: Survival Emotion and Free Will* (Sussex: Harvester Press, 1985).

485   Skolimowski, Henryk. "Freedom, Responsibility and the Information Society," *Vital Speeches* **50,** no. 16 (June 1, 1984), 493–497. A somewhat caustic critique of the idea of an "information society" with a call for the exercise of human responsibility in the use of computers.

486   Slezak, Peter. "Gödel's Theorem and the Mind," *British Journal for the Philosophy of Science* **33,** no. 1 (March 1982), 41–52. This is the most recent contribution in a discussion about the implications of Gödel's theorem for the possibility of artificial intelligence, a discussion that goes back to J. R. Lucas, "Minds, Machines and Gödel," *Philosophy* **36** (April–July 1961), 112–127. Slezak reviews the literature as well as adding his own argument.

487   Sloman, Aaron. *The Computer Revolution in Philosophy: Philosophy, Science and Models of Mind*. Atlantic Highlands, NJ: Humanities, 1978. Pp. xvi, 304. One of the strongest pro-AI books, by someone educated as a philosopher but now actually doing cognitive simulation research. Critical review by Kenneth Sayre, *Philosophy of Science* **46,** no. 4 (Dec. 1979), 651–652.

488   Stover, William James. *Information Technology in the Third World: Can I.T. Lead to Humane National Development?* Boulder, CO: Westview Press, 1984. Pp. xiii, 183. Overview of how information technologies can serve as agents of change in economic, social, and political aspects of Third World development. But centralized mass communications will not solve all problems and can be used either "to oppress through authoritarian controls, surveillance and propaganda" or "to increase freedom, democracy and human dignity by bringing together mass and elite, public and leaders, advanced and Third World countries in a common effort at human liberation."

489   Sudnow, David. *Pilgrim in the Microworld*. New York: Warner Books, 1983.
       Pp. viii, 227. A phenomenological biography of an encounter with computers by a
       descriptive sociologist. Compares video games with everything from pool halls and
       one-arm bandits to typewriters and pianos, and shows how such games demand
       skills that legitimately engage both hand and brain in complex coordinated activi-
       ties.

490   Taviss, Irene, ed. *The Computer Impact*. Englewood Cliffs, NJ: Prentice-Hall, 1970
       Pp. xi, 297. Broad collection focusing in turn on economics, politics, and culture,
       with over twenty articles by Paul Armer, H. A. Simon, Alan Westin, John Diebold,
       Patrick Suppes, Anthony G. Oetinger, Ithiel de Sola Pool, Bruce Mazlish, etc. The
       issues covered continue to involve the key questions.

491   Tomeski, Edward. *The Computer Revolution: The Executive and the New Informa-
       tion Technology*. New York: Macmillan, 1970. Pp. xii, 276. What an executive
       should know to keep up. Not profound, but symptomatic.

492   Torrance, Steve (ed.). *The Mind and the Machine; Philosophical Aspects of Artifi-
       cial Intelligence*. New York: John Wiley, 1984. Pp. 213. Proceedings from a confer-
       ence. Good collection of fifteen papers.

493   Turkle, Sherry. *The Second Self: Computers and the Human Spirit*. New York:
       Simon & Schuster, 1984. Pp. 362. "Technology catalyzes changes not only in what
       we do but in how we think. It changes people's awareness of themselves, of one
       another, of their relationship with the world. . . . Most considerations of the com-
       puter concentrate on the 'instrumental computer,' on what work the computer will
       do. But my focus here is on something different, on the 'subjective computer' . . .
       the computer as it affects the way that we think, especially the way we think about
       ourselves" (p. 13). Incorporates two earlier studies: "Computer as Rorschach,"
       Society 17, no. 2 (Jan.–Feb. 1980), 15–25, dealing with how people project onto the
       computer; and "The Subjective Computer: A Study in the Psychology of Personal
       Computation," *Social Studies of Science* 12, no. 2 (May 1982), 173–205, which ex-
       amines how human-computer interactions can be used to strengthen a self identity,
       construct a world which completely understands, articulate a political ideology, or
       experience a sense of wholeness otherwise absent from life.

494   Vester, Frederic. *Neuland des Denkens: Vom technokratischen zum kybernetischen
       Zeitalter* [New thinking: from technocratic to cybernetic age]. Stuttgart: Deutsche
       Verlags-Anstalt, 1980. Pp. 544. Based in part on Vester's 1974 book *Das kyberne-
       tische Zeitalter*.

495   Weizenbaum, Joseph. *Computer Power and Human Reason: From Judgment to
       Calculation*. San Francisco: W. H. Freeman, 1976. Pp. xii, 300. A remarkably deft
       critique by a computer scientist of the epistemological and moral limitations of
       artificial intelligence. Makes it clear that although he is speaking directly about
       computers, he is dealing indirectly with technology as a whole and "logicality itself –
       quite apart from whether logicality is encoded in computers or not" (p. 3). Chapter
       I distinguishes tools as extensions of man from automatic machines and argues that
       technology both closes as well as opens possibilities for human action. Chapters 2

and 3 give a good description of how computers work. The thesis of the last seven chapters is "first, that there is a difference between man and machine, and second, that there are certain tasks which computers *ought* not be made to do, independent of whether computers *can* be made to do them" (p. x). An earlier statement of this part of the argument can be found in Weizenbaum's "On the Impact of the Computer on Society," *Science* **176,** whole no. 4035 (May 12, 1972), 609–614. "I believe we are now at the beginning of . . . a crisis in the mental life of our civilization. . . . The possibility that the computer will, one way or another, demonstrate that, in the inimitable phase of one of my esteemed colleagues, 'the brain is merely a meat machine,' is one that engages academicians, industrialists, and journalists in the here and now. . . . The failure to make distinctions between descriptions, even those that 'work,' and theories accounts in large part for the fact that those who refuse to accept the view of man as machine have been put on the defensive" (pp. 610–611). For a subsequent statement, see Weizenbaum's contribution to Michael L. Dertouzos and Joel Moses (eds.), *The Computer Age* (1979), also included in Tom Forester (ed.), *The Microelectronic Revolution* (1980), both cited above. For sympathetic or moderately critical reviews of Weizenbaum's book, see Daniel Lang, untitled review, *New Republic* **174,** no. 10 (March 20, 1976), 29–30; Theodore Roszak, "The Computer – A Little Lower than the Angels," *Nation* **222,** no. 17 (May 1, 1976), 29–30; and N. S. Sutherland, "The Electronic Oracle," (London) *Times Literary Supplement,* whole no. 3881 (July 30, 1976), 957–959. For three very critical responses, see *Three Reviews of J. Weizenbaum's Computer Power and Human Reason* (Memo AIM-291, Stanford AI Laboratory, Nov. 1976), containing a piece by Bruce Buchanan said to be scheduled for the newsletter *Pharos* (Autumn 1976), another by Joshua Lederburg said to have been done for the *New York Times Book Review* (March 1, 1976) but never published, and John McCarthy's "An Unreasonable Book" reprinted from the ACM *SIGART Newsletter* and *Creative Computing* (Sept.–Oct. 1976), 84–88, where it appeared along with other critical reviews of Weizenbaum. For another McCarthy review, see *Physics Today* **30,** no. 1 (Jan. 1977), 68–69. One last critical review is Daniel D. McCracken, *Datamation* (April 1976). An interview and profile of Weizenbaum by Elisabeth Rosenthal can be found in *Science Digest* (August 1983), 94–97.

496   Westin, Alan F. *Privacy and Freedom*. New York: Atheneum, 1967. Pp. xvi, 487. Comprehensive study. Part one analyzes the historical origins of the theory and practice of privacy. Part two examines the "new tools for invading privacy" – with one chapter on "The Revolution in Information Collecting and Processing: Data Surveillance." Part three consists of five case studies of how American society has defended itself against threats to privacy. Part four outlines policy issues for the 1970s.

496a   Westin, Alan F. (ed.). *Information Technology in a Democracy*. Cambridge, MA: Harvard Univ. Press, 1970. Pp. x, 499. A collection of 51 articles organized into four general categories: "Descriptions of Developing Systems by Their Advocates," "Control Technology in a Democracy: The Broad Socio-Political Debate," "The Information Function in Organization Decision-Making," and "Emerging Information Systems: The Policy Debates." Good general introduction and special introductions to each section, extensive (although unannotated) bibliography, and excellent index. A necessary resource.

496b   Westin, Alan F. *Computers, Health Records, and Citizen Rights*. New York: Pet-
        rocelli Books, 1976. Pp. xviii, 381. A detailed case study with extensive documenta-
        tion of the problems of privacy created by medical record computerization. Two
        other detailed Westin studies are: with Michael A. Baker, *Databanks in a Free Soci-
        ety: Computers, Record Keeping, and Privacy* (New York: Quadrangle, 1972).
        Pp. xxi, 522. *Computer Science and Technology: Computers, Personnel Admini-
        stration, and Citizen Rights* (Washington, DC: U.S. Dept. of Commerce, National
        Bureau of Standards, 1979). Pp. xxiv, 439. The lead title is actually that of a series of
        which this volume is part.

497   Wicklein, John. *Electronic Nightmare: The Home Communications Set and Your
        Freedom*. New York: Viking, 1981. Pp. xv, 282. Paperback reprint, Boston:
        Beacon Press, 1982. Pp. xv, 282. A well-informed scare book. Good reporting on
        videotex in England, data control in Sweden, communication control in Brazil.
        Conclusion, "To Gain the Promise and Avoid the Threat," is more even-handed.
        By a reporter with wide experience.

498   Wiener, Norbert. *Cybernetics: or Control and Communication in the Animal and
        the Machine*. New York: John Wiley 1948. Pp. 194. 2nd edition, with sup-
        plementary chapter "On Learning and Self-Reproducing Machines," Cambridge,
        MA: MIT Press, 1961. Pp. 212. See also Wiener's article on "Cybernetics" in *En-
        cyclopedia Americana* (New York: Americana Corp., 1960).

498a   Wiener, Norbert. *The Human Use of Human Beings: Cybernetics and Society*. New
        York: Houghton Mifflin, 1950. Paperback reprint, Garden City, NY: Doubleday
        Anchor, 1954. Pp. 199. Second paperback reprint, with an "Afterword" by Walter
        A. Rosenblith, New York: Avon, 1967. Pp. 288. See also Wiener's "Some Moral
        and Technical Consequences of Automation," *Science* **131** (May 6, 1960), 1355–
        1358.

498b   Wiener, Norbert. *God and Golem, Inc: A Comment on Certain Points Where
        Cybernetics Impinges on Religion*. Cambridge, MA: MIT Press, 1964. Pp. 99. On
        what theories of machines which learn and machines which reproduce themselves
        imply for theological ideas about God's omniscience and omnipotence (Chapters
        2–4), and the relation between machine-man coordination and religious morality
        (Chapters 5–7).

499   Woodward, Kathleen (ed.). *The Myth of Information: Technology and Postindus-
        trial Culture*. Madison, WI: Coda Press, 1980. Pp. xxvi, 250. Reprinted, Blooming-
        ton: Indiana Univ. Press, 1984; and London: Routledge & Kegan Paul, 1984. Pro-
        ceedings of a two-part symposium at the University of Wisconsin at Milwaukee in
        1977 and the University of Nice in May 1978. Contents: Jean-Pierre Dupuy's
        "Myths of the Informational Society," Heinz von Foerster's "Epistemology of
        Communication," Magoroh Maruyama's "Information and Communication in
        Poly-Epistemological Systems," Eric Leed's " 'Voice' and 'Print': Master Symbols
        in the History of Communication," Oskar Negt's "Mass Media: Tools of Domina-
        tion or Instruments of Emancipation? Aspects of the Frankfurt School's Com-
        munications Analysis," Jack Zipes' "The Instrumentalization of Fantasy: Fairy

Tales and the Mass Media," Andrew Feenberg's "The Political Economy of Social Space," David Hall's "Irony and Anarchy: Technology and the Utopian Sensibility," Jean Baudrillard's "The Implosion of Meaning in the Media and the Implosion of the Social in the Masses," Andreas Huyssen's "The Hidden Dialectic: The Avant Garde – Technology – Mass Culture," Mikel Dufrenne's "Art and Technology: Alienation or Survival?" Kathleen Woodward's "Art and Technics: John Cage, Electronics, and World Improvement," Daniel Charles' "Music, Voice, Waves," Jack Burnham's "Art and Technology: The Panacea That Failed," Anthony Wilden's "Changing Frames of Order: Cybernetics and the Machina Mundi," and Jacques Ellul's "The Power of Technique and the Ethics of Non-Power."

5. AUTHOR INDEX

This index is to the bibliography only. Numbers indicate entries rather than page numbers.

Koh, T. T. B. under 469
Krueger, Myron W. 450

Lachman, Roy under 443
Lamborghini, Bruno under 431
Landauer, Rolf under 469
Landon, Michael de L. under 460
Lang, Daniel under 495
Larsen, Judith K. under 205
Lasswell, Harold D. 451, 451a
Laughery, Kenneth under 429
Laurenzo, Frederick E. under 460
Laver, F. J. Murray 452
Lederburg, Joshua under 495
Ledley, Robert S. under 402
Lee, Alfred M. 453
Leed, Eric under 499
Lefevre, Bruno under 430
Lehnert, Wendy under 478
Lenk, Klaus under 431
Lerner, Daniel 451a
Lerner, Eric J. under 444
Levin, Michael E. under 481b
Levinthal, Cyrus under 469
Levy, Steven 210
Lhamon, Deborah under 439
Licklider, J. C. R. under 422, under 434, under 476
Lihisse, Jean 454
Lindsay, Robert K. under 429
Loveland, D. W. under 429
Lowi, Theodore J. under 430
Lucas, J. R. under 401, under 481c, under 486
Lucky, Robert W. under 469

McAdam, Douglas under 440, 479
McCarthy, John 312, under 434, under 469, under 478, under 495
McClintock, Robert under 427
McCorduck, Pamela 211, 429a, under 469
McCracken, David D. under 495
McCulloch, Warren S. 305
McDermott, Drew under 436
McDermott, John under 478
McDonald, Stuart under 204
McHale, John 455

Machly, John W. under 423a
Mackay, D. M under 403, under 417, under 423a, 456, under 481c
Mackey, Robert 107
Mackingtosh, Ian M. under 430
McLuhan, Marshall under 419, under 427, under 446a, 457
Mander, Jerry under 416, 458
Marks, Sema 466
Marr, David 306, under 436
Martin, James 459
Martin, R. M. under 443
Maruyama, Magoroh under 499
Massey, J. L. under 417a
Mathews, Walter M. under 460
Mauchly, John W. under 434
Mazlish, Bruce under 476, under 490
Mead, Graham P. 404
Mead, Margaret under 423a, under 476
Megarry, Jacquetta 461
Melman, Seymour under 469
Merton, Thomas under 445
Metropolis, N. 212
Michael, Donald N. under 476
Mikulak, Maxim W. under 419
Miller, George A. under 434
Minc, Alain 464
Minsky, Marvin under 313, under 422, under 429, under 434, under 436, 462, under 469
Mitcham, Carl 107
Mooers, Gary R. under 460
Moore, Jennifer Mills 441
Moreau, Rene under 206
Morison, Elting E. under 434
Moses, Joel under 405, under 495
Mowshowitz, Abbe 463

Negroponte, Nicholas P. under 422
Negt, Oskar under 499
Neisser, Ulric under 417, under 419, under 423a, under 429
Neugroschel, Joachim under 428
Neumann, John von under 423a, under 476
Nevitt, Barrington under 455
Newell, Allen under 417, under 429, under 434, under 436, under 476, under 481c

NAME INDEX

The index includes proper names of all individuals referred to in the preface, introduction, and twenty papers of this collection. An "n" after the number indicates a footnote on the referenced page.

347

BOSTON STUDIES IN THE PHILOSOPHY OF SCIENCE

Editors:

ROBERT S. COHEN and MARX W. WARTOFSKY

(Boston University)

1. Marx W. Wartofsky (ed.), *Proceedings of the Boston Colloquium for the Philosophy of Science 1961–1962*. 1963.
2. Robert S. Cohen and Marx W. Wartofsky (eds.), *In Honor of Philipp Frank*. 1965.
3. Robert S. Cohen and Marx W. Wartofsky (eds.), *Proceedings of the Boston Colloquium for the Philosophy of Science 1964–1966. In Memory of Norwood Russell Hanson*. 1967.
4. Robert S. Cohen and Marx W. Wartofsky (eds.), *Proceedings of the Boston Colloquium for the Philosophy of Science 1966–1968*. 1969.
5. Robert S. Cohen and Marx W. Wartofsky (eds.), *Proceedings of the Boston Colloquium for the Philosophy of Science 1966–1968*. 1969.
6. Robert S. Cohen and Raymond J. Seeger (eds.), *Ernst Mach: Physicist and Philosopher*. 1970.
7. Milic Capek, *Bergson and Modern Physics*. 1971.
8. Roger C. Buck and Robert S. Cohen (eds.), *PSA 1970. In Memory of Rudolf Carnap*. 1971.
9. A. A. Zinov'ev, *Foundations of the Logical Theory of Scientific Knowledge (Complex Logic)*. (Revised and enlarged English edition with an appendix by G. A. Smirnov, E. A. Sidorenka, A. M. Fedina, and L. A. Bobrova.) 1973.
10. Ladislav Tondl, *Scientific Procedures*. 1973.
11. R. J. Seeger and Robert S. Cohen (eds.), *Philosophical Foundations of Science*. 1974.
12. Adolf Grünbaum, *Philosophical Problems of Space and Time*. (Second, enlarged edition.) 1973.
13. Robert S. Cohen and Marx W. Wartofsky (eds.), *Logical and Epistemological Studies in Contemporary Physics*. 1973.
14. Robert S. Cohen and Marx W. Wartofsky (eds.), *Methodological and Historical Essays in the Natural and Social Sciences. Proceedings of the Boston Colloquium for the Philosophy of Science 1969–1972*. 1974.
15. Robert S. Cohen, J. J. Stachel, and Marx W. Wartofsky (eds.), *For Dirk Struik. Scientific, Historical and Political Essays in Honor of Dirk Struik*. 1974.
16. Norman Geschwind, *Selected Papers on Language and the Brain*. 1974.
17. B. G. Kuznetsov, *Reason and Being: Studies in Classical Rationalism and Non-Classical Science*. (forthcoming).
18. Peter Mittelstaedt, *Philosophical Problems of Modern Physics*. 1976.
19. Henry Mehlberg, *Time, Causality, and the Quantum Theory* (2 vols.). 1980.
20. Kenneth F. Schaffner and Robert S. Cohen (eds.), *Proceedings of the 1972 Biennial Meeting, Philosophy of Science Association*. 1974.
21. R. S. Cohen and J. J. Stachel (eds.), *Selected Papers of Léon Rosenfeld*. 1978.

22. Milic Capek (ed.), *The Concepts of Space and Time. Their Structure and Their Development.* 1976.
23. Marjorie Grene, *The Understanding of Nature. Essays in the Philosophy of Biology.* 1974.
24. Don Ihde, *Technics and Praxis. A Philosophy of Technology.* 1978.
25. Jaakko Hintikka and Unto Remes. *The Method of Analysis. Its Geometrical Origin and Its General Significance.* 1974.
26. John Emery Murdoch and Edith Dudley Sylla, *The Cultural Context of Medieval Learning.* 1975.
27. Marjorie Grene and Everett Mendelsohn (eds.), *Topics in the Philosophy of Biology.* 1976.
28. Joseph Agassi, *Science in Flux.* 1975.
29. Jerzy J. Wiatr (ed.), *Polish Essays in the Methodology of the Social Sciences.* 1979.
30. Peter Janich, *Protophysics of Time.* 1985.
31. Robert S. Cohen and Marx W. Wartofsky (eds.), *Language, Logic, and Method.* 1983.
32. R. S. Cohen, C. A. Hooker, A. C. Michalos, and J. W. van Evra (eds.), *PSA 1974: Proceedings of the 1974 Biennial Meeting of the Philosophy of Science Association.* 1976.
33. Gerald Holton and William Blanpied (eds.), *Science and Its Public: The Changing Relationship.* 1976.
34. Mirko D. Grmek (ed.), *On Scientific Discovery.* 1980.
35. Stefan Amsterdamski, *Between Experience and Metaphysics. Philosophical Problems of the Evolution of Science.* 1975.
36. Mihailo Marković and Gajo Petrović, *Praxis. Yugoslav Essays in the Philosophy and Methodology of the Social Sciences.* 1979.
37. Hermann von Helmholtz, *Epistemological Writings. The Paul Hertz/Moritz Schlick Centenary Edition of 1921 with Notes and Commentary by the Editors.* (Newly translated by Malcolm F. Lowe. Edited, with an Introduction and Bibliography, by Robert S. Cohen and Yehuda Elkana.) 1977.
38. R. M. Martin, *Pragmatics, Truth, and Language.* 1979.
39. R. S. Cohen, P. K. Feyerabend, and M. W. Wartofsky (eds.), *Essays in Memory of Imre Lakatos.* 1976.
42. Humberto R. Maturana and Francisco J. Varela, *Autopoiesis and Cognition. The Realization of the Living.* 1980.
43. A. Kasher (ed.), *Language in Focus: Foundations, Methods and Systems. Essays Dedicated to Yehoshua Bar-Hillel.* 1976.
44. Trân Duc Thao, *Investigations into the Origin of Language and Consciousness.* (Translated by Daniel J. Herman and Robert L. Armstrong; edited by Carolyn R. Fawcett and Robert S. Cohen.) 1984.
46. Peter L. Kapitza, *Experiment, Theory, Practice.* 1980.
47. Maria L. Dalla Chiara (ed.), *Italian Studies in the Philosophy of Science.* 1980.
48. Marx W. Wartofsky, *Models: Representation and the Scientific Understanding.* 1979.
49. Trân Duc Thao, *Phenomenology and Dialectical Materialism.* 1985.
50. Yehuda Fried and Joseph Agassi, *Paranoia: A Study in Diagnosis.* 1976.
51. Kurt H. Wolff, *Surrender and Catch: Experience and Inquiry Today.* 1976.
52. Karel Kosík, *Dialectics of the Concrete.* 1976.
53. Nelson Goodman, *The Structure of Appearance.* (Third edition.) 1977.

54. Herbert A. Simon, *Models of Discovery and Other Topics in the Methods of Science.* 1977.
55. Morris Lazerowitz, *The Language of Philosophy. Freud and Wittgenstein.* 1977.
56. Thomas Nickles (ed.), *Scientific Discovery, Logic, and Rationality.* 1980.
57. Joseph Margolis, *Persons and Minds. The Prospects of Nonreductive Materialism.* 1977.
59. Gerard Radnitzky and Gunnar Andersson (eds.), *The Structure and Development of Science.* 1979.
60. Thomas Nickles (ed.), *Scientific Discovery: Case Studies.* 1980.
61. Maurice A. Finocchiaro, *Galileo and the Art of Reasoning.* 1980.
62. William A. Wallace, *Prelude to Galileo.* 1981.
63. Friedrich Rapp, *Analytical Philosophy of Technology.* 1981.
64. Robert S. Cohen and Marx W. Wartofsky (eds.), *Hegel and the Sciences.* 1984.
65. Joseph Agassi, *Science and Society.* 1981.
66. Ladislav Tondl, *Problems of Semantics.* 1981.
67. Joseph Agassi and Robert S. Cohen (eds.), *Scientific Philosophy Today.* 1982.
68. Władysław Krajewski (ed.), *Polish Essays in the Philosophy of the Natural Sciences.* 1982.
69. James H. Fetzer, *Scientific Knowledge.* 1981.
70. Stephen Grossberg, *Studies of Mind and Brain.* 1982.
71. Robert S. Cohen and Marx W. Wartofsky (eds.), *Epistemology, Methodology, and the Social Sciences.* 1983.
72. Karel Berka, *Measurement.* 1983.
73. G. L. Pandit, *The Structure and Growth of Scientific Knowledge.* 1983.
74. A. A. Zinov'ev, *Logical Physics.* 1983.
75. Gilles-Gaston Granger, *Formal Thought and the Sciences of Man.* 1983.
76. R. S. Cohen and L. Laudan (eds.), *Physics, Philosophy and Psychoanalysis.* 1983.
77. G. Böhme et al., *Finalization in Science,* ed. by W. Schäfer. 1983.
78. D. Shapere, *Reason and the Search for Knowledge.* 1983.
79. G. Andersson, *Rationality in Science and Politics.* 1984.
80. P. T. Durbin and F. Rapp, *Philosophy and Technology.* 1984.
81. M. Marković, *Dialectical Theory of Meaning.* 1984.
82. R. S. Cohen and M. W. Wartofsky, *Physical Sciences and History of Physics.* 1984.
83. E. Meyerson, *The Relativistic Deduction.* 1985.
84. R. S. Cohen and M. W. Wartofsky, *Methodology, Metaphysics and the History of Sciences.* 1984.
85. György Tamás, *The Logic of Categories.* 1985.
86. Sergio L. de C. Fernandes, *Foundations of Objective Knowledge.* 1985.
87. Robert S. Cohen and Thomas Schnelle (eds.), *Cognition and Fact.* 1985.
88. Gideon Freudenthal, *Atom and Individual in the Age of Newton.* 1985.
89. A. Donagan, A. N. Perovich, Jr., and M. V. Wedin (eds.), *Human Nature and Natural Knowledge.* 1985.
90. C. Mitcham and A. Huning (eds.), *Philosophy and Technology II.* 1986.
91. M. Grene and D. Nails (eds.), *Spinoza and the Sciences.* 1986.
92. S. P. Turner, *The Search for a Methodology of Social Science.* 1986.
93. I. C. Jarvie, *Thinking About Society: Theory and Practice.* 1986.
94. Edna Ullmann-Margalit (ed.), *The Kaleidoscope of Science.* 1986.
95. Edna Ullmann-Margalit (ed.), *The Prism of Science.* 1986.
96. G. Markus, *Language and Production.* 1986.